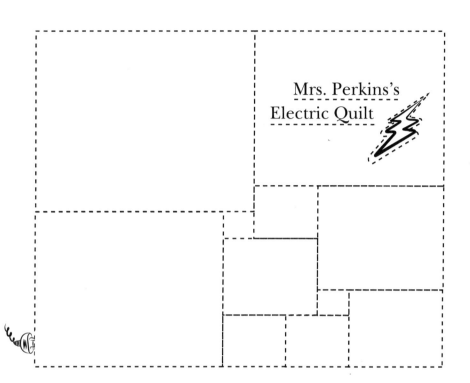

Looking for all the world like little kids at recess, two of the twentieth century's greatest physicists (both won the Nobel Prize) watch a spinning tippy-top in fascination during a break at the 1954 inauguration of the Institute of Physics, Lund, Sweden. Wolfgang Pauli (1900–1958), on the left, was a deep mathematical theoretician, while Niels Bohr (1885–1962) was more of an intuitionist, yet the physics of the everyday schoolyard top straddled the purely mathematical and the experimental to embrace the imaginations of both men. Photograph courtesy of the AIP Emilio Segrè Visual Archives, the Margrethe Bohr Collection.

Mrs. Perkins's Electric Quilt

and Other Intriguing Stories of Mathematical Physics

PAUL J. NAHIN

Princeton University Press
Princeton and Oxford

Copyright © 2009 by Princeton University Press

Published by Princeton University Press,
41 William Street, Princeton, New Jersey 08540

In the United Kingdom: Princeton University Press,
6 Oxford Street, Woodstock, Oxfordshire OX20 1TW

All Rights Reserved

Library of Congress Cataloging-in-Publication Data
Nahin, Paul J.
Mrs. Perkins's electric quilt: and other intriguing
stories of mathematical physics / by Paul J. Nahin.
p. cm.
Includes index.
ISBN 978-0-691-13540-3 (hardcover: alk. paper)
1. Mathematical physics. I. Title.
QC20.N34 2009
530.15-dc22   2008044856

British Library Cataloging-in-Publication Data is
available

This book has been composed in ITC New Baskerville
Printed on acid-free paper. ∞
press.princeton.edu
Printed in the United States of America

10 9 8 7 6 5 4 3 2 1

For Patricia Ann

who, unlike all the other young ladies I wrote letters to when I went off to college more than a half-century ago, took pity on me and wrote back

(she took in stray cats then, too).

Contents

For the Reader	xi
Preface	xiii

1   Three Examples of the Mutual Embrace   1
    1.1 Unphysical Laws   1
    1.2 When Math Goes Wrong   6
    1.3 Math from Physics   13

2   Measuring Gravity   18
    2.1 First, a Little Theory   18
    2.2 Out in the Author's Garage   21

3   Feynman's Infinite Circuit   24
    3.1 An Infinity of Resistors   24
    3.2 An Infinity of Reactances, and Recursion   27
    3.3 Convergence—or Not?   32
    3.4 Three More Infinite, All-Resistor Networks   36

4   Air Drag—A Mathematical View   44
    4.1 Air Drag Treated Broadly   44
    4.2 Air Drag Treated with Some Detail   51

5 — Air Drag—A Physical View — 62
5.1 The Quadratic Force Law — 62
5.2 Long Falls through a Real Atmosphere — 70

6 — *Really* Long Falls — 82
6.1 Falling into the Sun — 82
6.2 Falling from Heaven to Hell — 86

7 — The Zeta Function—and Physics — 94
7.1 A Curious Double Integral — 94
7.2 Fourier Series and the Zeta Function — 95
7.3 The Zeta Function in Physics — 100

8 — Ballistics—With No Air Drag (Yet) — 107
8.1 Shooting a Cannon in a Vacuum — 107
8.2 What Makes a Champion Shot-Putter? — 112
8.3 Another Cannon Question — 116

9 — Ballistics—With Air Drag — 120
9.1 Thin Air *Cannot* Be Ignored! — 120
9.2 Air Drag and Baseball — 126

10 — Gravity and Newton — 136
10.1 The Beginnings of Modern Gravity — 136
10.2 Newton's Superb Theorems — 140
10.3 The Moon Test and Blowing-Up Planets — 148
10.4 A Surprising Gravity Calculation — 152
10.5 Gravitational Contraction — 157

11 — Gravity Far Above the Earth — 170
11.1 Kepler's Laws of Planetary Motion — 170
11.2 Weighing the Planets — 175

12   Gravity Inside the Earth   186
    12.1   Newton's Experiment   186
    12.2   Gravity Inside the Earth   191
    12.3   Pressure at the Center of the Earth   200
    12.4   Travel Inside the Earth   203
    12.5   Epilogue   209

13   Quilts & Electricity   215
    13.1   Recreational Mathematics   215
    13.2   Electric Quilts   220
    13.3   Three Impossibility Proofs   225

14   Random Walks   233
    14.1   Ronald Ross and the Flight of Mosquitoes   233
    14.2   Karl Pearson Formulates a Famous Problem   236
    14.3   Gambler's Ruin   241
    14.4   The Monte Carlo Method   245

15   Two More Random Walks   261
    15.1   Brownian Motion   261
    15.2   Shrinking Walks   269

16   Nearest Neighbors   285
    16.1   Cannibals Can Be Fun!   285
    16.2   Neighbors Beyond the Nearest   291
    16.3   What Happens When We Have Lots of Cannibals   294
    16.4   Serious Physics   296

17   One Last Random Walk   299
    17.1   Resistor Mathematics   299
    17.2   Electric Walks   301

	17.3 Monte Carlo Circuit Simulation	305
	17.4 Symmetry, Superposition, and Resistor Circuits	313
18	**The Big Noise**	321
	18.1 An Interesting Textbook Problem	321
	18.2 The Polar Equations of the Big-Noise Flight	322
	18.3 The Acceleration on a Big-Noise Flight Path	328
	SOLUTIONS TO THE CHALLENGE PROBLEMS	333
	SPECIAL BONUS DISCUSSION	371
	Warning: Do Not Read before Reading Disscussion 17	373
19	**Electricity in the Fourth Dimension**	373
	19.1 The Tesseract	373
	19.2 Connecting a Tesseract Resistor Cube	376
	Acknowledgments	385
	Index	387

For the Reader

This book is organized as a series of chapters that, because of the informal style of writing I've adopted, I have called *discussions*. Some of the discussions have light cross-referencing to others, but in the main, each is a stand-alone presentation. The discussions are of subjects typically encountered in high school and lower-level college physics courses but are here taken to the next level. That is, they include analyses and calculations that are often skipped over, either because of lack of time or because the mathematical techniques required have not yet been developed in those classes. Each discussion has at least one Challenge Problem for the reader (a total of thirty-five problems). Complete, detailed solutions to nearly all of the Challenge Problems are provided at the end of the book (there are two exceptions, Challenge Problems 3.4 and 17.4).

The choice of what to include in these discussions is of course a highly personal one. *I* think they are all incredibly interesting—certainly interesting enough for me, in several cases, to have devoted huge chunks of time and energy to trying to understand what is going on so I could write about them. I can only hope that you'll agree with my selections, but if not, well, listen to the late Ricky Nelson's wonderful 1972 song, "Garden Party."

The assumed mathematical and physics background is that of a student who has finished his or her first year in a technical major at a good American college or university. And understood it. If you think, as I recently heard a TV commentator claim, that you'll save gas by driving as fast as possible because you'll get where you are going sooner, then there is simply no hope for you as an analyst, and you shouldn't buy this book—save your money for the extra gas you'll use!

The intended audience of this book includes students of mathematical physics, starting with very bright high school seniors who have taken (or are taking) AP calculus and physics. Also targeted are physicists and engineers who are interested in mathematical techniques (absolutely pure experimentalists, covered with vacuum pump grease and proud of it, need not apply). I am also writing for mathematicians who are interested in real, physical applications of their subject (absolutely pure mathematicians, e.g., those who lust *only* for a one-page analytical proof of the four-color theorem, or of Fermat's Last Theorem, can go stand in the corner with the absolutely pure, greasy experimentalists). *All* readers of this book will be those who yearn for the mutual embrace of physics and mathematics.

Preface

Philosophy is written in this grand book—I mean the universe—which stands continually open to our gaze, but it cannot be understood unless one first learns to comprehend the language and interpret the characters in which it is written. It is written in the language of mathematics, and its characters are triangles, circles, and other geometrical figures, without which it is humanly impossible to understand a single word of it; without these, one is wandering about in a dark labyrinth.
— Galileo Galilei, *The Assayer* (1623)

Mathematics may be compared to a mill of exquisite workmanship, which grinds you stuff of any degree of fineness; but, nevertheless, what you get out depends upon what you put in; and as the grandest mill in the world will not extract wheat-flour from peascod, so pages of formulae will not get a definite result out of loose data.
— Thomas H. Huxley, "Geological Reform" (1869)

Lord Rayleigh's solution . . . is most valuable . . . I ought to have known it, but . . . one does not expect to find the [solution to] a biometric problem provided in a memoir on sound.
— Karl Pearson, in a 1905 letter to *Nature*

Just past the midpoint of the nineteenth century, the Scottish mathematical physicist James Clerk Maxwell achieved the *second* great unification in the history of physics: he showed how to mathematically connect the previously separate subjects of electricity and magnetism. A link between the two had already been achieved by the great English experimenter Michael Faraday, but it was Maxwell who made the theoretical connection. (The first unification was Isaac Newton's mathematical description of gravity, near the end of the seventeenth century, by the inverse square law and his extension of that law from mere earthly confines to all of the universe: the universal law of gravity unified earthly and celestial physics. The third unification was, of course, Albert Einstein's connection of space and time at the beginning of the twentieth century, and the fourth, yet to come, will be the eventual connection of quantum mechanics with electrodynamics and gravity.) So taken by his unification of electricity and magnetism was Maxwell, so enthused was he by the intimate interaction of electric and magnetic phenomenon in his mathematical theory of the electromagnetic field, a connection described by what would soon be the world-famous Maxwellian equations, that he called the electric–magnetic interaction a *mutual embrace*.

In this book I have tried to illustrate Maxwell's mutual embrace, broadly interpreted to include all the interactions of physics and mathematics, with specific, detailed case studies of problems that I think show the astonishing way the two can combine to become something vastly more powerful and more useful than either of the subjects is alone. As the quotations from Galileo, Huxley, and Pearson that stand at the head of this preface show, this is an observation that was made long ago. And yet even today the connection of the two still amazes. The modern mathematical physicist Freeman Dyson wrote of this wonderfully curious and fruitful interaction of physics and mathematics in his 1979 book, *Disturbing the Universe*. There, when discussing the famous Millikan oil drop experiment from the early 1900s to determine the size of the electronic charge, he wrote how that experiment had

> brought home to me as nothing else could the truth of Einstein's remark, "One may say the eternal mystery of the world is its comprehensibility."

Here I was, sitting at my desk for weeks on end, doing the most elaborate and sophisticated calculations to figure out how an electron should behave. And here was the electron on my little oil drop, knowing quite well how to behave without waiting for the result of my calculation. How could one seriously believe that the electron really cared about my calculation, one way or the other? And yet experiments... showed that it did care. Somehow or other, all this complicated mathematics I was scribbling established rules that the electron on the oil drop was bound to follow. We know that this is so. Why it is so, why the electron pays attention to our mathematics, is a mystery that even Einstein could not fathom.

The great twentieth-century mathematical physicist Paul Dirac (famous in physics for his work in quantum mechanics, and famous in mathematics for his delta or impulse function) thought he knew how to explain that mystery: with *beauty*. To mention just one of his many proclamations on this, in a May 1963 *Scientific American* essay ("The Evolution of the Physicist's Picture of Reality"), he asserted that "it is more important to have beauty in one's equations than to have them fit experiment." The thesis of this book denies Dirac's claim.[1] The thesis of this book is that beauty comes precisely from our equations fitting reality (that is, experiment).

Not everybody, however, agrees with Galileo, Huxley, Pearson, Dyson—or me, for that matter—especially poets! Some people have (I know this is hard to believe) actually argued that an appreciation of reality is spoiled by the introduction of mathematics and physics! One such person was the famed essayist Charles Lamb at the so-called "Immortal Dinner," a party given on December 28, 1817, at the home of the English painter Benjamin Haydon. In attendance at what Haydon modestly described as "a night worthy of the Elizabethan age... with Christ hanging over us like a vision" were such luminaries as the poets Wordsworth and Keats. That evening Lamb toasted a portrait containing the image of Isaac Newton with words describing Newton as "a fellow who believed nothing unless it was as clear as the three sides of a triangle, and who had destroyed all the poetry of the rainbow by reducing it to the prismatic colors." Lamb was described by Haydon as having been "delightfully merry" just before he made his toast, which I interpret to mean he was thoroughly drunk. Certainly an intelligent man like Lamb wouldn't have made such a silly statement— I would hope—if sober.

One of Lamb's younger dinner companions was apparently greatly influenced by that toast, however, as three years later John Keats repeated the sentiment in his poem *Lamia*, where we find the words

> ... Do not all charms fly
> At the mere touch of cold philosophy?
> There was an awful rainbow once in heaven:
> We know her woof, her texture; she is given
> In the dull catalogue of common things.
> Philosophy will clip an Angel's wings,
> Conquer all mysteries by rule and line,
> Empty the haunted air, and gnomed mine—
> Unweave a rainbow. ...

I suspect neither Lamb or Keats knew much (if anything) about mathematics or physics, and so, harsh as it might seem, I dismiss both as being ignorant of what they spoke. Their words are beautiful in sentiment, yes, but empty of truth.[2]

Much better, I think, and in the spirit in which I've written this book are the following words by the English poet William Wordsworth (written in 1802, years before he attended the Immortal Dinner):

> My heart leaps up when I behold
>   A rainbow in the sky;
> So was it when my life began;
> So is it now I am a man;
> So be it when I shall grow old;
>   Or let me die!
> The Child is father of the Man;
> And I could wish my days to be
> Bound each to each by natural piety.

That I like.

The best reason I know of for why we need mathematics to understand physics was given by Vincent Icke (a professor of astrophysics at the University of Leiden and of cosmology at the University of Amsterdam) in his 1995 book, *The Force of Symmetry*. There he wrote, "Physics is not difficult; it's just weird.... Physics is weird because intuition is [usually] false." Physics needs mathematics because, unlike our intuitions about how things ought to work (but all too often don't), mathematics has no false preconceptions. This amazing power of mathematics was expressed by the Hungarian-born physicist Eugene Wigner, winner of the 1963 Nobel Prize in Physics, when he wrote (in his famous 1960 essay "The Unreasonable Effectiveness of Mathematics in the Natural Sciences") "The miracle of the appropriateness of the language of mathematics for the formulation of the laws of physics is a wonderful gift which we neither understand nor deserve."

To give you a quick example of what I have in mind for the discussions in this book, let me tell you a little story. A priest, a rabbi, and a minister are in a famous Boston bar (I don't have to tell you their names—everybody there knows their names). Suddenly a tall, elegantly outfitted, model-thin stork dressed to the nines walks through the door on the arm of a rhinoceros who is wearing a $2,000 business suit and an enormous gold Rolex watch. The rhino is big and very tough-looking, but badly needs a shave. There is an ominous, gun-shaped bulge under his monogrammed, meticulously tailored coat. He's either a vicious gangster or a dashing, heroic private eye who walks the mean streets, but who knows for sure? The rabbi turns to his companions and says—oops, wait, wrong story. That tale will need a *lot* more than physics and mathematics to explain it. Math and physics can't do everything, you know. Sorry! Let me start again.

A computer scientist (Professor C), a mathematician (Professor M), and a physicist (Professor P) are at the snack table in a large classroom as the weekly all-college seminar is about to begin. Each is wolfing down stale crackers layered with once sharp but now a bit flat cheddar cheese and drinking cheap sherry from cracked plastic cups. Budget cuts, you know. Suddenly Professor Grunderfunk, the seminar speaker, an already joyfully tipsy member of the Psychology Department, lurches into their midst and points a dirty finger at them (my heavens, are those nasty-looking cuts on his finger rat bites, bites from starving rodents driven half-mad from running diabolically fiendish mazes for hours on end in the Psych 101 lab?) and shouts, "Aha! Our infamous trio of college technicians, our disciples of the false belief that science and mathematics will save the world! Think you're so smart, do you? Well, we'll just see about that. I'll bet none of you can prove *this* without looking it up in Maor's book!"[3] And with that he tosses a crumpled piece of filthy paper on the floor, stuffs two cheese-laden crackers in his mouth, washes them down with one swift backward toss of a cracked sherry cup, and staggers off into a crowd of professors, who have seen far worse at previous weekly seminars.

"Should be an interesting talk," says P, as he bends over and retrieves the paper. "I wonder if he'll make it all the way through, though—you can hear the sherry sloshing when he walks. Oh, well, I wonder what Grunderfunk has left us." He smooths the paper out on the snack table, and C, P, and M look at what has so excited their colleague.

$$\text{Prove: } \lim_{n \to \infty} \left(1 + \frac{x}{n}\right)^n = e^x.$$

C is the first to speak, beginning with a slightly superior chuckle of amusement that both P and M find annoying. "This is trivial," says C, as he reaches inside his tweedy coat pocket and pulls out a flat, gleaming little black box covered with tiny buttons and a postcard-sized flat-screen display. "Behold the ByteBlaster 1000, the latest in portable programmable calculators. I never go anywhere without it. Well, *naturally* I don't take it into the *shower*, of course...." C stammers to a stop as both P and M grin at the image of C in the shower with his ByteBlaster 1000.

C blushes, then recovers. "Okay, I'll start by rewriting the problem as having to prove

$$\lim_{n \to \infty} \left\{ \left(1 + \frac{x}{n}\right)^n - e^x \right\} = 0.$$

"I'm stunned, C," says P with a grin. "That's a totally unexpected step to take. It's so unobvious. How *did* you think of it?"

C ignores P and his sarcasm, as by now his fingers are flying over the keyboard's tiny buttons. "See, what I'm doing is programming the calculator to accept any input value of x, and then to calculate $(1+\frac{x}{n})^n - e^x$ for $n = 1, 2, 3, 4$, and so on until I stop it. For example, let's put in x equal to 2 and see what we get." As P and M watch they see the numbers -4.3891, -3.3891, -2.7594, -2.3266, -2.0108, and more flash by on the screen.

"There," exclaims C. "The numbers *are* approaching zero. And that will happen no matter what I put in for x. Case closed." C places his calculator back in his coat pocket, picks up another cracker, and waits for his friends to pat him on the back.

"My lord, C," says M. "That's the ugliest so-called 'proof' I've ever seen. Why, it's even uglier than that awful computer proof, years ago, of the four-color map theorem. What you just did doesn't *prove* anything!"

"The proper way to prove the desired result," continues M, ignoring C's embarrassment while quickly falling into the professorial lecture mode that has induced numbness in his classes for decades, "is first to establish that the limit even exists in the first place. We can do that by expanding $(1+\frac{x}{n})^n$ with the binomial theorem and then applying the well-known result from real analysis that every monotonic increasing sequence that has an upper bound does indeed approach a limit. Once we have done that we can—"

"Stop, stop, for God's sake, stop, M," yells P, "you'll kill us all with that mathematician's mumbo jumbo! Let me show you how a *physicist* can do this problem in one-tenth the time all your formal axioms and theorems take."

"Well, then, go ahead," sniffs M, who really is badly hurt by his friend's harsh tone. "I'm sure we'll be entertained by how a physicist thinks about a mathematical question. And I use the word *think* loosely."

"Now, now, M, don't be angry. Perhaps I was a bit out of line there. I apologize." P is feeling generous because a beautiful way to attack the problem has just occurred to him. "I assume you two are familiar with the first law of motion?" P looks at his two friends expectantly.

"Well, yes, I think so," mumbles a still red-faced C, and "Oh, isn't that something Newton once said about how things blow around in the wind? Or something like that?" says M.

"Yes, something like that," replies P, suppressing a laugh, "but somewhat more accurately the law says every mass will continue in its state of rest, or of uniform motion in a straight line, except insofar as it is compelled to change that state by impressed forces."

"Yes, yes, of course," says M. "*That's* what I meant."

"Of course you did, M," says P, "of course. Now, pay attention." And with that P pulls a piece of chalk out of a pocket of his tweedy pants (Lord, how his wife hates it when chalk goes through the washing machine!) and strides over to a nearby blackboard.

"Suppose first that we have a particle moving along the positive x-axis at the constant speed $v = v_0$, says P, "and that at time $t = 0$ the particle is at the origin. There are no forces acting on the particle. Then, at any time $t \geq 0$ the particle will be distance s from the origin on the positive x-axis, where $s = v_0 t$."

Both C and M node in agreement; that's pretty easy to understand.

"So," continues P, "let's now suppose that the particle experiences a retarding, *drag* force proportional to the square of its speed. The speed is still v_0 at $t = 0$, but for $t > 0$ the speed will clearly be less and continually decreasing with time. In any case, I can write the following, where k is some constant of proportionality."

P writes the following on the blackboard:

$$\frac{d^2s}{dt^2} = \frac{dv}{dt} = -kv^2.$$

Both C and M now look perplexed.

P explains: "Here I've used Newton's second law of motion, which says applied force equals mass times acceleration. All my equation says is that the acceleration of a particle—which of course is actually a *de*celeration—is proportional to the retarding force, and the minus sign is the mathematical way of expressing that the force is indeed a *retarding* force."

"But P," says M, "why are you using the square of the speed? How do you know that? Why not a retarding force that is linear with speed, or even as the square root of the speed?"

"Good question, M," replies P. "It's because that is the way a retarding force actually works, according to hydrodynamics, as long as the speed is big enough, but not too big. But you'll soon see that the actual functional form of the drag force is really not all that important, and I'm just using the speed squared because that particular assumption will let me easily do the mathematics. I'll come back to this in just a bit."

M and C brighten up at that, and P continues.

"I can now rewrite my equation, by flipping it upside down, as

$$\frac{dt}{dv} = -\frac{1}{kv^2},$$

or as

$$dt = -\frac{1}{kv^2}\, dv,$$

and then by the obvious integration we have

$$t = -\int \frac{1}{kv^2}\, dv + W = \frac{1}{kv} + W,$$

where W is the constant of indefinite integration."

"Yes, yes," agrees M with enthusiasm, "I see *that*."

"Um, yeah, okay," says C with almost as much enthusiasm. "That looks right to me, too."

"Now," continues P, "since $v = v_0$ at $t = 0$, we see that $W = -\frac{1}{kv_0}$, and so

$$t = \frac{1}{kv} - \frac{1}{kv_0} = \frac{1}{k}\left(\frac{1}{v} - \frac{1}{v_0}\right).$$

Or, since $v = \frac{ds}{dt}$,

$$kt = \frac{1}{\left(\frac{ds}{dt}\right)} - \frac{1}{v_0}.$$

Then, with a little algebra...," and as P hums and mumbles to himself, he ends with a flourish and writes

$$\frac{ds}{dt} = \frac{v_0}{1 + kv_0 t}.$$

"Okay, okay, you're good at algebra," says M, "but just where is all this going, P?"

"Just this, M," replies P, "we can now easily integrate this last equation. See?" And P writes

$$s = \ln\{(1+kv_0 t)^{1/k}\} + Z.$$

"What's Z?," asks C, who has been thinking things over.

"It's just another constant of indefinite integration, C," says M. "Come on, man, pay attention."

"Well, you're a fine one to be telling *me* to pay attention, M," huffs C. "You as good as just admitted you don't know what P is doing, either!"

"Okay, okay, both of you, calm down," says P. "I'm almost done. Notice first that, since $s = 0$ at $t = 0$, we must have $Z = 0$. That is,

$$s = \ln\{(1+kv_0 t)^{1/k}\}.$$

Now comes the *coup d'etat!*"

This last exclamation is delivered with such a thunderous voice that not only are C and M startled, so is everybody else in the room. All eyes fasten on P's racing piece of chalk.

"Imagine now that the retarding force *vanishes*, that is, we let $k \to 0$. This means that our earlier assumption of a retarding force varying as the square of the speed is not really important, because we're letting that force go to *zero*. Then we approach our original case of a particle moving freely at constant speed, where we saw that $s = v_0 t$. That is, it must be the case that

$$\lim_{k \to 0} \ln\{(1+kv_0 t)^{1/k}\} = v_0 t.$$

If I now write $n = 1/k$, then as $k \to 0$ we have $n \to \infty$, and if I write $x = v_0 t$, we then have

$$\lim_{n \to \infty} \ln\left\{\left(1 + \frac{x}{n}\right)^n\right\} = x."$$

And before P can continue, M sees the last step, grabs the chalk from P, walks to the blackboard, and boldly writes the following while saying,

"Now, raising e to the power of both sides, we have

$$\lim_{n\to\infty} \left(1 + \frac{x}{n}\right)^n = e^x.$$

QED. That's *quod erat demonstrandum*, C, Latin for 'which was to be shown,' " says M, who likes to show off his knowledge of Latin phrases. Flushed with success, P grins with pleasure, C claps his hands together in an appreciative applause, and then the room follows C's lead and erupts as one into a thunderous ovation.

Physics has derived a mathematical theorem![4]

Even Grunderfunk appears to be impressed, although by now he is so loaded with cheap sherry and stale cheese that it's anybody's guess what's really going on inside his head. So excited are M and P at the result on the blackboard that they fall into each other's arms and a mutual embrace—and then, realizing all are watching, they just as suddenly jump back from each other. Quickly giving each other a chest bump, followed by the famous arm-pulling sports animation of "starting the lawnmower," and then finally a loud, smacking high five, each shouts at the other, "How about them Patriots!" "How about those Red Sox!" "Yeah, man, when we get outa here let's grab a bruski at Cheers!" "Wanna go duck huntin' on Saturday?" "Sure, we'll blast us some birds!"

"Excellent work, excellent work, gentlemen," yells the seminar chairperson over the still tumultuous excitement that fills the room. "Now, if you will all please be seated, let's welcome this week's speaker, Professor Grunderfunk of the Psychology Department, who will speak to us on his prize-winning research into the cognitive nervous system of the common sewer rodent. Or, as he has charmingly titled his talk, 'What's It Like to Be a Rat?' "

And so we leave P, M, and C as Grunderfunk begins his talk on rats, and we prepare to plunge into our next example of the mutual embrace of mathematics and physics. But, before we do that, here's a quick rundown on what I'm assuming on your part. It's really pretty straightforward—just that you have a math and physics background equivalent to the first year or so of college. I'll fill in a lot of the technical details as we go along, but it all will flow much more easily if, for example, acceleration as the first derivative of velocity, the differentiation and integration of simple functions, and the conservation laws of energy and momentum are already familiar concepts to you. In addition, familiarity with Ohm's law,[5] the alternating current (ac) impedances of inductors and capacitors, and with Kirchhoff's electrical circuit laws will help smooth the way for the electronic circuit simulation package (*Electronics Workbench*) that I'll use to physically confirm some important results from pure mathematics in one of the discussions.

The style will be informal (perhaps you've already guessed that), in an attempt to avoid a situation illustrated by a hilarious example from a mathematical physicist (the well-known Russian V. I. Arnold) of the differences between mathematicians of old and modern times. As he wrote in his marvelous 1989 book, *Huygens & Barrow, Newton & Hooke*, while a mathematician of yesteryear might write "Bob washed his hands," today's pure mathematician might instead say, "There is a $t_1 < 0$ such that the image Bob(t_1) of the points t_1 under the natural mapping $t \to$ Bob(t) belongs to the set of people having dirty hands and a t_2 of the half-open interval (t_1, 0] such that the image of the point t_2 under the same mapping belongs to the complement of the set concerned when the point t_1 is considered." Holy cow! Well, there won't be any of *that* sort of thing here. Indeed, to be mathematically obscure is not at all hard. Here's a classic example of that, one popular among physicists, mathematicians, and engineers, to amuse you, as well as to serve as an illustration of the level of mathematics you should feel comfortable with to be able to read this book.

CP. P1:

The following line of mathematics expresses a great truth; what is it? You can find the answer as the first entry in the section titled Solutions to the Challenge Problems, at the end of the book. But don't peek until you've tried!

$$\ln[\lim_{x \to \infty}(1+x^{-1})^x] + [\sin^2(y) + \cos^2(y)] = \sum_{n=0}^{\infty} \frac{\cosh(z)\sqrt{1-\tanh^2(z)}}{2^n}.$$

As another example, if you find the following story funny, then you'll have no problem with any of the math in this book. I've taken it from a book review[6] written by an evolutionary biologist who was describing the resistance to mathematics he once encountered from a journal editor: "I had particular difficulties with my paper on [mammalian] gaits, which may have been the first genuine 'optimization' paper in biology. To find the optimal gait, the one that would minimize energy expenditure at a given speed, I wrote down a differential equation: $(dw/dj = 0)$. The editor, offended by this attempt to sully a respectable biological journal with mathematics, asked, "Why don't you cancel the d's?' I fear that story will amuse only some readers, but I want to get it on record."

Let me put your mind at ease, however: the mathematics in this book is not all that highbrow. I liked to tell my students, when I was still teaching in the classroom, a little observation[7] made by the electrical engineer Richard Hamming during a talk he gave to a group of mathematicians: "[F]or more than 40 years I have claimed that if whether an airplane would fly or not depended on whether some function that arose in its design was Lebesgue *but not Riemann* integrable, then I would not fly in it. Would you? Does Nature recognize the difference? I doubt it!"

A book of this sort, a finite collection of individual problems selected from a truly vast number of possible candidates, is bound to be reflective of the author's particular tastes. I see no way to avoid that and will simply admit that what you'll find on the following pages is the result of a subjective, certainly *not* an objective, process of selection. With that said, let me also say that I have made a real effort to write discussions that illustrate fundamental issues in physics or mathematics, or that have a surprising twist. Best of all, of course, are those discussions that have both features—which may be different ones for different readers. Each discussion ends with at least one challenge problem for readers, with solutions at the back of the book.

CP. P2:

As your first real challenge problem in pure math applied to a real-life physics problem, consider the following impressive-looking integral:

$$I(t) = \int_0^{2\pi} \{\cos(\phi) + \sin(\phi)\} \sqrt{\frac{1 - \sqrt{t}\sin(\phi)}{1 - \sqrt{t}\cos(\phi)}} d\phi.$$

This integral occurred in a 1937 paper in *The Physical Review*, in a quantum mechanical analysis of the helium atom. As part of that analysis, it was important to show that $I(t) \neq 0$ for $0 \leq t \leq 1$. The author of that paper (a University of Illinois physicist on sabbatical at the Institute for Advanced Study in Princeton, New Jersey) apparently had little luck in formally attacking this problem and in the end was reduced to simply evaluating $I(t)$ numerically for five different values of t distributed over the interval zero to one (from 0.09 to 0.95, in particular). Finding that all five values for $I(t)$ are positive, he then argued that $I(t) > 0$ (and so $\neq 0$) for every value of t over the entire interval. Such an argument wouldn't cut much ice with mathematicians, of course (although Professor C would probably like it); the interesting thing for us is that there actually is an elementary analytic demonstration, requiring only high school algebra and trigonometry. Can you find it? The solution is the second entry in Solutions to the Challenge Problems at the end of the book.

So, here we go.

Notes and References

1. To be fair to Dirac, it is only right that I should not (as have other writers) take his words out of context. Dirac actually did elaborate and offer some explanation for why he felt as he did, in the same *Scientific American* essay. I quote:

> I might tell you a story I heard from [Erwin] Schrödinger, of how, when he first got the idea for [the Schrödinger wave equation of quantum mechanics], he immediately applied it to the behavior of the electron in the hydrogen atom, and then he got results that did not agree with experiment. The disagreement arose because

at that time it was not known that the electron has a spin. That, of course, was a great disappointment to Schrödinger, and it caused him to abandon the work for some months. Then he noticed that if he applied the theory in a more approximate way, not taking into account the refinements required by relativity, to this rough approximation his work was in agreement with observation. He published his first paper with only this rough approximation, and in that way Schrödinger's wave equation was presented to the world. Afterward, of course, when people found out how to take into account correctly the spin of the electron, the discrepancy between the results of applying Schrödinger's relativistic equation and the experiments was completely cleared up.... If Schrödinger had been more confident of his work, he could have published it some months earlier.... If there is not complete agreement between the results of one's work and experiment, one should not allow oneself to be too discouraged, because the discrepancy may well be due to minor features that are not properly taken into account and that will get cleared up with further developments of the theory.

2. A very erudite (but, in my opinion, misguided) assertion that the rise of science is the cause for a decline in poetry—really, I am not making this up—is made by the book by Max Eastman, *The Literary Mind: Its Place in an Age of Science* (New York: Charles Scribner's Sons, 1932). The tone of that work is immediately established when we read (on pp.3-4)

> Scientists are of course afflicted with a kind of professional vanity, and they occasionally get into a state of rapture over their ability to explain something mechanically, or mathematically, or .... You will see them at such times step out of their laboratories with a very unscientific light in their eyes, blow a horn on the street corner, assemble a few gullible people, and try to make a great impression on them.... This never works very well, and need not be taken more seriously than philosophy or any of the other frailties of human nature.

There is another exhortation against rational analysis that has incorrectly become famous among mathematicians. In his 1953 book *Mathematics in Western Culture*, Morris Kline quotes St. Augustine (from his *On Genesis Literally Interpreted*), as follows: "The good Christian should beware of mathematics.... The danger already exists that mathematicians have made a covenant with the Devil to darken the spirit and confine man in the bonds of Hell." Kline misquotes the fourth-century A.D saint, however, as Augustine used the word *mathematici* (which Kline translated as *mathematicians*), which actually means astrologers.

3. The reference is, of course, to Eli Maor's *e, the Story of a Number* (Princeton, N.J.: Princeton University Press, 1994, 1998).

4. In the first discussion of this book I'll show another example of this. My little story of P, M, and C is a Huge fictional embellishment of a brief technical note published long ago by Robert E. Shafer, "A Proof of Euler's Limit from a Well Known Physical Principle" (*Mathematics Magazine*, March–April 1959, pp. 211–212). My sincere apologies to Shafer and all surviving family relatives if he (they) should read what I've done to his work.

5. Here's an example of an elementary electrical problem I expect you to find "easy," as well as having the virtue of illustrating the difference between an inductive and a deductive mathematical proof. Suppose we have $n \geq 2$ ordinary resistors, of arbitrary values $R_1, R_2, \ldots, R_n$, and we connect them in parallel. Prove that the equivalent resistance of the result is less than the smallest of the n individual resistors. I'll do this two ways, first by induction. With R_e as the equivalent resistance of the n resistors in parallel, an application of Kirchhoff's current law at the common node (the sum of all the currents into any node is zero, which is of course simply a statement of the conservation of electric charge) quickly gives the textbook result of

$$\frac{1}{R_e} = \frac{1}{R_1} + \frac{1}{R_2} + \cdots + \frac{1}{R_n}.$$

Now, suppose $n = 2$. Let $R_s = \text{minimum}(R_1, R_2)$, and so the other resistor has value $R_b = R_s + \epsilon$, $\epsilon > 0$. (The subscripts s and b denote *small* and *big*, respectively.) Thus,

$$\frac{1}{R_e} = \frac{1}{R_s} + \frac{1}{R_s + \epsilon} = \frac{2R_s + \epsilon}{R_s(R_s + \epsilon)},$$

or

$$R_e = \frac{R_s(R_s + \epsilon)}{2R_s + \epsilon}.$$

The claim is that $R_e < R_s$; that is, the claim is that

$$\frac{R_s(R_s + \epsilon)}{2R_s + \epsilon} = \frac{R_s^2 + R_s \epsilon}{2R_s + \epsilon} < R_s,$$

which is clearly true because $R_s^2 < 2R_s^2$, i.e., $1 < 2$. So, we have shown that R_e is less than the minimum of R_1 and R_2. Suppose next that we have $n = 3$ resistors in parallel. We can take R_1 and R_2 in parallel as equivalent to R, and so our new problem is simply that of R in parallel with R_3. But our result for the $n=2$ case says the equivalent resistance is smaller than the smaller of R (which is smaller than either R_1 or R_2) and R_3, which means the equivalent resistance of three resistors in parallel is smaller than the smallest of the three. And so on for any n.

The deductive proof is much faster. As before, we have

$$\frac{1}{R_e} = \sum_{k=1}^{n} \frac{1}{R_k}.$$

Let R_s be the smallest of the n resistors, where s is the subscript of that resistor. Then,

$$\frac{1}{R_e} = \sum_{k=1}^{n} \frac{1}{R_k} = \frac{1}{R_s} + \sum_{k=1, k \neq s}^{n} \frac{1}{R_k}, n \geq 2.$$

Since every term in the sum on the right is positive, then the sum is positive, and therefore

$$\frac{1}{R_e} > \frac{1}{R_s},$$

and so, immediately, $R_e < R_s$, and this is true for all $n \geq 2$, because we made no assumptions about the value of n (other than it is at least 2).

6. John Maynard Smith, "Genes, Memes, & Minds" (*The New York Review of Books*, November 30, 1995). Smith's story was part of a review of Daniel C. Dennett's book, *Darwin's Dangerous Idea: Evolution and the Meaning of Life*.

7. R. W. Hamming, "Mathematics on a Distant Planet" (*American Mathematical Monthly*, August–September 1998, pp. 640–650).

Three Examples of the Mutual Embrace

> With arithmetic the Creator adjusted the World to unity, with geometry he balanced the design to give it stability. . . .
> — Nicholas of Cusa, *Of Learned Ignorance* (1440)

> I omit their [astronomers'] vain disputes about Eccentricks, Concentricks, Epicycles, Retrogradations, Tripidations, accessus, recessus, swift motions and circles of motion, as being the works neither of God nor Nature, but the Fiddle-Faddles and Trifles of Mathematicians.
> — Henry Cornelius Agrippa, *Of the Vanity and Uncertainties of Arts and Sciences* (1531)

1.1 Unphysical Laws

As the epigraphs illustrate, the role of mathematics as an aid to understanding the world has been evaluated differently by different philosophers. (I side with Nicholas, not Henry, as you have probably guessed.) The central thesis of this book is that physics needs mathematics, but the converse is often true, too. In this opening discussion I'll show you three examples of what I'm getting at.[1] In the first example you'll see a physical problem that is not difficult to state, and for which it is only very slightly more difficult to actually calculate a solution. It all comes about so smoothly, in fact, that it is easy to convince yourself that the job is done. But it is not, because that solution makes no *physical* sense, and it will be elementary mathematical arguments that show us that. The second example illustrates the converse: a physical problem that is easy to state but that clearly has a nonsensical *mathematical* solution. And it will now be physics that shows us the way out. The final example mirrors my earlier story of C, M, and P, that is, it will show you yet another example of physics "deriving" a mathematical theorem—probably the most famous theorem in all of mathematics, in fact.

For my first example, imagine a one-kilogram mass that we are going to accelerate straight down the positive x-axis, starting from rest at $x = 1$. We'll accelerate the mass by applying the benign-looking position-dependent force $F = x^2$. So, for example, at time $t = 0$ we have the mass motionless at $x = 1$ with an initial applied force of $F = 1$ newton directed toward increasing x. We now ask what appears to be a simple question: where on the x-axis is the mass at time $t = 5$ seconds? The astonishing (I think) answer is, *there is no answer!* The mathematics seems easy enough—I'll show it to you starting right now—but the problem is simply *physical* nonsense. Here's why.

From Newton's second law of motion (net applied force equals mass times acceleration), we have

$$F = x^2 = \frac{d^2x}{dt^2}, \tag{1.1}$$

where our initial conditions are

$$\left.\frac{dx}{dt}\right|_{t=0} = 0, x(t=0) = 1.$$

To integrate (1.1), begin by defining

$$v = \frac{dx}{dt}.$$

Then (1.1) becomes

$$x^2 = \frac{dv}{dt},$$

or, because of the chain rule from calculus,

$$\frac{dv}{dt} = \frac{dv}{dx} \cdot \frac{dx}{dt} = v\frac{dv}{dx},$$

and so we have
$$x^2 = v\frac{dv}{dx}.$$
That is,
$$x^2 dx = v dv,$$
which integrates by inspection to give
$$\frac{1}{3}x^3 = \frac{1}{2}v^2 + C,$$
where C is the constant of indefinite integration. From the initial conditions, $x(t=0) = 1$ while $v(t=0) = 0$, we see that $C = 1/3$, and so
$$\frac{1}{3}x^3 = \frac{1}{2}v^2 + \frac{1}{3},$$
or
$$x^3 = \frac{3}{2}v^2 + 1 = \frac{3}{2}\left(\frac{dx}{dt}\right)^2 + 1,$$
or
$$\frac{dx}{dt} = v = \sqrt{\frac{2}{3}(x^3 - 1)}. \tag{1.2}$$

This result tells us the speed of the accelerating mass as a function of its location. In particular, for any finite value of x the speed, too, is finite.

Now, (1.2) integrates immediately to give us our second formal answer, which provides a direct relationship connecting x and t (w, of course, is simply a dummy variable of integration):

$$t = \sqrt{\frac{3}{2}} \int_1^x \frac{dw}{\sqrt{w^3 - 1}}. \tag{1.3}$$

In (1.3) we have a formula that, for a given x, gives us the time in which the mass reaches x, and with (1.2) we can calculate how fast the mass is moving at that location. So, with these two results we are done, right? Well, some readers might wonder about the lower limit on the integral because, at $w = 1$, the denominator of the integrand is zero, and so there might be concern over the very existence of the integral.

However, as I'll show you next, the integral does indeed exist, so that is not a concern. Ironically, however, that very existence is a *problem*, because the integral exists (i.e., t has a finite value) even when the upper limit (x) is infinity! That is,

$$\int_1^\infty \frac{dw}{\sqrt{w^3-1}} < \infty. \tag{1.4}$$

This is disastrous, because that means our one-kilogram mass has moved infinitely far in finite time. Okay, you think about that for a bit while I demonstrate the above claim of existence for (1.4).

We can write the integral in (1.3), for an infinite upper limit, as

$$\int_1^\infty \frac{dw}{\sqrt{w^3-1}} = \int_1^\infty \frac{dw}{\sqrt{w-1}\sqrt{w^2+w+1}} < \int_1^\infty \frac{dw}{\sqrt{w-1}\sqrt{w^2}}.$$

That is,

$$\int_1^\infty \frac{dw}{\sqrt{w^3-1}} < \int_1^\infty \frac{dw}{w\sqrt{w-1}}.$$

Next, change variable to $q = w - 1$. Then,

$$\int_1^\infty \frac{dw}{\sqrt{w^3-1}} < \int_0^\infty \frac{dq}{(q+1)\sqrt{q}} = \int_0^1 \frac{dq}{q^{3/2}+q^{1/2}} + \int_1^\infty \frac{dq}{q^{3/2}+q^{1/2}}.$$

We make the inequality even stronger by replacing each of the two integrals on the right by even larger ones:

$$\int_1^\infty \frac{dw}{\sqrt{w^3-1}} < \int_0^1 \frac{dq}{q^{1/2}} + \int_1^\infty \frac{dq}{q^{3/2}}.$$

That is,

$$\int_1^\infty \frac{dw}{\sqrt{w^3-1}} < (2q^{1/2})|_0^1 - (2q^{-1/2})|_1^\infty = 2 + 2 = 4.$$

Thus, our one-kilogram mass has moved infinitely far away in less than $\sqrt{\frac{3}{2}} \cdot 4 = 4.899$ seconds.[2] Where the mass is at $t = 5$ seconds, therefore, simply has no answer (can anything be further away than infinity?) Before discussing just what is going on to cause this outrageous result, let me show you another seemingly faultless mathematical analysis that results in an even more outrageous conclusion (as hard as that may be to conceive at this moment).

Imagine that we now have a speed-dependent force, e.g., $F = v^3$. I'll further imagine that, as before, we have a one-kilogram mass to be accelerated by this force down the positive x-axis. But now we'll have it start from the origin $x = 0$ with an initial speed of 1 m/s. Then, $x(t = 0) = 0$ and $v(t = 0) = 1$, and

$$F = v^3 = \frac{d^2x}{dt^2} = \frac{dv}{dt}.$$

This integrates immediately to

$$t = -\frac{1}{2v^2} + C,$$

where C is an arbitrary constant. Since $v(t = 0) = 1$, we have $C = 1/2$, and so

$$\frac{1}{2v^2} = \frac{1}{2} - t,$$

or

$$v = \frac{dx}{dt} = \frac{1}{\sqrt{1 - 2t}}. \tag{1.5}$$

So, the mass approaches infinite speed as t approaches 1/2 second. This looks bad, alright, but what makes this result really bad is *where* the mass is when its speed becomes arbitrarily large. It is not infinitely far down the x-axis but rather pretty close to where it started from! This is easy to demonstrate.

Integrating (1.5), we have

$$\int_0^x du = \int_0^t \frac{du}{\sqrt{1 - 2u}} = \left\{ -(1 - 2u)^{\frac{1}{2}} \right\} \Big|_0^t,$$

or
$$x = 1 - \sqrt{1-2t}. \qquad (1.6)$$

That is, as $t \to 1/2$, we have $x \to 1$, which means that our mass achieves infinite speed in just 1/2 second after moving just one meter! After $t = 1/2$ second and $x = 1$ meter, our solution offers no answers to either how fast the mass is moving or where it is. The mass leaves any realm of reality *right in front of our eyes!*[3]

This is, of course, absolutely absurd. The problem with both of these examples is that we have assumed *unphysical* forces that have no bound, as well as used physics that is not relativistically correct as the speed of the mass approaches the speed of light. The initial physical situations initially appear to be so benign, however, that our sense of well-being is not threatened or alerted—or at least not until we work through the detailed mathematics. Then we can see the inherent absurdity of the physics and observe just how quickly things spin out of control. Now, let me reverse the situation and give you a situation where it is the mathematics of an analysis that leads us into silliness and physics that both identifies the origin of the difficulty and removes the confusion.

1.2 When Math Goes Wrong

I don't think there has been a calculus textbook written that doesn't have a problem, either as an example or as an end-of-chapter question, that goes something as follows. Imagine, as shown in Figure 1.1, that a pole of length L is leaning against a wall at angle $\theta_0 = \theta(t=0)$. By some means the bottom end is pulled away from the wall (all the while remaining in contact with the floor) at the constant speed v_0. How fast is the other end sliding down the wall? The intent of such a question in a calculus textbook is purely mathematical, simply to provide a physical situation in which one has to calculate derivatives. From the geometry of the problem we have, from the Pythagorean theorem,

$$x^2 + y^2 = L^2. \qquad (1.7)$$

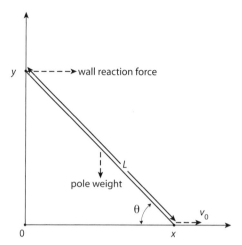

Figure 1.1. A sliding pole.

Then, differentiating (1.7) with respect to time,

$$2x\frac{dx}{dt} + 2y\frac{dy}{dt} = 0,$$

or

$$x \cdot v_x + y \cdot v_y = 0,$$

where v_x and v_y are, respectively, the speed of the floor end and the speed of the wall end of the pole. Now, from Figure 1.1 we see that $x = L\cos(\theta)$ and $y = L\sin(\theta)$, so (as $v_x = v_0$) we have the answer to the textbook question:

$$v_y = -\frac{v_0 L \cos(\theta)}{L \sin(\theta)} = -v_0 \cot(\theta). \tag{1.8}$$

The minus sign in (1.8) means that the wall end of the pole is sliding in the direction of decreasing y, i.e., that end is sliding *down* the wall. Notice that, as $\theta \to 0$, the magnitude of the speed of the wall end continually increases; that is, the falling end of the pole is accelerating. This says not only that θ is becoming more and more negative as the fall progresses but also that the rate of increase in the "negativeness" of θ is increasing; that is, $\frac{d^2\theta}{dt^2} < 0$, or $-\frac{d^2\theta}{dt^2} > 0$.

This is generally all that calculus textbook writers expect students to get from this calculation, but there is an implication in the result that bothers attentive readers. At the instant the falling end of the pole has slid all the way down to the floor, at the moment $\theta = 0$, (1.8) indicates that $v_y = -\infty$. That's right, our analysis says that the left end of the pole is moving infinitely fast at the moment it hits the floor! That can't be right. (Invoking relativistic physics isn't the answer, either: before the speed becomes infinite, the math says the speed at some finite time exceeds the speed of sound, which ought to raise an eyebrow, too: do falling ladders generate sonic booms?) But where did the mathematics go wrong? *That* is the question we have to answer. The origin of the problem lies in (1.7). That's not to say that the Pythagorean theorem is wrong but rather that it simply does not *physically* apply over the entire duration of the fall. At some point during the slide, at some intermediate angle $\theta = \theta_c$, (1.7) ceases to describe what is happening. That means, for $\theta < \theta_c$, it is no longer true that the wall end of the pole remains in contact with the wall. At the instant when θ becomes less than θ_c, the pole will "break away" from the wall. Mathematics alone cannot tell us what θ_c is, but physics *and* mathematics can. Here's how.

Figure 1.1 shows the pole and all the forces acting on it. There is the weight of the pole (mg, where m is the mass of the pole and g is the acceleration of gravity at the Earth's surface) acting at the pole's center of mass, which for a uniform pole is at the midpoint of the pole, directed perpendicular to the floor; there is the reaction force of the wall (F_w) acting on the wall end of the pole perpendicular to the wall; there is the applied force at the floor end of the pole, acting parallel to the floor, and it is the direct cause of the constant speed v_0; and there are, in general, friction wall and floor forces acting at the two ends of the pole. At this point, however, I'll make the simplifying assumption that the wall is frictionless (whether the floor has friction or not won't matter). Then, using Figure 1.1 as a guide, we can write Newton's *rotational* analogue of the second law of motion (net applied torque equals moment of inertia times angular acceleration) as

$$mg\cos(\theta)\cdot\frac{1}{2}L - F_w\sin(\theta)\cdot L = -I\frac{d^2\theta}{dt^2}, \tag{1.9}$$

where I is the moment of inertia of a uniform pole rotating about one end (the floor end). (If the minus sign on the right-hand side of (1.9) puzzles you, look back at my comments just after (1.8).) As shown in any good freshman physics textbook,

$$I = \frac{1}{3}mL^2.$$

Now, a few words about each term in (1.9). Since a *torque* in this problem is the product of a force acting perpendicular to the pole times the *moment arm* of that force (that is, the distance of the application point of the force from the axis of rotation, which is the floor end of the pole), we can see what is going on with the left-hand side of (1.9). The first term is the normal (to the pole) component of the pole's weight times half the length of the pole, and the second term is the normal (to the pole) component of the wall reaction force times the full length of the pole. It is clear that these two torques tend to rotate the pole in opposite senses, and (1.9) follows the convention that a positive torque causes a counterclockwise rotation, which a look at Figure 1.1 shows is in the sense of decreasing θ. The right-hand side of (1.9) is simply moment of inertia times angular acceleration, and as you'll soon see, during the initial part of the slide, when the pole is in contact with the wall, $-I\frac{d^2\theta}{dt^2}$ is indeed positive (because of the net positive torque), just as it should be for a pole that is rotating counterclockwise, i.e., that has its wall end sliding downward toward the floor.[4]

From Figure 1.1 we have $x = L\cos(\theta)$, and so

$$\frac{dx}{dt} = v_0 = -L\sin(\theta) \cdot \frac{d\theta}{dt}, \qquad (1.10)$$

or

$$\frac{d\theta}{dt} = -\frac{v_0}{L\sin(\theta)}. \qquad (1.11)$$

Differentiating (1.10) and using $\frac{dv_0}{dt} = 0$, we have

$$0 = \sin(\theta)\frac{d^2\theta}{dt^2} + \frac{d\theta}{dt}\cos(\theta)\frac{d\theta}{dt},$$

or

$$-\sin(\theta)\frac{d^2\theta}{dt^2} = \cos(\theta)\left(\frac{d\theta}{dt}\right)^2.$$

Using (1.11) in this last result, we have
$$-\frac{d^2\theta}{dt^2} = \frac{v_0^2 \cos(\theta)}{L^2 \sin^3(\theta)},$$
which is, as claimed, positive over the entire interval $0 \leq \theta \leq \theta_0 < \pi/2$. Thus, (1.9) becomes
$$\frac{1}{2}mgL\cos(\theta) - F_w L \sin(\theta) = \frac{1}{3}\frac{mv_0^2 \cos(\theta)}{\sin^3(\theta)},$$
or, solving for F_w,
$$F_w = \frac{\frac{1}{2}mgL\cos(\theta)}{L\sin(\theta)} - \frac{mv_0^2 \cos(\theta)}{3L\sin^4(\theta)}.$$

That is,
$$F_w = \frac{1}{2}mg\cot(\theta) - \frac{\frac{1}{2}mgv_0^2 \cot(\theta)}{3 \cdot \frac{1}{2} \cdot Lg\sin^3(\theta)},$$
or
$$F_w = \frac{1}{2}mg\cot(\theta)\left[1 - \frac{2v_0^2}{3Lg\sin^3(\theta)}\right]. \quad (1.12)$$

The important observation to make with (1.12) is that, as long as the wall end of the pole is leaning against the wall, we'll have $F_w \geq 0$, as it is physically meaningless to talk of $F_w < 0$. After all, the wall can't pull on the pole! The instant $F_w < 0$ is the instant the wall end of the pole loses contact with the wall.

We can normalize (1.12) by not talking of F_w or v_0^2 directly but of their ratios with the naturally occurring force and speed (squared), mg and gL, respectively. That is, let's write (1.12) as
$$\frac{F_w}{mg} = \frac{1}{2}\cot(\theta)\left[1 - \frac{\frac{2}{3} \cdot \frac{v_0^2}{gL}}{\sin^3(\theta)}\right]. \quad (1.13)$$

When the pole is simply leaning against the wall, and we have not yet begun to pull the lower end, (1.13) tells us that for $v_0^2 = 0$ the normalized reaction force of the wall is
$$\frac{F_w}{mg} = \frac{1}{2}\cot(\theta), \, v_0 = 0.$$

Once we start pulling on the lower end, however, what happens depends, as (1.13) shows, on whether v_0^2/gL is less than or greater than 3/2.

For the first case, that is, for $v_0^2/gL < 3/2$, (1.13) says the normalized wall reaction force does not drop to zero until

$$1 - \frac{\frac{2}{3} \cdot \frac{v_0^2}{gL}}{\sin^3(\theta)} = 0.$$

That is, the wall end of the pole stays in contact with the wall until it reaches the "breakaway" angle of θ_c, where

$$\theta_c = \sin^{-1}\left\{ \sqrt[3]{\frac{2}{3} \cdot \frac{v_0^2}{gL}} \right\}. \tag{1.14}$$

If the initial pole angle $\theta_0 < \theta_c$, then the wall end of the pole breaks away from the wall as soon as we start pulling the floor end at speed v_0. And that is precisely what happens in the second case of $v_0^2/gL > 3/2$ no matter what v_0 may be: the pole breaks away from the wall as soon as we start pulling the floor end at *any* speed.

Suppose we concentrate from this point on the first case, of $v_0^2/gL < 3/2$, and assume $\theta_0 > \theta_c$. From (1.14) we have

$$\cot(\theta_c) = \frac{\sqrt{1 - \left(\frac{2}{3} \cdot \frac{v_0^2}{gL}\right)^{2/3}}}{\left(\frac{2}{3} \cdot \frac{v_0^2}{gL}\right)^{1/3}},$$

and so from (1.8) we see that the magnitude of the speed v_{ba} with which the wall end of the pole is sliding downward is, at the breakaway moment,

$$v_{ba} = v_0 \sqrt{\left(\frac{3}{2} \cdot \frac{gL}{v_0^2}\right)^{2/3} - 1}.$$

We can write this in a more transparent form as

$$v_{ba} = \sqrt{\left(\frac{3gL}{2}\right)^{2/3} \cdot \frac{v_0^2}{v_0^{4/3}} - v_0^2} = \sqrt{\left(\frac{3gL}{2}\right)^{2/3} \cdot v_0^{2/3} - v_0^2},$$

or

$$v_{ba} = \sqrt{\alpha v_0^{2/3} - v_0^2}, \quad \alpha = \left(\frac{3gL}{2}\right)^{2/3}. \tag{1.15}$$

It is clear from (1.15) that there is some value of v_0 in the interval 0 to $\alpha^{3/4}$ for which v_{ba} reaches its maximum value. We can find that value for v_0 by setting the derivative of (1.15) to zero, which results in the equation

$$\frac{1}{2}(\alpha v_0^{2/3} - v_0^2)^{-1/2}\left\{\frac{2}{3}\alpha v_0^{-1/3} - 2v_0\right\} = 0,$$

or

$$2v_0 = \frac{\frac{2}{3}\alpha}{v_0^{1/3}},$$

or

$$v_0^{4/3} = \frac{\alpha}{3}.$$

This says that

$$v_0^{2/3} = \left(\frac{\alpha}{3}\right)^{1/2}$$

and also that

$$v_0^2 = \left(\frac{\alpha}{3}\right)^{3/2}. \tag{1.16}$$

Substituting these last two results back into (1.15), we get the maximum value for v_{ba}, what I'll call v_{bam}, to be

$$v_{bam} = \sqrt{\left(\frac{3gL}{2}\right)^{2/3}\left(\frac{1}{3}\right)^{1/2}\left(\frac{3gL}{2}\right)^{1/3} - \left(\frac{1}{3}\right)^{3/2}\left(\frac{3gL}{2}\right)}$$

$$= \sqrt{\frac{3gL}{2}}\sqrt{\frac{1}{\sqrt{3}} - \frac{1}{3\sqrt{3}}} = \sqrt{\frac{3gL}{2}}\sqrt{\frac{2}{3\sqrt{3}}} = \sqrt{\frac{gL}{\sqrt{3}}}$$

$$= \frac{1}{\sqrt[4]{3}}\sqrt{gL} = 0.76\sqrt{gL}.$$

And finally, from (1.16), we can write

$$\frac{v_0^2}{gL} = \frac{1}{gL}\cdot\left(\frac{\alpha}{3}\right)^{3/2} = \frac{1}{gL}\cdot\frac{1}{3^{3/2}}\cdot\left\{\left(\frac{3gL}{2}\right)^{2/3}\right\}^{3/2} = \frac{1}{gL}\cdot\frac{1}{3^{3/2}}\cdot\frac{3gL}{2} = \frac{1}{2\sqrt{3}}.$$

Inserting this result into (1.14), we find the pole angle θ at the instant the pole breaks away from the wall, for $v_{ba} = v_{bam}$, to be

$$\theta_c = \sin^{-1}\left\{\sqrt[3]{\frac{2}{3} \cdot \frac{1}{2\sqrt{3}}}\right\} = \sin^{-1}\left\{\frac{1}{\sqrt[3]{3^{3/2}}}\right\} = \sin^{-1}\left\{\frac{1}{\sqrt{3}}\right\} = 35.3 \text{ degrees}.$$

We could do a lot more similar calculations, but this is enough to make my point: the original calculus textbook solution of (1.8) is only the beginning of what can be done if physics is brought into the analysis, along with the mathematics of calculating derivatives.

1.3 Math from Physics

In the Preface I told you a story in which physics "derived" a mathematical theorem. What I'll show you next is another example of this amazing entanglement of mathematics and physics. When I was a freshman in Physics 51 at Stanford more than fifty years ago (1958), I took a lot of examinations, but one in particular sticks in my memory. One of the questions on that test described a physical situation in which, at the end, the problem was to calculate how far up a glass tube capillary action would draw a fluid. It was a gift question, one the professor had put on the exam to get everybody off to a good start; to answer it all you had to do was remember a formula that had been derived in lecture and in the course text and that we had used on at least a couple of homework assignments. All the exam required was plugging numbers into the formula. The professor had kindly provided all the numbers, too. Unfortunately, I couldn't remember the formula, and so no gift points for me.

Later, back in the dorm, I was talking with a friend in the class, who was most grateful for that gift question; he wasn't doing well in the course, and the "free" points were nice. "So, you remembered the formula, right?" I asked. "Nope, but you didn't have to. I nailed that one, anyway," he replied. "What do you mean, you didn't have to remember the formula?" I asked, a sinking feeling in my stomach. "All you had to do," my friend grinned back, "was just take all the different

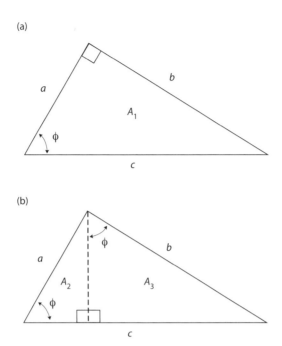

Figure 1.2. Deriving the Pythagorean theorem.

numbers the prof gave us and try them in different ways until the units worked out as a *length*, the unit of *distance up the tube*." "But, but," I sputtered, "that's, that's ... *cheating!*"

But of course, it wasn't cheating. I was just angry at myself for not being sharp enough to have thought of the same idea my friend had. It was my first (painful) introduction to the honorable technique of dimensional analysis. Here, then, as another example of that idea, is how a physicist might derive the Pythagorean theorem using dimensional analysis.

In the upper half of Figure 1.2 I've drawn a right triangle with perpendicular sides of lengths a and b and a hypotenuse of length c. One of the interior acute angles of the triangle is ϕ. I think it obvious that the triangle is *determined* once we know the values of c and ϕ. That is, for a given c and a given ϕ, the other side lengths (a and b) and the remaining interior acute angle all have unique values. And certainly, then, the *area* A_1 of the triangle is also determined. Since area has units of length squared, and since ϕ is dimensionless, it must be that the area depends on the square of c. So, let's write the area of our triangle as

$$A_1 = c^2 f(\phi), \tag{1.17}$$

where $f(\phi)$ is some function of ϕ. (we do not, as you'll soon see, have to actually know the detailed nature of $f(\phi)$!)

Now, draw the perpendicular line from the right angle to the hypotenuse of the triangle, as shown in the lower half of Figure 1.2. This divides the triangle, into two smaller right triangles, one with area A_2, an acute angle ϕ, and hypotenuse a, and another with area A_3, an acute angle ϕ, and hypotenuse b. Thus, just as in (1.17), we can write

$$A_2 = a^2 f(\phi)$$

and

$$A_3 = b^2 f(\phi).$$

Since $A_1 = A_2 + A_3$, then $c^2 f(\phi) = a^2 f(\phi) + b^2 f(\phi)$, and so the unknown function $f(\phi)$ cancels away (that's why we don't have to know what it is) and we suddenly have, dramatically and seemingly out of thin air, the well-known

$$a^2 + b^2 = c^2.$$

That's it. Pretty nifty, don't you think?

CP. P1.1:

Looking back at the sliding pole problem, assume $\theta_0 > \theta_c$. What is the acceleration of the wall end of the sliding pole at the instant of breakingaway? *Hint*: The answer is independent of v_0 and L and is simply of the form kg, where k is a particular number.

Notes and References

1. The examples discussed here are based on the papers by A. John Mallinckrodt, "The Pathological Kinematics of Unphysical Force Laws"

(*American Journal of Physics*, March 1992, pp. 238–241), and M. Freeman and P. Palffy-Muhoray, "On Mathematical and Physical Ladders" (*American Journal of Physics*, March 1985, pp. 276–277).

2. Matters are actually worse than this. A numerical evaluation of (1.3)—I simply typed the following MATLAB Symbolic Toolbox command line, sqrt(3/2)*double(int(1/sqrt(x^3-1),1,inf))—almost instantly produced the value 2.97447742··· seconds for how long it takes the mass to move out to infinity. (The command double is MATLAB's "double precision" command.) That is, in less than three seconds it becomes meaningless to ask where the mass is.

3. As I wrote the line in the text describing the disappearing mass, I couldn't help but be reminded of a wonderful passage from H. G. Wells's classic novella, *The Time Machine* (1895). When the story opens, the Time Traveller has invited some friends to his home for dinner, and afterward, over drinks, tells them of his time machine. Soon after he demonstrates a working model of the machine by sending it on a journey through time, in a fantastic scene described by one of the invited friends:

> We all saw the lever turn. I am absolutely certain there was no trickery. There was a breath of wind, and the lamp flame jumped. One of the candles on the mantel was blown out, and the little machine suddenly swung around, became indistinct, was seen as a ghost for a second perhaps, as an eddy of faintly glittering brass and ivory; and it was gone—vanished! Save for the lamp the table was bare. Every one was silent for a minute. Then Filby said he was damned.

4. You can see now that both the friction (if any) force and the "pulling" force at the floor end of the pole don't appear in (1.9) because those forces have *zero-length* moment arms, and so neither of them produces a torque. If we *did* have a non-zero friction wall force, then that force, acting on the wall end of the downward-sliding pole, would produce a clockwise torque. That complication would not change the character of the results but would complicate the mathematics. Hence my assumption of frictionless surfaces. There is also a perpendicular contact force of the floor on the floor end of the pole, but it too has a zero-length moment arm and so produces no torque. As a final note, observe that all torques are taken about a *moving* point (the floor end of the pole), which is legitimate if the moving point is not accelerating. Since the floor end is moving at the constant speed v_0, this condition is satisfied. For a nice tutorial discussion of this issue, see Fredy R. Zypman, "Moments to Remember: The Conditions for Equating Torque and Rate of

Change of Angular Momentum" (*American Journal of Physics*, January 1990, pp. 41–43). One reviewer has observed that, since the floor is frictionless, to maintain a constant speed for the floor end of the pole, one would actually have to push to the left rather than pull to the right. I have put the word "pulling" in quotes, but perhaps this comment may help make the physics of what is happening more transparent.

5. The definitive modern work on the Pythagorean theorem is the book by Eli Maor, *The Pythagorean Theorem: A 4,000-Year History* (Princeton, N.J.: Princeton University Press, 2007). The dimensional analysis derivation in this discussion is, however, not in Maor's book; I came across it while browsing in a book by A. B. Migdale, *Qualitative Methods in Quantum Theory* (W. A. Benjamin, New York: 1977 [published originally in Russian in 1975], p. 2). Dimensional analysis has been around in mathematical physics for some time. For example, it is used several times in Rayleigh's classic 1877 work, *The Theory of Sound*. (Lord Rayleigh will appear again in this book, most prominently in Discussion 14.)

Measuring Gravity

"Kids, you should do this at home!"

2.1 First, a Little Theory

In a number of the discussions to follow in this book, we'll find that we will need to know the numerical value of the gravitational acceleration at the Earth's surface, usually denoted, as in the first discussion, by g (pronounced *gee*). The experimental determination of the value of g is, in fact, a classic experiment performed every year in thousands of college freshman physics labs worldwide. I remember well when I did it as a freshman in Physics 51 at Stanford (1958). I remember it as a clunky, uninspiring experiment that required watching a high-speed, pulsed spark generator burn holes through a strip of falling wax paper (I recall that even the graduate student teaching assistant looked like she would rather have been somewhere else). That was followed by the measurement of the distances between adjacent burn holes to eventually arrive, with some arcane intermediate calculations, at a value for g.

Here's a better way—faster and pedagogically superior—to measure g with a simple two-step experiment. All you'll need in the way of equipment is a yardstick, a bouncy rubber ball, and a stopwatch.[1] You don't need an expensive and mysterious (to most college freshmen, anyway) spark generator. You do need to be able to follow a little elementary physics and some simple high school algebra. Then you can measure g where you live in less than 60 seconds.

Begin by imagining that you are holding a nice round bouncy rubber ball in your hand, about hip high, over a concrete floor. You let the ball fall. What happens? Well, of course, it eventually hits the floor and then bounces up, then falls again, then bounces up again, then falls. ... With each new bounce after the first one the ball doesn't go quite as high as it did on the previous bounce, because energy is being dissipated during this process. Every time the ball hits the concrete floor, for example, both the floor and the ball get a little hotter, and of course the impact noise itself is energy traveling away from the impact point at the speed of sound. After a while the ball stops bouncing and rolls away to a stop somewhere.

Physicists model the decreasing bounce heights with the aid of what is called the *coefficient of restitution*. In our case, we look at the speed of the ball just before it hits the floor (call that speed v_b) and then just after the bounce (call that rebound speed v_a). The coefficient of restitution is then defined to be

$$c = \frac{v_a}{v_b},$$

where, because of energy dissipation on impact, we know $v_a \leq v_b$, and so $c \leq 1$. If we suppose that we initially drop our ball (with mass m) from height h_0, then conservation of energy says

$$\frac{1}{2}mv_b^2 = mgh_0,$$

which is simply equating the kinetic energy of the ball just before the first impact with the potential energy at the start of the first fall. (By doing this you should realize that we are ignoring resistive air drag on the moving ball as being an important energy dissipation mechanism.[2]) Thus,

$$v_b = \sqrt{2gh_0},$$

and so

$$v_a = c\sqrt{2gh_0}.$$

The kinetic energy of the ball just after the start of the first bounce is therefore

$$\frac{1}{2}mv_a^2 = mc^2gh_0,$$

and so, when the ball reaches the top of the first bounce, all that energy will be potential energy. Therefore, the height of the first bounce (call it h_1) is given by

$$mgh_1 = mc^2 gh_0.$$

Thus, $h_1 = c^2 h_0$, or

$$c = \sqrt{h_1/h_0}, \tag{2.1}$$

and this gives us a quick way to measure c for our ball and floor. All we have to do is position a yardstick vertically upright, hold the ball in front of and at the top of the yardstick ($h_0 = 36$ inches), and then let the ball fall. If we then observe how high up the yardstick the ball rebounds (h_1), we can calculate c from (2.1), and that's the first half of our experiment to measure g. Now, what do we do with c?

Begin by noticing that if the initial drop height is h_0, then (from above) the first rebound height is $h_1 = c^2 h_0$, as well as the next drop height, and so the second rebound height is $h_2 = c^4 h_0$. But that says the third rebound height is $h_3 = c^6 h_0$, and so on. Now, let's call the first impact of the ball with the floor the $n=0$ impact. The second impact will be the $n=1$ impact, and so on. Starting with the $n=0$ impact, let's write ΔT_1 as the time interval until the $n=1$ impact, ΔT_2 as the time interval between the $n=1$ impact and the $n=2$ impact, and so on. For example, ΔT_1 is the time it takes the ball to rise from the floor at speed v_a up to the height h_1 and then to fall back down to the floor, where it will again be moving at speed v_a just before the second impact. Since the rise and fall times are equal (remember, we are assuming zero air drag—see note 2 again), and since the fall time for a ball to fall through distance h_1 when starting from rest (at the height of the first rebound) is given by $\sqrt{2h_1/g}$ (remember, the distance traveled from rest with a constant acceleration g during a time interval of duration t is $\frac{1}{2}gt^2$), we see that

$$\Delta T_1 = 2\sqrt{2h_1/g}.$$

In the same way

$$\Delta T_2 = 2\sqrt{2h_2/g},$$

$$\Delta T_3 = 2\sqrt{2h_3/g},$$

and in general
$$\Delta T_n = 2\sqrt{2h_n/g}$$
where ΔT_n is the time interval between the $n-1$st and the nth impact.

If we write T_n as the total time from the $n=0$ impact (the first impact) until the nth impact (i.e., n more bounces after the first bounce), then

$$T_n = \Delta T_1 + \Delta T_2 + \Delta T_3 + \cdots + \Delta T_n$$
$$= 2\sqrt{\frac{2}{g}}\left(\sqrt{h_1} + \sqrt{h_2} + \sqrt{h_3} + \cdots + \sqrt{h_n}\right),$$

or

$$T_n = 2\sqrt{\frac{2}{g}}\left(\sqrt{c^2 h_0} + \sqrt{c^4 h_0} + \sqrt{c^6 h_0} + \cdots + \sqrt{c^{2n} h_0}\right)$$
$$= 2\sqrt{\frac{2h_0}{g}}(c + c^2 + c^3 + \cdots + c^n).$$

The expression in parentheses is a geometric series, easily summed to give

$$T_n = 2\sqrt{\frac{2h_0}{g}} \cdot c \cdot \frac{1-c^n}{1-c}.$$

Solving for g gives us the result we are after:

$$g = \frac{8h_0 c^2}{T_n^2} \cdot \left(\frac{1-c^n}{1-c}\right)^2. \tag{2.2}$$

2.2 Out in the Author's Garage

From the first part of the analysis we already know c for our ball, so all we need to do now to measure g is to drop the ball from a known height and, beginning with the first bounce, start a timer and measure how long it takes for n more bounces. What could be easier? And so, taking a yardstick, a stopwatch, and a bouncy rubber ball, I went out to my garage (and its concrete floor) one cold New England winter evening and measured the value of g in Lee, New Hampshire. First, with $h_0 = 36$ inches (the midpoint of the ball), I observed a rebound height of thirty inches, so (2.1) says

$$c = \sqrt{\frac{30}{36}} = 0.913.$$

Then, again using $h_0 = 36$ inches, I again dropped the ball and, at the instant of the first impact, I started my stopwatch. I counted off eight *more* bounces ($n = 8$) and, at the last impact, I stopped the watch. Observing $T_8 = 4.6$ seconds, the formula for g, (2.2), gave a value of

$$g = \frac{8 \cdot 36 \cdot (0.913)^2}{4.6^2} \cdot \left(\frac{1 - 0.913^8}{1 - 0.913}\right)^2 = 401 \text{ in.}/\text{s}^2 = 33.4 \text{ ft/s}^2.$$

The typical textbook value for g is 32.2 ft/s² (9.81 m/s²), and so I made a less than 4% error. Not too shabby for a sixty-six-year-old guy—my age, when I did this experiment—with poor eyesight and questionable reflexes, to say nothing about doing it while freezing in my garage. (This had to be luck! As I recall, my Physics 51 lab report was far less impressive in its estimate of g.)[3]

CP. P2.1:

Here are a couple of little challenge calculations—they are easy!—for your amusement. What is the total distance, up *and* down, traveled by our bouncing ball as it goes through an infinity of bounces? How long does it take the ball, from the instant you release it until it has completed that infinity of bounces, to come to rest? *Hint*: Neither answer is infinity.

Notes and References

1. The analysis in this discussion is based entirely on the work of J. G. Dodd, "Determination of g by a Bouncing Ball" (*American Journal of Physics*, April 1958, p. 268). It may amuse you to know that, all the while I was writing this discussion, the words of Bobby Vee's 1960 smash pop hit "Rubber Ball"

reverberated through my mind. Here are a few of the opening lines to get your feet tapping:

> (*Key of A major*)
> Rubber ball, I come bouncin' back to you,
> rubber ball, I come bouncin' back to you
> hoo-ah-ooh-ooh.
> I'm like a—rubber ball, baby that's all
> that I am to you (bouncy, bouncy) (bouncy, bouncy)
> Just a rubber ball 'cuz you think you
> can be true to two (bouncy, bouncy) (bouncy, bouncy).
> You bounce my heart around (You don't even put her down)
> And like a rubber ball, I come bouncin' back to you
> rubber ball, I come bouncin' back to you.

Ah, yes, one of the deep philosophical chants from the dawning of the Age of Aquarius. Undeniably dumb, it was, but very hard to sing correctly and very popular in its day.

2. In later discussions in this book, I have much more to say on the mathematical physics of air drag.

3. The measurement of T_n can be made more accurate using electronics (which, of course, makes the experiment more complicated): see G. Guercio and V. Zanetti, "Determination of the Gravitational Acceleration Using a Rubber Ball" (*American Journal of Physics*, January 1987, pp. 59–63). A nice discussion of the experimental errors and of the theoretical assumptions inherent in this experiment is presented by the authors. Dodd's much earlier (1958) paper—see note 1 again—is not cited, however.

Feynman's Infinite Circuit

> In physics, caution when taking infinite limits is often considered as a useless subtlety by students (not to speak of their teachers). Many interesting physical properties can however be missed because of the improper use of mathematical techniques.
> — H. Krivine and A. Lesne, "Phase Transition-Like Behavior in a Low-Pass Filter" (2003)

3.1 An Infinity of Resistors

Have you heard of the famous mathematical hotel that has an infinite number of rooms? It has the wonderful feature—unlike any real hotel I've ever tried to check into in a strange city at midnight without a reservation—of never being full. Even if you think it's full, there is always room for more. Here's why. Let's suppose it *is* full, i.e., there is an infinity of people in the hotel, with each individual person enjoying a private room and with each and every room having a person in it. That night, an infinity *more* people arrive at the front desk, each wanting his or her own private room, too. No problem! The front desk clerk simply moves the original person in room 1 to room 2, the original person in room 2 to room 4, the original person in room 3 to room 6, ... the original person in room n to room $2n$, and so on. All the original infinity of people have then been moved to the even-numbered rooms, and all the odd-numbered rooms (of which there is an infinite number) are now empty, available to that rather large crowd at the front desk.

There is nothing to prevent doing this again, either. That is, every single night, from tonight to the end of time, a new infinity of people

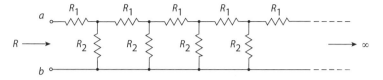

Figure 3.1. An infinite resistive ladder.

could arrive in town and, using the above procedure, all could be given a room of their very own! This little story should, by itself, be enough to convince you that infinity is not anything at all like an ordinary number. (The physicist George Gamow in his famous 1947 book *One, Two, Three ... Infinity* attributes this wonderful illustration of infinity to the great German mathematician David Hilbert [1862–1943].) Indeed, mathematicians do not count infinity as a number but rather recognize it as a (potentially rather slippery) concept. And it is not true that only mathematicians need to use great care when manipulating infinities; physicists need to be careful, as well. Let me give you a concrete example of this, from electric circuit theory.

Consider, first, the so-called *infinite ladder* circuit of Figure 3.1. The ladder is constructed from an infinite number, each, of two generally different-valued resistors, R_1 and R_2. (Ladder networks built from resistors, as well as other components, are important in electrical engineering, occurring in a multitude of useful electronic circuits.) The question here is, what is the net input resistance R "seen" when "looking into" the two terminals marked a and b at the left? Clearly, we won't make much progress with this question by applying Kirchhoff's circuit laws to each of the loops and nodes of the ladder—there is an *infinity* of each!

There is, however, a very clever way of actually taking advantage of the infinite extent of the circuit and making that work for us rather than against us. We start by redrawing Figure 3.1 as Figure 3.2. Because the ladder is infinite to the right from terminals a and b, it is also infinite to the right from the terminals marked a' and b'. Therefore, if we disconnect the leftmost stage of the ladder and then "look into" the terminals marked a' and b' we will still "see" an infinite ladder with input resistance R. In his famous three-volume work, *Lectures on Physics*, the great American physicist Richard Feynman (1918–1988) treated

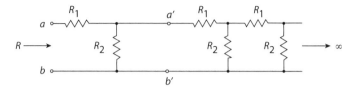

Figure 3.2. Figure 3.1 again.

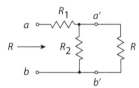

Figure 3.3. The *infinite* ladder is one stage, followed by itself.

this very problem with the explanation, "Surely, if we add one more section to an infinite network it is still the same infinite network."[1] So, our infinite ladder can be redrawn one last time, as in Figure 3.3.

From Figure 3.3 we can use the well-known formulas for combining series and parallel resistors to write

$$R = R_1 + R_2 \parallel R = R_1 + \frac{RR_2}{R+R_2}, \tag{3.1}$$

or

$$R^2 + RR_2 = RR_1 + R_1R_2 + RR_2,$$

or

$$R^2 - R_1R - R_1R_2 = 0,$$

or, using the quadratic formula to solve for R,

$$R = \frac{R_1 \pm \sqrt{R_1^2 + 4R_1R_2}}{2} = R_1 \frac{1 \pm \sqrt{1 + 4\frac{R_2}{R_1}}}{2}. \tag{3.2}$$

We know physically that the input resistance must be positive, and so we use the plus sign and not the negative one (that choice would give, incorrectly, a negative input resistance). Thus, at last,[2]

$$R = R_1 \frac{1 + \sqrt{1 + 4\frac{R_2}{R_1}}}{2}. \tag{3.3}$$

It is interesting to note that if $R_1 = R_2$, then $R = R_1 \frac{1+\sqrt{5}}{2}$, where $\frac{1+\sqrt{5}}{2} = 1.60803\cdots$ is the famous irrational number often called the *golden ratio*.

3.2 An Infinity of Reactances, and Recursion

Suppose now that, instead of building our infinite ladder from just resistors, we replace every R_1 with an inductance of value L and replace every R_2 with a capacitance of value C. Let the input signal to the ladder now be an ac voltage source with angular frequency ω radians per second (rad/s) (at the power outlets in your home, for example, the frequency—if you live in the United States—is $f = 60$ Hz, which is $\omega = 2\pi f = 377$ rad/s). The ac *impedance* (measured in ohms, just as with resistors) of each L and C is then given by $Z_1 = i\omega L$ and $Z_2 = 1/i\omega C$, respectively, where $i = \sqrt{-1}$. Substituting these expressions[3] into (3.2)—using the plus sign—we get, for the input impedance Z of our new infinite ladder made from just inductors and capacitors,

$$Z = \frac{i\omega L + \sqrt{-\omega^2 L^2 + 4\frac{L}{C}}}{2}. \tag{3.4}$$

Notice that as long as the expression under the radical in (3.4) is negative—as long as $-\omega^2 L^2 + 4\frac{L}{C} < 0$, which means $\omega > 2/\sqrt{LC}$—then Z is pure imaginary. This makes sense, too, as every component in the ladder has a pure imaginary impedance, and so it seems logical that the input impedance should be pure imaginary as well. But what if $\omega < 2/\sqrt{LC}$? Then Z is *complex* with a *positive* real part, i.e., Z behaves, in part, just like an ordinary resistor!

The reason I ended the last sentence with an exclamation mark is because the behavior of Z for $\omega < 2/\sqrt{LC}$ ought to strike you as quite odd. Resistors dissipate energy in electrical circuits, a fact easy to demonstrate by simply touching them: they are warm, even hot. Inductors and capacitors, however, do not dissipate electrical energy as heat energy but rather (temporarily) store it in either a magnetic field (inductors) or an electric field (capacitors). The electrical energy stored in such fields can always be recovered later, as electrical energy. So, there's our puzzle: how can a network, made only of inductors and capacitors, possibly have a real, positive, energy-dissipating component to its impedance? Notice carefully that if we had taken the *negative* root of the positive radical in (3.4), then matters would be even worse, as a negative resistance *generates* energy, and such a circuit, made only of inductors and capacitors, could exist only in Alice's wonderland.

As a start to explaining this puzzle, we might change our language to say resistors *transform* electrical energy in a way that makes it no longer available to us—ever—*as electrical energy*. This interpretation allows our original view of resistors turning electrical energy into heat energy, of course, but it now allows another possibility, too. With that little twist in mind, here's what Feynman says, specifically:

> But how can the circuit continuously absorb energy, as a resistance does, if it is made only of inductances and capacitances? *Answer*: Because there is an infinite number of inductances and capacitances, so that when a source is connected to the circuit, it supplies energy to the first inductance and capacitance, then to the second, to the third, and so on. In a circuit of this kind, energy is continually absorbed from the generator at a constant rate and flows constantly out into the network, supplying energy which is stored in the inductances and capacitances down the line.

It is the absorption of energy, *endlessly* (because the ladder is spatially *infinite*), that causes the mathematics to end up with a positive real part to the input impedance if $\omega < 2/\sqrt{LC}$. Feynman's explanation is a physically satisfying one, at least at first glance. When I was still teaching I would bring in the infinite *reactive* (so called because inductance and capacitance are lumped together, in tech-speak, as *reactances*) *ladder* and give my students Feynman's explanation. But one day I came across a fascinating paper that stopped me in my

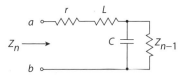

Figure 3.4. A finite-length resistive ladder.

tracks and made me rethink the problem.[4] The thesis of that paper was essentially this: let's not take our ladder as infinite right off the bat, as did Feynman, but rather start with a ladder having a finite number of stages (say, n), calculate the input impedance Z_n, and *then* let $n \to \infty$, i.e., calculate $Z = \lim_{n \to \infty} Z_n$. Further, let's not assume that each inductance is perfect, but instead let's add a bit of realism and insert a "small" resistance r in series with each L. We'll get Feynman's case when $r = 0$. (We should do the same for the capacitances—insert a "large" resistance in parallel with each C—but it turns out that would have no effect on our results but would significantly add to the pain of the mathematics. So, I won't do that.) Now, as shown in Figure 3.4, it is not difficult to calculate Z_n.

We have, from the usual parallel/series formulas for impedances,

$$Z_n = r + i\omega L + \frac{\frac{1}{i\omega C} Z_{n-1}}{\frac{1}{i\omega C} + Z_{n-1}}, \quad Z_1 = r + i\left(\omega L - \frac{1}{\omega C}\right),$$

or

$$Z_n = r + i\omega L + \frac{Z_{n-1}}{1 + i\omega C Z_{n-1}}.$$

Z_n is, of course, a complex quantity in general, so let's explicitly write its real and imaginary parts as R_n and I_n, respectively. Then, $Z_n = R_n + i I_n$, and so

$$\begin{aligned} Z_n &= r + i\omega L + \frac{R_{n-1} + i I_{n-1}}{1 + i\omega C (R_{n-1} + i I_{n-1})} \\ &= r + i\omega L + \frac{R_{n-1} + i I_{n-1}}{1 - \omega C I_{n-1} + i\omega C R_{n-1}}. \end{aligned}$$

Thus,
$$Z_n = r + i\omega L + \frac{(R_{n-1} + iI_{n-1})(1 - \omega C I_{n-1} - i\omega C R_{n-1})}{(1 - \omega C I_{n-1})^2 + \omega^2 C^2 R_{n-1}^2}. \quad (3.5)$$

If we equate imaginary parts on the left- and right-hand sides of (3.5), we get the recursion
$$I_n = \omega L + \frac{(1 - \omega C I_{n-1})I_{n-1} - \omega C R_{n-1}^2}{(1 - \omega C I_{n-1})^2 + \omega^2 C^2 R_{n-1}^2}, \quad I_1 = \omega L - \frac{1}{\omega C}. \quad (3.6)$$

And if we equate real parts on the left- and right-hand sides of (3.5), we get the recursion
$$R_n = r + \frac{(1 - \omega C I_{n-1})R_{n-1} + \omega C I_{n-1} R_{n-1}}{(1 - \omega C I_{n-1})^2 + \omega^2 C^2 R_{n-1}^2},$$

or
$$R_n = r + \frac{R_{n-1}}{(1 - \omega C I_{n-1})^2 + \omega^2 C^2 R_{n-1}^2}, \quad R_1 = r. \quad (3.7)$$

We can significantly reduce the complexity of the expressions for R_n and I_n if we normalize them. As a start on doing this, let's define the frequency $\omega_0 = 1/\sqrt{LC}$, and write an arbitrary frequency as $\omega = h\omega_0$ where h is a dimensionless number ($h \geq 0$). Now we can write[5]

$$I_1 = h\omega_0 L - \frac{1}{h\omega_0 C} = h\frac{1}{\sqrt{LC}}L - \frac{1}{h\frac{1}{\sqrt{LC}}C} = h\sqrt{\frac{L}{C}} - \frac{1}{h\sqrt{\frac{C}{L}}}$$

$$= h\sqrt{\frac{L}{C}} - \frac{1}{h}\sqrt{\frac{L}{C}} = \sqrt{\frac{L}{C}}\frac{h^2 - 1}{h}.$$

If we restrict ourselves for now to Feynman's case and set $r = 0$, then we also have $R_1 = 0$. Equation (3.7) then tells us immediately that, for Feynman's case, $R_n = 0$ for all n and for all ω. That is, our recursive formulation of the pure reactive ladder (remember, we have set $r = 0$) does not have a positive real part to the input impedance, no matter how long we make the ladder. Mathematically, (3.7) is saying $\lim_{n\to\infty} R_n \neq$ real part of Z as given by (3.4) for $\omega < 2/\sqrt{LC}$. Something is clearly not right here, and if we look at (3.6), for Feynman's case of $r = 0$, we get an even more dramatic indication of that.

Since $R_n = 0$, as we just showed, (3.6) reduces to

$$I_n = \omega L + \frac{I_{n-1}}{1 - \omega C\, I_{n-1}}, \quad I_1 = \sqrt{\frac{L}{C}}\frac{h^2 - 1}{h}.$$

Since $\omega = h\omega_0$, we then have

$$I_n = h\omega_0 L + \frac{I_{n-1}}{1 - h\omega_0 C\, I_{n-1}} = h\frac{1}{\sqrt{LC}}L + \frac{I_{n-1}}{1 - h\frac{1}{\sqrt{LC}}C\, I_{n-1}}$$

$$= h\sqrt{\frac{L}{C}} + \frac{I_{n-1}}{1 - h\sqrt{\frac{C}{L}}I_{n-1}} = \frac{h\sqrt{\frac{L}{C}} - h^2 I_{n-1} + I_{n-1}}{1 - h\sqrt{\frac{C}{L}}I_{n-1}},$$

or

$$I_n = \frac{h\sqrt{\frac{L}{C}} - (h^2 - 1)I_{n-1}}{1 - h\sqrt{\frac{C}{L}}I_{n-1}}, \quad I_1 = \sqrt{\frac{L}{C}}\frac{h^2 - 1}{h}. \tag{3.8}$$

So, for example,

$$I_2 = \frac{h\sqrt{\frac{L}{C}} - (h^2 - 1)I_1}{1 - h\sqrt{\frac{C}{L}}I_1} = \frac{h\sqrt{\frac{L}{C}} - (h^2 - 1)\sqrt{\frac{L}{C}}\frac{h^2-1}{h}}{1 - h\sqrt{\frac{C}{L}}\sqrt{\frac{L}{C}}\frac{h^2-1}{h}} = \sqrt{\frac{L}{C}} \cdot \frac{h - \frac{(h^2-1)^2}{h}}{1 - (h^2 - 1)}.$$

That is, just as $I_1 = k_1\sqrt{L/C}$, we see that $I_2 = k_2\sqrt{L/C}$, where k_1 and k_2 are functions of just h. In general, in fact, you should be able to see that we can write $I_n = k_n\sqrt{L/C}$. Inserting this into (3.8), we have

$$k_n\sqrt{\frac{L}{C}} = \frac{h\sqrt{\frac{L}{C}} - (h^2 - 1)k_{n-1}\sqrt{\frac{L}{C}}}{1 - h\sqrt{\frac{C}{L}}\sqrt{\frac{L}{C}}k_{n-1}} = \sqrt{\frac{L}{C}} \cdot \frac{h - (h^2 - 1)k_{n-1}}{1 - hk_{n-1}}.$$

Or, cancelling the $\sqrt{L/C}$ factors on both sides, we have the remarkably simple recursion that $I_n = k_n\sqrt{L/C}$, where

$$k_n = \frac{h - (h^2 - 1)k_{n-1}}{1 - hk_{n-1}}, \quad k_1 = \frac{h^2 - 1}{h}. \tag{3.9}$$

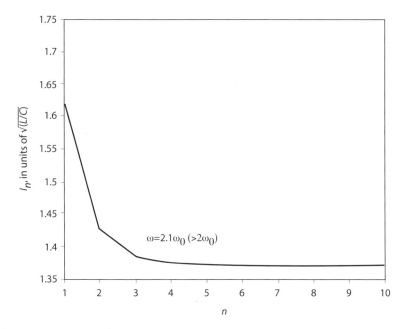

Figure 3.5. Feynman's $r = 0$. Convergence? Yes!

This recursion is very easy to code for a computer, and it is quite interesting to see what (3.9) says. Remember, (3.9) is strictly for Feynman's case of a ladder constructed from pure reactances ($r = 0$), with an input impedance of $Z_n = iI_n$. That is, $R_n = 0$, where n is the number of stages in a *finite* ladder (which is the only difference we have from Feynman's result of (3.4), which is for $n = \infty$). The result produced by (3.9) gives I_n in units of $\sqrt{L/C}$ for any frequency we wish, simply by specifying the value of h.

3.3 Convergence—or Not?

Feynman's analysis makes physical sense, as stated earlier, if $\omega > 2\omega_0$, in that (3.4) gives a pure imaginary result for the input impedance. And, indeed, if we plot I_n versus n at any frequency $\omega > 2\omega_0$, we'll get a plot like that shown in Figure 3.5 (where $h = 2.1$). There we see I_n converging to about $1.37\sqrt{L/C}$. In fact, if you insert $\omega = h\omega_0$ into (3.4),

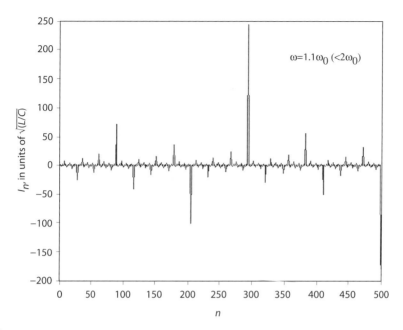

Figure 3.6. Feynman's $r = 0$. Convergence? No!

you should be able to show that Feynman's input impedance is

$$Z = i\sqrt{\frac{L}{C}} \cdot \frac{h + \sqrt{h^2 - 4}}{2},$$

which, for $h = 2.1$, says $I = 1.3701562\sqrt{L/C}$. So, in this case, the recursive formulation of Z_n *does* converge to Feynman's result. The convergence is fast, too; $k_{10} = 1.3701584$. But what if $\omega < 2\omega_0$? Figure 3.6 shows I_n versus n for $\omega = 1.1\omega_0$ ($h = 1.1$), and we see that now I_n is all over the place and certainly is *not* converging. We could have seen this sort of behavior without writing any computer code at all if, instead of setting $\omega = 1.1\omega_0$, I had set $\omega = \omega_0$ ($h = 1$). Then (3.9) easily gives $k_1 = 0$, $k_2 = 1$, $k_3 = \infty$, $k_4 = 0$, $k_5 = 1$, and so on, endlessly. So, for $\omega = \omega_0$ we see that Feynman's claim that "adding one more stage to an infinite ladder makes no difference" can't be right. Each new stage, at least for $\omega = \omega_0$, makes a lot of difference in what we see for the input impedance of the ladder. And Figure 3.6 strongly suggests that this lack of convergence is not confined to just $\omega = \omega_0$;

any frequency $\omega < 2\omega_0$ will result in a failure of the I_n recursion to converge.

What are we to make of this? Was the great Feynman *wrong*? Well, sort of, but not really. When actual ladder networks are made by cascading many LC stages (built from actual electronic components) together, the behavior derived by Feynman is indeed what is actually observed. The explanation of this apparent contradiction will show you the power of mathematics in a dramatic way. What we need to do is to go back to the full-blown, general recursions of (3.6) and (3.7) and *not* set $r = 0$. Instead, let's set r to some small, non-zero value, say $r = s\sqrt{L/C}$, where $s \leq 0.01$ (see note 5 again). I'll let you do the easy algebra, but if you repeat the analysis that I did above for Feynman's $r = 0$ case for the $r > 0$ case, you should be able to convince yourself that $R_n = g_n\sqrt{L/C}$, and thus arrive at the two coupled recursions

$$k_n = h + \frac{(1 - hk_{n-1})k_{n-1} - hg_{n-1}^2}{(1 - hk_{n-1})^2 + h^2 g_{n-1}^2}, \quad k_1 = \frac{h^2 - 1}{h} \qquad (3.10)$$

and

$$g_n = s + \frac{g_{n-1}}{(1 - hk_{n-1})^2 + h^2 g_{n-1}^2}, \quad g_1 = s. \qquad (3.11)$$

These recursions are just as easy to code as was (3.9), and Figure 3.7 shows that now, with $r \neq 0$, I_n and R_n do converge to values close to Feynman's result of (3.4):

$$Z = \frac{ih\omega_0 L + \sqrt{-h^2 \omega_0^2 L^2 + 4\frac{L}{C}}}{2} = \frac{ih\frac{1}{\sqrt{LC}}L + \sqrt{-h^2 \frac{1}{LC}L^2 + 4\frac{L}{C}}}{2}$$

$$= \frac{ih\sqrt{L/C} + \sqrt{L/C}\sqrt{4 - h^2}}{2} = \sqrt{L/C}\left(\frac{1}{2}\sqrt{4 - h^2} + i\frac{1}{2}h\right)$$

$$= \sqrt{L/C}(R + iI).$$

That is,

$$R = \frac{1}{2}\sqrt{4 - h^2}$$

and

$$I = \frac{1}{2}h.$$

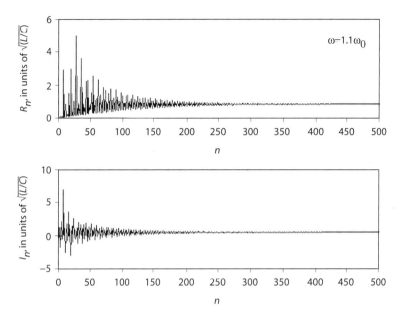

Figure 3.7. $r = 0.01\sqrt{L/C}$.

With $h = 1.1$ (as in Figure 3.6), for example, Feynman's impedance is given by $R = 0.8352$ and $I = 0.55$. Figure 3.7 is for $r = 0.01\sqrt{L/C}$ ($s = 0.01$, which gives $g_{500} = 0.8372$ and $k_{500} = 0.5449$), and Figure 3.8 repeats the calculations for the smaller value of $r = 0.005\sqrt{L/C}$ ($s = 0.005$, which gives $g_{500} = 0.7793$ and $k_{500} = 0.493$). We still have convergence using the smaller r, but it takes a larger n to achieve it. In the limit as $r \to 0$, we approach the case of Figure 3.6 and $r = 0$, but *at $r = 0$ convergence will not occur.*

Feynman's thesis that "adding one more stage to an infinite ladder makes no difference" works for an all-resistor ladder because there we have just one limiting operation, that of $n \to \infty$. But what we have encountered in the pure reactive ladder problem are *two* limiting operations, $n \to \infty$ and $r \to 0$. The lesson to be taken from our results is that *it makes a difference* in which order we take these two limits, i.e., the two limiting operations do not commute.[6] A mathematician would express this fact by writing

$$\lim_{n \to \infty} \lim_{r \to 0} Z_n \neq \lim_{r \to 0} \lim_{n \to \infty} Z_n.$$

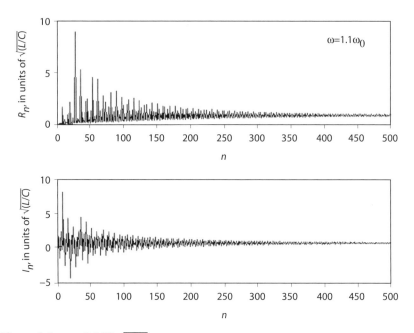

Figure 3.8. $r = 0.005\sqrt{L/C}$.

Feynman's original, not always correct analysis is the first double limit on the left, while the always correct sequence of limiting is the second one, on the right. Feynman, therefore, was not correct from a purely mathematical point of view. Even Homer (as historians like to put it) nodded off now and then. I don't think Feynman would have been much concerned about this. As the mathematical physicist Freeman Dyson said of his friend (in his 1972 Gibbs Lecture to the American Mathematical Society), "Mathematical rigor is the last thing Feynman was ever concerned about." And, if we remember that any actual inductance always has *some* resistance, Feynman was, as engineers like to put it, correct "for all practical purposes."

3.4 Three More Infinite, All-Resistor Networks

Although Feynman's infinite resistor ladder network has a straightforward explanation, and we didn't run into real trouble until we added reactances to the circuit, this is not to say infinite connections of just

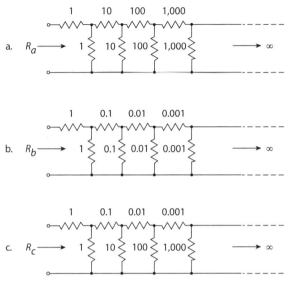

Figure 3.9. Three infinite resistor ladders.

resistors alone can't have their difficulties, too. I'll show you here three curious circuits that raise curious questions, two with answers and one that, well, you'll see!

Consider the three infinite resistor ladder networks in Figure 3.9. They look somewhat the same, as all have the same connection pattern—what electrical engineers call the *topology* of the network—and all have a first section at the far left made from two one-ohm resistors. In Figure 3.9a the resistor values thereafter all increase by a factor of 10 with each new section as we move to the right, and in Figure 3.9b the resistor values all decrease by a factor of 10 with each new section. In Figure 3.9c, however, we have a sort of mixture of the first two ladders. In this third ladder the horizontal resistor values decrease by a factor of 10 but the vertical resistor values increase by a factor of 10.

The first two ladders can still be easily analyzed, with Feynman's trick, to compute the input resistances R_a and R_b: in the first ladder we have the first section followed by an infinite ladder with an input resistance of $10R_a$, and in the second ladder we have the first section followed by an infinite ladder with an input resistance of $\frac{1}{10}R_b$. This is because of a general theorem in circuit theory that says if we have an all-resistor ladder with an input resistance of R, and if we then multiply every resistor value in the ladder by the same factor k, then the new input resistance is kR. (Think about this as you read on. The proof is not difficult, and it's a Challenge Problem at the end of this discussion.) If you use this theorem, you should easily be able to show that

$$R_a = \frac{19 + \sqrt{401}}{20} = 1.95 \text{ ohms}$$

and

$$R_b = -4 + \sqrt{26} = 1.099 \text{ ohms}.$$

The theorem doesn't help us with the computation of R_c, however, as *all* the resistor values in that ladder are *not* multiplied by the same factor.[7] What do we do now?

Suppose the ladder for R_c had all those horizontal resistors (except for the first one) replaced with zero resistance, that is, all the vertical resistors are now simply in parallel. Does it seem reasonable to you that the new input resistance would then be *smaller* than R_c? Thus, R_c must be greater than the leftmost horizontal resistance of 1 ohm in series with the parallel combination of all the vertical resistors. Mathematically, then,

$$R_c > 1 + 1 \parallel 10 \parallel 100 \parallel 1000 \parallel \cdots = 1 + \frac{1}{1 + \frac{1}{10} + \frac{1}{100} + \frac{1}{1000} + \cdots}$$

$$= 1 + \frac{1}{\frac{10}{9}} = 1 + \frac{9}{10} = 1.9 \text{ ohms}.$$

Also, does it seem reasonable to you that if all we keep of the original ladder is the first two one-ohm resistors, that is, if we disconnect all the rest of the ladder that is in parallel with the first vertical resistor, then R_c must be *less* than those first two resistors (now in series)? That is, it must be that

$$1.9 \text{ ohms} < R_c < 2 \text{ ohms}.$$

This double inequality bounds R_c, but what *is* R_c? Well, you think about that; it's another Challenge Problem!

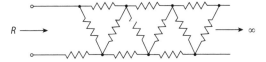

Figure 3.10. A challenge circuit.

CP. P3.1:

Find, for the infinite circuit (made entirely from one-ohm resistors) shown in Figure 3.10, the input resistance R. If you look at the circuit long enough, you may be able to spot a certain symmetry to it that will let you derive the answer without having to write even a single equation.

CP. P3.2:

In the theory of quantum electrodynamics (far, far beyond what we will talk about in this book), physicists run into a very curious mathematical problem in which a physical quantity (that means it has a *finite* value) ends up expressed as the difference of two individually *infinite* quantities. The mathematical technique used to handle this situation has been given the special name of *renormalization*, which sounds very elegant and elite, but in fact the very same sort of thing occurs in a much less esoteric setting, at the level of this book. Imagine an infinitely long straight wire with a uniform electrical charge density of q coulombs per meter, with the wire defining the z-axis of a three-dimensional coordinate system. The electric potential at every point in space around the wire is then shown in elementary electromagnetic field theory courses to be given by

$$V(x,y,z) = \int_{-\infty}^{\infty} \frac{q}{4\pi \epsilon_0 \sqrt{x^2+y^2+(z-z')^2}} \, dz' + C,$$

where ϵ_0 is a physical constant called the electric permittivity of space and C is an arbitrary constant. We can adjust C to be anything we wish, depending on how we define the *surface of zero potential* in three-dimensional space. Suppose we take that surface to be the cylindrical surface centered on the wire with radius R, that is, as the surface defined by $x^2 + y^2 = R^2$. Then

$$C = -\int_{-\infty}^{\infty} \frac{q}{4\pi\epsilon_0 \sqrt{R^2 + (z-z')^2}} dz'.$$

Writing $x^2 + y^2 = r^2$ in general, we have

$$V(x, y, z) = \frac{q}{4\pi\epsilon_0} \left[\int_{-\infty}^{\infty} \frac{dz'}{\sqrt{r^2 + (z-z')^2}} - \int_{-\infty}^{\infty} \frac{dz'}{\sqrt{R^2 + (z-z')^2}} \right].$$

You can see that the integrand in each integral behaves, as $|z'| \to \infty$, like $\frac{1}{|z'|}$. That means each integral is individually logarithmically divergent, yet their difference, $V(x, y, z)$, is finite. Prove this claim by direct evaluation of $V(x, y, z)$. (Historical note: This very same issue puzzled Isaac Newton at the end of the seventeenth century, when he wondered about the distribution of stars throughout the universe. If there is a finite number of stars in a finite volume of space, then there would, of course, have to be stars at the outermost reaches of that volume. Those remote stars would "feel" an imbalance of gravitational pull from the other stars that would tend to pull the outer stars inward. Eventually, such an arrangement would collapse. The only way out, in Newton's mind, was an infinity of stars in an infinite universe. That way each star would "feel" an infinite pull in every direction, and all those infinities would exactly cancel, to result in a finite pull of zero.)

CP. P3.3:

Prove, for an infinite resistor ladder circuit, with each resistor of any value, that if *every* resistor is multiplied by the same factor k, then the input resistance is also multiplied by the same factor.

CP. P3.4:

What is the exact value of R_c, the input resistance of the infinite ladder of Figure 3.9c?

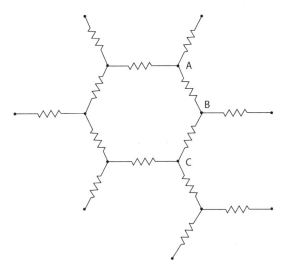

Figure 3.11. An infinite resistor honeycomb network.

CP. P3.5:

In Figure 3.11 a portion of an infinite resistor network is shown that is very different from a ladder. It is an infinite honeycomb of hexagons, with each edge of each hexagon a one-ohm resistor. What is the resistance between nodes A and B? Between A and C? This problem requires a different sort of approach than we've used in our discussion of ladders, but see what you can do with it. If you have trouble, wait until you get to Discussion 17, where I'll show you how to attack this sort of infinite network.

Notes and References

1. R. P. Feynman, R. B. Leighton, and M. L. Sands, *The Feynman Lectures on Physics*, vol. 2, (New York: Addison-Wesley, 1964, chap. 22, p. 13). As you'll soon see, it was with this apparently benign, "obviously true" observation that Feynman stumbled. Feynman took several electrical engineering courses while an undergraduate at MIT (1935–1939), and perhaps it was then that he first came across this problem.

2. This trick is certainly quite old; I have traced it back to at least as early as 1887, when it appeared in a paper on the analysis of telephone cables by the English electrical engineer Oliver Heaviside (see my biography, *Oliver*

Heaviside, [Baltimore: Johns Hopkins University Press, 2002, pp. 230–232]). The mathematically sophisticated way to analyze the infinite ladder (also called a *periodic circuit*, because of its spatially repetitive structure) is to write Kirchhoff's laws for an arbitrary three consecutive stages (say, the $n-1$st, the n th, and the $n+1$st), which leads to a second-order difference equation. Then, with the physically appropriate boundary conditions (here's what that means: to measure the input resistance we'll apply an ideal [zero internal resistance] one-volt dc source to the input terminals, and of course the node voltages and loop currents at "infinity" are clearly zero), that equation is solved with the aid of what mathematicians and electrical engineers alike call the z-*transform*. This is completely analogous to the use of the Laplace transform to solve differential equations. When I was a first-year graduate student in electrical engineering at Caltech (1962–1963) the infinite resistor ladder network was on one of my exams, with the professor's intent being that students use the z-transform. I was familiar (since high school) with the trick used in the text, however, and in the interest of saving time for other questions that were giving me more trouble, I used it to quickly answer the question. I recall my surprise at the professor being surprised by what I did, even though, even as I took my exam, his Caltech colleague Professor Feynman—see note 1—had already given a lecture on it to the sophomore class. (My professor was a nice guy. Even though I didn't solve the problem the "right" way, he still gave me full credit.)

3. We have $i = \sqrt{-1}$ in the expressions for the ac impedances of the inductors and the capacitors because the ac voltage drop across an inductor or a capacitor is 90 degrees out of phase with the ac current through the component. If you remember the geometrical meaning of i, then you'll appreciate that i, is just what we need, mathematically, to represent a 90-degree phase difference. And if you don't remember the geometrical meaning of i, see any one of my three books, *An Imaginary Tale* (Princeton, NJ: Princeton University Press, 1998, 2007), *The Science of Radio* (New York: Springer, 2001), or *Doctor Euler's Fabulous Formula* (Princeton, NJ: Princeton University Press, 2006) which will tell you everything you need to know.

4. S. J. van Enk, "Paradoxical Behavior of an Infinite Ladder Network of Inductors and Capacitors" (*American Journal of Physics*, September 2000, pp. 854–856).

5. Notice that $\sqrt{L/C}$ has the units of ohms (that is, it is an impedance, which of course I_1 is). This is most easily seen by simply observing that, in the case of $r = 0$, $Z_1 Z_2 = (i\omega L)(1/i\omega C) = L/C$, and of course $Z_1 Z_2$ has the units of ohms squared. So, $\sqrt{L/C}$ has the units of ohms. The radical $\sqrt{L/C}$ therefore

serves as a natural scale of impedance in our ladder, and I'll (somewhat arbitrarily) take any impedance (e.g., the inductor resistance r) that is less than $0.01\sqrt{L/C}$ to be "small."

6. H. Krivine and A. Lesne, "Phase Transition-Like Behavior in a Low-Pass Filter" (*American Journal of Physics*, January 2003, pp. 31–33).

7. See, for example, Armen H. Zemanian, *Infinite Electrical Networks* (New York: Cambridge University Press, 1991).

Air Drag— A Mathematical View

> My opinion is, therefore, that under the circumstances which occur in nature, the acceleration of any body falling from rest reaches an end, and that the resistance of the medium finally reduces its speed to a constant value which is thereafter maintained.
> — Galileo (1638), introducing the concept of *terminal speed*

4.1 Air Drag Treated Broadly

One of the many anecdotal stories told to illustrate the mathematical genius of the Hungarian-born American mathematician John von Neumann (1903–1957) is the following:

> Two bicyclists are 20 miles apart and head toward each other at 10 miles per hour each. At the same time a fly traveling at a steady 15 miles per hour starts from the front wheel of the northbound bicycle. It lands on the front wheel of the southbound bicycle, and then instantly turns around and flies back, and after next landing instantly flies north again. Question: What total distance did the fly cover before it was crushed between the two front wheels?

This problem was allegedly put to von Neumann at a cocktail party, supposedly to amuse onlookers, who wanted to watch the great man dance from foot to foot (an eccentricity said to have been his habit) as he thought through an analysis. The story then goes on to say

> The slow way of answering is to calculate the distance that the fly travels on its first trip to the southbound wheel, then the distance it travels on its next trip to the northbound wheel, and finally to sum the infinite series so obtained. It is extraordinary how many mathematicians can be fooled into doing that long sum. The short way is to note that the bicycles will meet in exactly one hour, by which time the 15-miles-per-hour fly must have covered 15 miles. When the question was put to [von Neumann] he answered immediately, "15 miles." "Oh, you've heard the trick before," said the disappointed questioner. "What trick?" asked the puzzled [von Neumann]. "I simply summed the infinite series."[1]

I do like this story, whether true or not, because it makes the point that it is not always necessary (or even desirable) to immediately apply high-powered mathematics to a problem. Often a little preliminary thought will do as well, or even better. With that in mind, here's a physics question as a more challenging example than the von Neumann fly/bicycle problem. A stone with mass m is projected upward from the ground with an initial vertical speed of v_i. Question 1: When the stone falls back to Earth, does it hit the ground with final speed $v_f = v_i$? Question 2: Is the ascent time t_{up} equal to, more than or less than the descent time t_{down}? For both questions I'll assume there is air drag, but I'll make no assumptions about its precise nature other than being *physically plausible*.[2] I'll also assume that the entire trajectory of the stone is sufficiently local that we can take the acceleration of gravity as everywhere constant. We could, of course, start by writing the differential equations of motion for a mass moving up (and then down) in a constant gravitational field, but here's a better, *physical* way to quickly arrive at the answers without writing even a single equation.

At the start of its upward motion the energy of the stone is entirely kinetic (I'm taking the ground as the zero reference level for potential energy). The stone on its ascent loses energy because of the resistive air drag. It therefore arrives (with zero speed) at its maximum height with all its energy now as potential energy, and that energy is *less* than the initial kinetic energy the stone was given at launch. On its return trip it loses even more energy, again because of air drag. Now, when it hits the ground the stone has, once again, *zero potential* energy. Since its energy is once again all kinetic energy (which, it is now clear, is less than its initial kinetic energy), then (obviously!) $v_f < v_i$. That's it, the entire analysis for Question 1, and it shows that the answer $v_f < v_i$ is true no matter how the air drag may vary with the stone's instantaneous speed (just as long as that variation is physically plausible). Notice—no equations! Not one.

Question 2 is just as easy to argue. First, notice the (obvious, I hope) fact that the distance traveled upward by the stone equals the distance traveled downward. Now, if you look at the analysis for Question 1, you should see that the potential energy of the stone at any given height is the same whether the stone is going up or coming down, and this is so whether there is air drag or not. Thus, as the total energy of the stone at any given height is always *less* coming down than going up, it follows that the kinetic energy coming down, at every given height, must be less than when going up; this simply says that the *speed* of the stone is, at any height, less coming down than when going up. So, we have a stone traveling the same distance up as it does down, but moving slower at every point coming down than it is when going up. Thus, coming down must take longer than going up, i.e., $t_{down} > t_{up}$. Again, this conclusion holds no matter how the air drag may vary (see note 2 again) with the stone's instantaneous speed, and again, no equations.

With that settled, let me add a tiny new twist to things. Instead of comparing the times of ascent and descent when there is air drag in both cases, let's compare the ascent time when there is *zero* air drag ($t_{up/z}$) with the ascent time when there *is* air drag ($t_{up/d}$). We'll assume that the upward launch speed is the same in both cases. I will need to write some mathematics now but, perhaps surprisingly, not very much. I'll start by asking you to accept that the maximum height reached by the stone when there is zero air drag (h_z) is greater than when there is air drag (h_d), i.e., that $h_z > h_d$. I think this should be immediately obvious, but if not, just remember the earlier argument about air drag as a mechanism of energy loss. So, here's Question 3: Is $t_{up/z}$ less than, equal to, or greater than $t_{up/d}$? What makes this question particularly interesting is that while $h_z > h_d$, it is also true that with zero air drag the stone will, at every height on the upward trip, be moving faster than it would if there were air drag. That is, with no air drag the stone has farther to travel, but it is also always moving faster (compared to the case of air drag), so it might now seem that we must know something about how the air drag varies with the stone's speed to determine which effect, increased distance or increased speed, dominates. But, as it turns out, the answer is again, for any physically plausible drag force law, independent of such details. Here's why.

Let's write $v_d(t)$ and $v_z(t)$ for the speeds of the stone with and without air drag, respectively, m for the mass of the stone, and $f(t)$ for the air drag force on the stone. I'll use the convention that positive forces are in the downward direction. (Remember, during an ascent the gravitational and air drag forces are in the same direction—downward—while during descent they are in opposite directions.) From Newton's second law of motion we have the net force on the stone as the time-rate of change of the stone's momentum.[3] So, for the two cases of zero air drag and air drag we have, respectively, for an upward-moving object,

$$mg = -\frac{d}{dt}(mv_z)$$

and

$$mg + f = -\frac{d}{dt}(mv_d).$$

The minus signs on the right in these two expressions are there because the left-hand sides are positive, but the rate of change of momentum during an ascent is negative. As $f \to 0$, of course, the two equations become one and the same. Integrating, we have

$$\int_0^{t_{up/z}} mg\, dt = -\int_0^{t_{up/z}} d(mv_z(t)) = \text{total momentum change with zero drag}$$

and

$$\int_0^{t_{up/d}} (mg + f)\, dt = -\int_0^{t_{up/d}} d(mv_d(t)) = \text{total momentum change with drag}.$$

Now, while even though $v_z(t) \neq v_d(t)$ in general, it must be true that

$$-\int_0^{t_{up/z}} d(mv_z(t)) = -\int_0^{t_{up/d}} d(mv_d(t)),$$

because each integral is the *total change* in the momentum of the stone from initial launch (zero height, with maximum momentum) to maximum height (where $v_z(t) = v_d(t) = 0$ and so zero momentum), which is the same whether there is air drag or not. From this it immediately follows that

$$\int_0^{t_{up/z}} mg\,dt = \int_0^{t_{up/d}} (mg+f)\,dt,$$

or

$$mgt_{up/z} = mgt_{up/d} + \int_0^{t_{up/d}} f\,dt,$$

or

$$t_{up/z} = t_{up/d} + \frac{1}{mg} \cdot \int_0^{t_{up/d}} f\,dt.$$

Since it is clear that $\int_0^{t_{up/d}} f\,dt > 0$ for any physically plausible drag force law (see note 2 again), it follows that $t_{up/z} > t_{up/d}$. That is, even though the stone is always moving faster if there is zero air drag than if air drag is present, the increased upward distance the stone travels is enough to dominate and result in an *increase* in the required upward travel time to reach maximum height.

By now you are probably beginning to think *any* question we might ask about our up-and-down stone is going to turn out to be independent of the specific details of the air drag force law. Well, that's not so. To show you that, suppose we now ask Question 4: How does the *total* up-and-down time for the case of zero air drag compare to the *total* up-and-down time when there is air drag? Again, we'll take the initial launch energy (launch speed) to be the same in both cases. The answer, as you'll see, does depend on the air drag force law. By assuming two different force laws, both physically plausible, we can have the zero drag case for either the larger *or* the smaller total up-and-down time. Here's how we can show this, and now I will have to write a fair amount of mathematics.

When there is zero air drag, matters are simple. The total time for the stone's motion when there is zero air drag is simply $t_{up/z} + t_{down/z}$, where it should be clear that $t_{up/z} = t_{down/z}$. Since $t_{up/z} = v_i/g$ when there is zero air drag, then the total up-and-down time for the stone in this case is $2v_i/g$. Matters when air drag is present are just a bit more complicated. When we launch the stone upward, the equation of motion (as the stone rises), when there is air drag, is

$$m\frac{dv}{dt} = -mg - f(v), 0 \le t \le t_{up/d},$$

and when the stone is falling the equation of motion when there is air drag is

$$m\frac{dv}{dt} = mg - f(v), t_{up/d} \le t \le t_{up/d} + t_{down/d}.$$

That is,

$$dt = \frac{dv}{-g - \frac{1}{m}f(v)}, 0 \le t \le t_{up/d} \quad (4.1)$$

and

$$dt = \frac{dv}{g - \frac{1}{m}f(v)}, t_{up/d} \le t \le t_{up/d} + t_{down/d}. \quad (4.2)$$

Integrating (4.1), remembering that v goes from v_i to 0 as t goes from 0 to $t_{up/d}$, gives

$$\int_0^{t_{up/d}} dt = \int_{v_i}^0 \frac{dv}{-g - \frac{1}{m}f(v)},$$

or

$$t_{up/d} = \int_0^{v_i} \frac{dv}{g + \frac{1}{m}f(v)}.$$

And integrating (4.2), remembering that v goes from 0 to v_f as t goes from $t_{up/d}$ to $t_{up/d} + t_{down/d}$, gives

$$\int_{t_{up/d}}^{t_{up/d}+t_{down/d}} dt = \int_0^{v_f} \frac{dv}{g - \frac{1}{m}f(v)},$$

or

$$t_{down/d} = \int_0^{v_f} \frac{dv}{g - \frac{1}{m}f(v)}.$$

The total time for the stone's motion is $t_{up/d} + t_{down/d}$, and so

$$\text{total time with drag} = \int_0^{v_i} \frac{dv}{g + \frac{1}{m}f(v)} + \int_0^{v_f} \frac{dv}{g - \frac{1}{m}f(v)}.$$

Let's define the ratio of the total time with drag to the total time with zero drag as τ, that is,

$$\tau = \frac{g}{2v_i} \left\{ \int_0^{v_i} \frac{dv}{g + \frac{1}{m}f(v)} + \int_0^{v_f} \frac{dv}{g - \frac{1}{m}f(v)} \right\},$$

or

$$\tau = \frac{1}{2v_i} \left\{ \int_0^{v_i} \frac{dv}{1 + \frac{1}{mg}f(v)} + \int_0^{v_f} \frac{dv}{1 - \frac{1}{mg}f(v)} \right\}. \quad (4.3)$$

If $\tau > 1$, then the stone's total motion, up and down, takes longer with drag than it does with no drag, and if $\tau < 1$, the stone's total motion up and down takes less time with drag than it does with no drag. I'll soon show you two different (but both physically plausible) drag force laws, one giving $\tau > 1$ and the other giving $\tau < 1$. We need one more technical result before I can do that, however.

We can write another expression that links v_i and v_f by remembering that, no matter what the drag force law may be, the distance traveled up equals the distance traveled down. Since differential distance is

$ds = v\,dt$, and using (4.1) and (4.2) for dt, we have

$$\int_{v_i}^{0} \frac{v\,dv}{-g - \frac{1}{m}f(v)} = \int_{0}^{v_f} \frac{v\,dv}{g - \frac{1}{m}f(v)},$$

or

$$\int_{0}^{v_i} \frac{v\,dv}{1 + \frac{1}{mg}f(v)} = \int_{0}^{v_f} \frac{v\,dv}{1 - \frac{1}{mg}f(v)}. \qquad (4.4)$$

One technical point, now, before continuing with the mathematics, about the integral on the right-hand side of (4.4). As the stone falls ever faster toward the ground, we know $f(v)$ monotonically increases (see note 2 again). But this cannot continue indefinitely, because eventually $f(v)$, which is directed upward during a fall, will become equal to the gravitational force mg (which is, of course, always directed downward), and so the net force on the stone will become zero. By Newton's second law (see note 3 again), that means the stone's acceleration will approach zero and its falling speed will approach a *constant speed*, called the *terminal speed* (v_T).[4] In any finite fall, of course, these are limits that are approached but never actually reached, and so the denominator in the integral on the right-hand side of (4.4) is always positive, i.e., $1 - \frac{1}{mg}f(v) > 0$ over the entire interval of integration, and we do not have to worry about a division-by-zero problem.

4.2 Air Drag Treated with Some Detail

We can now evaluate (4.3) and (4.4) by assuming various air drag force laws. For example, suppose we assume that $f(v) = kv^2$, $k > 0$. Then, writing $C = k/mg$, we have from (4.4) that

$$\int_{0}^{v_i} \frac{v\,dv}{1 + Cv^2} = \int_{0}^{v_f} \frac{v\,dv}{1 - Cv^2}, \; C > 0.$$

The constant C has a very simple physical interpretation. From the previous paragraph we have $f(v) = kv^2 = mg$ when $v = v_T$, the terminal speed. That is, $v_T^2 = mg/k = 1/C$, which says that $C = 1/v_T^2$. So far no mention has been made of units, but now we can see that if we measure all speeds in units of the terminal speed of the stone, then we'll have $v_T = 1$ and so $C = 1$. With this convention we can write

$$\int_0^{v_i} \frac{v\,dv}{1+v^2} = \int_0^{v_f} \frac{v\,dv}{1-v^2},$$

where it is understood that while $0 \leq v_i < \infty$ (we are ignoring the speed of light, and what relativity has to say about such matters!), we will have $0 \leq v_f < 1$ (when falling from rest the stone can at most only approach terminal speed). These two integrals are easily evaluated, either by looking in integral tables or by making the obvious change of variable $u = 1 \pm v^2$, and the result is (you should verify this!)

$$v_f = \frac{v_i}{\sqrt{1+v_i^2}}. \tag{4.5}$$

Notice, by the way, that (4.5) says $v_f < v_i$, a result that we already know holds for *any* physically plausible force law, not just the quadratic law we are explicitly working with here. Notice, too, that for *any* $v_i \geq 0$ we have $0 \leq v_f < 1$ just as we argued, above, must be so.

Inserting $f(v) = kv^2$ into (4.3) gives

$$\tau = \frac{1}{2v_i} \left\{ \int_0^{v_i} \frac{dv}{1+\frac{k}{mg}v^2} + \int_0^{v_f} \frac{dv}{1-\frac{k}{mg}v^2} \right\}$$

$$= \frac{1}{2v_i} \left\{ \int_0^{v_i} \frac{dv}{1+Cv^2} + \int_0^{v_f} \frac{dv}{1-Cv^2} \right\}$$

or, again because we have scaled all speeds to be in units of the terminal speed, we have $C = 1$ and so

$$\tau = \frac{1}{2v_i} \left\{ \int_0^{v_i} \frac{dv}{1+v^2} + \int_0^{v_f} \frac{dv}{1-v^2} \right\}.$$

And, again, both of these integrals are in standard math tables and the result is

$$\tau = \frac{\tan^{-1}(v_i) + \tanh^{-1}(v_f)}{2v_i}. \tag{4.6}$$

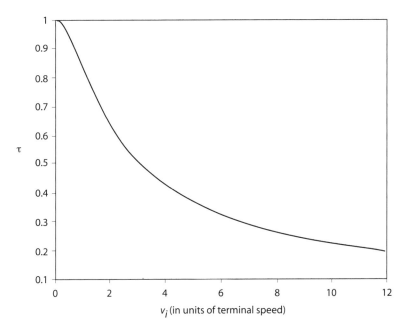

Figure 4.1. How τ varies with a quadratic force.

These two equations, (4.5) and (4.6), simply cry out for numerical evaluation. Given any launch speed v_i, we can use (4.5) to first calculate the associated landing speed v_f and then, inserting these two values of v_i and v_f into (4.6), we have τ. Doing that for a large number of values for v_i results in the plot of τ versus v_i shown in Figure 4.1. Notice that $\tau < 1$ always. That is, for a quadratic air drag force law the stone always takes less total time to return to Earth, no matter what the launch speed, as compared to the time the stone takes to travel up and down in the absence of air drag (given the same launch speed). The larger the launch speed, the more the difference in "air time." From Figure 4.1, for example, we see that if $v_i = 6$ (times the terminal speed) that $\tau = 0.33$ (the stone is in the air three times longer when there is no air drag as it is when there is air drag).

To finish my stated objective, I must now show that there does exist a physically plausible air drag force law that gives the opposite result of $\tau > 1$. Such a law is, in fact, $f(v) = k\sqrt{v}$. Inserting this law into (4.4) gives

$$\int_0^{v_i} \frac{v\,dv}{1+\frac{k}{mg}\sqrt{v}} = \int_0^{v_f} \frac{v\,dv}{1-\frac{k}{mg}\sqrt{v}}.$$

As before, I'll write v_T as the terminal falling speed of the stone, where $mg = f(v_T) = k\sqrt{v_T}$. Thus, $\sqrt{v_T} = mg/k = 1/C$, where $C = k/mg$, just as before. And, as before, if we take $C = 1$, then v_T becomes our unit speed, i.e., $v = 1$ means $v = v_T$. With this convention in mind, we have

$$\int_0^{v_i} \frac{v\,dv}{1+\sqrt{v}} = \int_0^{v_f} \frac{v\,dv}{1-\sqrt{v}}.$$

We can convert these integrals into something we can find in integral tables by making the change of variable $v = u^2$ (and so $dv = 2u\,du$). If you do this, you'll get

$$\int_0^{u_i} \frac{u^3\,du}{1+u} = \int_0^{u_f} \frac{u^3\,du}{1-u}, \qquad (4.7)$$

where $u_i = \sqrt{v_i}$ and $u_f = \sqrt{v_f}$. From standard integral tables we have the (perhaps surprisingly) complicated result

$$\int \frac{x^3}{ax+b}dx = \frac{(ax+b)^3}{3a^4} - \frac{3b(ax+b)^2}{2a^4} + \frac{3b^2(ax+b)}{a^4} - \frac{b^3}{a^4}\ln(ax+b).$$

On the left-hand side of (4.7) we use $a = b = 1$, and on the right-hand side we use $a = -1, b = 1$. So, (4.7) becomes

$$\left\{\frac{(u+1)^3}{3} - \frac{3(u+1)^2}{2} + 3(u+1) - \ln(u+1)\right\}\Big|_0^{u_i}$$

$$= \left\{\frac{(1-u)^3}{3} - \frac{3(1-u)^2}{2} + 3(1-u) - \ln(1-u)\right\}\Big|_0^{u_f}.$$

Or, expanding out all terms on both sides,

$$\left\{\frac{1}{3}u^3 - \frac{1}{2}u^2 + u - \ln(1+u)\right\}\Big|_0^{u_i} = \left\{-\frac{1}{3}u^3 - \frac{1}{2}u^2 - u - \ln(1-u)\right\}\Big|_0^{u_f},$$

or, at last, we arrive at the initially daunting transcendental equation

$$\frac{1}{3}u_i^3 - \frac{1}{2}u_i^2 + u_i - \ln(1+u_i) = -\frac{1}{3}u_f^3 - \frac{1}{2}u_f^2 - u_f - \ln(1-u_f), \quad (4.8)$$

where $u_i = \sqrt{v_i}$ and $u_f = \sqrt{v_f}$. It is probably clear to you that (4.8) isn't going to let us solve for v_f as an explicit function of v_i as we did in (4.5) for a quadratic force law. But that proves to be a mere shadow obstacle because, with a modern computer at hand, we can in a flash numerically solve[5] for the value of v_f that goes with any given value of v_i. And, if you think about it for a few seconds, that is all we really used (4.5) for, anyway—to give us the numerical value of v_f for any given value of v_i.

Continuing, from (4.3) and remembering that we are scaling all speeds such that when $v = v_T$, then $v = 1$, we have

$$\tau = \frac{1}{2v_i}\left\{\int_0^{v_i}\frac{dv}{1+\sqrt{v}} + \int_0^{v_f}\frac{dv}{1-\sqrt{v}}\right\},$$

or, again with $v = u^2$,

$$\tau = \frac{1}{2u_i^2}\left\{\int_0^{u_i}\frac{2u\,du}{1+u} + \int_0^{u_f}\frac{2u\,du}{1-u}\right\},$$

which reduces to (with the substitutions of $x = 1+u$ and $x = 1-u$ in the first and second integrals, respectively)

$$\tau = \frac{1}{u_i^2}\left\{\int_1^{1+u_i}\left(1-\frac{1}{x}\right)dx + \int_{1-u_f}^{1}\left(\frac{1}{x}-1\right)dx\right\}.$$

These two integrals are both well-known from freshman calculus, and so we have, by inspection, that

$$\tau = \frac{u_i - \ln(1+u_i) - \ln(1-u_f) - u_f}{u_i^2},$$

where u_i and u_f are as in (4.7) and (4.8). Thus,

$$\tau = \frac{\sqrt{v_i} - \ln(1+\sqrt{v_i}) - \ln(1-\sqrt{v_f}) - \sqrt{v_f}}{v_i}. \quad (4.9)$$

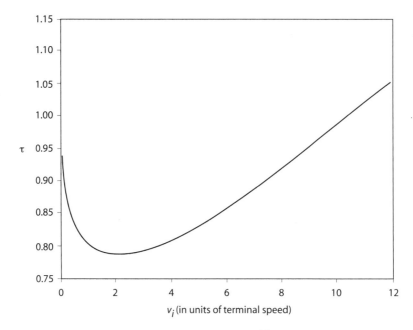

Figure 4.2. How τ varies for the force $f(v) = k\sqrt{v}$.

Using (4.8) and (4.9), Figure 4.2 shows τ plotted versus v_i for $0 \leq v_i \leq 12\, v_T$, and while τ at first does decrease as v_i increases, you can see that once v_i exceeds about twice the terminal speed, then τ begins to increase and exceeds unity for v_i greater than a little bit more than ten times the terminal falling speed.

I'll finish this discussion[6] by showing you that something you may have (perhaps unconsciously) concluded—that it is air drag that makes the difference between "going up" and "coming down"—is not true in general. That is, I'm going to show you a simple physical situation, with zero air drag, that nevertheless is asymmetrical in the up and down phases. Specifically, imagine a rocket that, from rest on the ground, accelerates straight up with constant acceleration a over the time interval $0 \leq t \leq t_1$, then coasts upward until its speed is reduced to zero by gravity, and then falls back to the surface. Let t_{up} be the time from blastoff to maximum altitude and t_{down} be the time from maximum altitude to impact with the surface. All of this is to occur in a vacuum (so we are not on Earth—perhaps it's the moon—but I'll still use g to denote the acceleration of gravity at the surface) and so there is no air drag. Finally, I'll assume that the entire rocket motion is always "close enough" to the surface that we can take g to be the acceleration of gravity experienced by the rocket at all times. How does t_{up} compare with t_{down}?

The speed of the rocket at the end of its upward acceleration is $v = at_1$, and so the rocket coasts upward for an additional time of at_1/g. That is,

$$t_{up} = t_1 + \frac{at_1}{g} = t_1\left(1 + \frac{a}{g}\right). \tag{4.10}$$

The maximum altitude reached by the rocket is H, where

$$H = \frac{1}{2}at_1^2 + \frac{1}{2}g\left(\frac{at_1}{g}\right)^2 = \frac{1}{2}at_1^2 + \frac{1}{2}\frac{a^2t_1^2}{g}.$$

Thus, the time to fall back to the surface is

$$t_{down} = \sqrt{\frac{2H}{g}} = \sqrt{\frac{at_1^2}{g} + \frac{a^2t_1^2}{g^2}} = t_1\sqrt{\frac{a}{g}\left(1 + \frac{a}{g}\right)}. \tag{4.11}$$

Now, from the expressions for t_{up} and t_{down} in (4.10) and (4.11), respectively, it is easy to show that $t_{up} > t_{down}$, always, for *any* positive values of t_1, a, and g. Here's how.

It is certainly true that

$$1 + \frac{a}{g} > 0,$$

and so

$$1 + 2\frac{a}{g} + \left(\frac{a}{g}\right)^2 > \frac{a}{g} + \left(\frac{a}{g}\right)^2,$$

and so

$$\left(1 + \frac{a}{g}\right)^2 > \frac{a}{g}\left(1 + \frac{a}{g}\right),$$

and so

$$t_1^2\left(1 + \frac{a}{g}\right)^2 > t_1^2\frac{a}{g}\left(1 + \frac{a}{g}\right),$$

or, at last, taking the square root of both sides of the inequality,

$$t_1\left(1 + \frac{a}{g}\right) = t_{up} > t_1\sqrt{\frac{a}{g}\left(1 + \frac{a}{g}\right)} = t_{down}.$$

That is, it *always* takes longer for the rocket to go up than it does to come down—and this is with no air drag present. I think this is perhaps not such an obvious result.

CP. P4.1:

Here's a challenge calculation for you. The analysis I just showed you is all mathematical. See if you can construct an all-physics (that means no equations) explanation for why $t_{up} > t_{down}$.

CP. P4.2:

Assuming Stokes's linear drag force law (see note 2 again), i.e., $f(v) = kv$, find an expression for τ. Is τ less than, equal to, or greater than one? *Hint*: You will not need a computer to do this.

Notes and References

1. The version I relate here is from the biography by Norman Macrae, *John von Neumann* (New York: Random House, 1992, pp. 10–11). My skepticism about this tale is caused, just a little bit, because I've heard more than one version of it ("a bee races back and forth between two onrushing locomotives..."), and I doubt von Neumann suffered through this stunt more than once.

2. Let v be the speed of a mass through a resistive medium that opposes the motion according to some force law $f(v)$. We'll take any drag force law as being physically plausible if (1) $f(v) > 0$ for $v > 0$, (2) $f(v)$ is monotonic increasing with v, and (3) $f(v) = 0$ for $v = 0$. Matters are actually much more involved than this, however. At the next level of complexity, physicists write the drag force law as $f(v) = \frac{1}{2} C_d \rho A v^2$, where C_d is the so-called drag coefficient (a catchall fudge factor that incorporates all the nasty fluid mechanical details of how the medium flows around the moving mass, e.g., is the flow laminar or is it turbulent?), ρ is the density of the medium in which the mass is moving, and A is the mass's cross-sectional area (if the mass is a sphere with radius R then $A = \pi R^2$). The next step is to define the dimensionless Reynolds number—named after the English engineer and physicist Osborne Reynolds (1842–1912)—as $Re \triangleq \frac{\rho l v}{\eta}$, where l is a characteristic length of the mass (for a sphere, e.g., l is the sphere's diameter), and η is the medium's viscosity. Then, over the range $0 < Re < 2 \cdot 10^5$, it is experimentally found that, for smooth

spheres,

$$C_d(Re) \approx \frac{24}{Re} + \frac{6}{1+\sqrt{Re}} + 0.4.$$

If, for example, the spherical mass moves very slowly, then Re will be small. If, to take this further, we say that v is so small that $Re < 1$, then in the expression for C_d the first term dominates the other two, and we have

$$C_d \approx \frac{24}{Re} = \frac{24}{\rho l v/\eta} = \frac{24\eta}{\rho l v} = \frac{24\eta}{\rho 2 R v} = \frac{12\eta}{\rho R v},$$

and so the drag force law (for v small) is, for a smooth sphere,

$$f(v) = \frac{1}{2} \cdot \frac{12\eta}{\rho R v} \cdot \rho \cdot \pi R^2 \cdot v^2 = 6\pi \eta R v.$$

This result, linear in speed, is known as *Stokes's law*, after the English mathematical physicist George Stokes (1818–1903), and played a central role in the Millikan oil drop experiment (see Freeman Dyson's comments in the Preface), in which electrically charged oil drops were observed as they fell very slowly through air while in an electric field that *almost* negated the gravitational force on the drops. Stokes's law would also apply, for another example, to a grain of sand settling *slowly* through water, or to the tiny water droplets in a cloud. Stokes's linear force law does not apply to massive objects moving rapidly through air, such as a thrown stone or a subsonic bullet fired from a gun. For objects with speeds large enough that Re is in the interval 10^3 to $2 \cdot 10^5$ (beyond the lower value the flow of the medium transitions from laminar, in which the drag force is due to skin friction between the moving object and the medium, to turbulent, in which the drag force is due to momentum transfer from the moving object to the medium via induced eddy currents in the medium), the first and second terms in the expression for C_d can be ignored compared to the third, and then we have $C_d \approx 0.4$. The force law then becomes, for v large,

$$f(v) = \frac{1}{2} \cdot 0.4 \cdot \rho \cdot \pi R^2 \cdot v^2 = 0.2 \rho \pi R^2 v^2.$$

That is, $f(v) = kv^2$, a quadratic force law (where k is $0.2\rho\pi R^2$, a particular constant for a given object). Additional interesting reading can be found in two papers in the *American Mathematical Monthly* by Lyle N. Long and Howard Weiss: "The Velocity Dependence of Aerodynamic Drag: A Primer for Mathematicians" (February 1999, pp. 127–35), and "How Terminal Is Terminal Velocity?" (October 2006, pp. 752–55).

3. Writing "net force=$\frac{d}{dt}$(momentum)" is the modern interpretation of the second law. Newton himself did not so express matters this way in his *Principia* (1687). See A. B. Arons and Alfred M. Bork, "Newton's Laws of Motion and the 17th Century Laws of Impact" (*American Journal of Physics*, April 1964, pp. 313–317).

4. The concept of a terminal speed for a body falling through a resistive medium can be found in Galileo's famous last work, *Dialogues Concerning Two New Sciences*, published in Leyden in 1638. It was from *Two Sciences* that I took the opening quotation to this discussion (from the English translation by Henry Crew and Alfonso De Salvio, published by Macmillan [New York: 1914, p. 94]).

5. Here's how to numerically solve (4.8) for the value of v_f given a value for v_i. We can write (4.8) as a function of u_f, that is, as $g(u_f)$, as follows:

$$g(u_f) = \frac{1}{3}u_f^3 + \frac{1}{2}u_f^2 + u_f + \ln(1 - u_f) + \left\{\frac{1}{3}u_i^3 - \frac{1}{2}u_i^2 + u_i - \ln(1 + u_i)\right\} = 0.$$

(Remember, $v_i = u_i^2$ and $v_f = u_f^2$, and so once we have u_f, we simply compute $v_f = u_f^2$.) The classic technique of numerically solving the equation $g(x) = 0$ is the iterative Newton-Raphson formula, which simply says that if x_n is the nth approximation to the actual solution, then the next, $n + 1$st approximation, is given by

$$x_{n+1} = x_n - \frac{g(x_n)}{g'(x_n)},$$

where $g' = dg/dx$. You can find the derivation of this formula in any good book on numerical analysis, or see my book, *When Least Is Best* (Princeton, N.J.: Princeton University Press, 2004 [corrected ed. 2007], pp. 120–23). Now,

$$\frac{dg}{du_f} = u_f^2 + u_f + 1 + \frac{-1}{1 - u_f} = -\frac{u_f^3}{1 - u_f}.$$

Remembering that the expression in the braces in the equation for $g(u_f)$ is a *constant*, K, once we have specified the value of $u_i = \sqrt{v_i}$, we have the iterative equation that numerically solves $g(u_f) = 0$ to be

$$u_{f_{n+1}} = u_{f_n} + \frac{\frac{1}{3}u_{f_n}^3 + \frac{1}{2}u_{f_n}^2 + u_{f_n} + \ln(1 - u_{f_n}) + K(u_i)}{\frac{u_{f_n}^3}{1 - u_{f_n}}}$$

$$= \frac{\frac{2}{3}u_{f_n}^4 - \frac{1}{6}u_{f_n}^3 - \frac{1}{2}u_{f_n}^2 + (1 - u_{f_n})\ln(1 - u_{f_n}) + u_{f_n}(1 - K(u_i)) + K(u_i)}{u_{f_n}^3}.$$

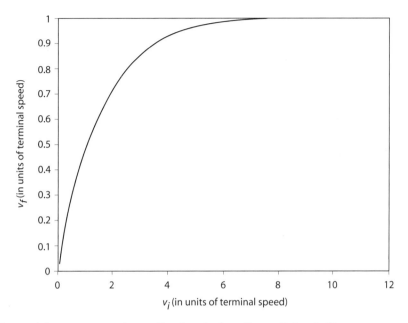

Figure 4.3. v_f compared to v_i for the air drag force $f(v) = k\sqrt{v}$.

All you need to do is start with an initial guess for u_f and continue the iterative process until you think you're "close enough" (I used the condition $|u_{f_{n+1}} - u_{f_n}| < 0.000001$ to be the signature of "close enough"). All this is quite easy to code, and a plot of v_f versus v_i is shown in Figure 4.3 This same code was at the heart of the code that produced Figure 4.2, too, for τ versus v_i with the air drag force law $f(v) = k\sqrt{v}$. The convergence is so fast that, even using a thousand values of v_i from 0 to 12, the code generated a plot in less than a second.

6. The inspiration for much of the mathematics of this discussion came from a paper by John Lekner, "What Goes Up Must Come Down: Will Air Resistance Make It Return Sooner, or Later?" *(Mathematics Magazine,* January 1982, pp. 26–28).

Air Drag— A Physical View

> It is common to find in textbooks a linear drag law of the form kv, where k is independent of speed. This yields an equation of motion that is readily solved at the expense of the solution being largely irrelevant.
> — Craig F. Bohern, "Dimensional Analysis, Falling Bodies, and the Fine Art of *Not* Solving Differential Equations"

5.1 The Quadratic Force Law

In the first discussion on air drag I wasn't terribly concerned about the actual applicability of our results to the real world. That is, while I limited myself to physically plausible air drag force laws, I didn't take the next step of asking which ones are the force laws that are actually observed. I have already opened the door to more than one law, of course; Stokes's linear law and the quadratic law. Can *both* apply in real life? Yes, but in entirely different situations. So, to be absolutely clear from the start, in this discussion we'll be interested in massive objects, such as iron balls and people, that are falling straight down through the atmosphere of Earth toward the ground. To start, I discuss falls that take place fairly close to the surface of the planet, but later on we'll see what happens when we extend our analyses to include very long falls from high altitude. Curious things indeed do happen on such falls. We will in every case assume that these falls begin from rest, at time $t = 0$. And, as you'll soon see, it is the quadratic law that will be the one of interest.

If we take the simplest possible case of masses falling through a vacuum, where there is no drag at all, then it is easy to calculate the answers to such questions as (1) how *far* has the mass fallen at time t? and (2) how *fast* is the mass moving at time t? Starting from rest, and with an acceleration of 1 g in a vacuum, a falling object descends a distance $y = \frac{1}{2}gt^2$ and is moving at speed $v = gt$, t seconds after the fall begins. This means, since $t = \sqrt{2y/g}$, that t seconds after the fall begins, the object is moving at speed $v = \sqrt{2yg}$. For example, after the object has fallen just *one foot*, it is already moving at the speed

$$v = \sqrt{2 \cdot 1 \cdot 32.2} \text{ ft/s} = \sqrt{64.4} \text{ ft/s} = 8.02 \text{ ft/s}.$$

Now, for purposes of seeing how the numbers work (and here I'll switch to the metric system), let's suppose we have a spherical object of radius R falling at speed v through air. For air, at sea level and room temperature—what physicists and chemists call standard temperature and pressure, or STP—the density and viscosity are, respectively,

$$\rho_a = 1.3 \cdot 10^{-3} \text{g/cm}^3$$

and

$$\eta = 1.7 \cdot 10^{-4} \text{g/cm} \cdot \text{s}.$$

The Reynolds number[2] for this fall is then

$$Re = \frac{\rho_a l v}{\eta} = \frac{1.3 \cdot 10^{-3} 2Rv}{1.7 \cdot 10^{-4}} = (15.3 \text{ s/cm}^2)Rv,$$

where R is in centimeters (cm) and v is in centimeters per second (cm/s). Remember, Re itself is a dimensionless number, which you can verify by working through the units. For the quadratic drag force law to apply we require that $Re > 10^3$, and for $R = 1$ cm (a steel ball bearing a little less than an inch in diameter, for example), this translates into a speed of at least $v = 65$ cm/s. Assuming the fall is in a vacuum, the time to reach this minimum speed is $t = v/g = 65/981$ seconds $= 0.066$ seconds. And the distance traveled in that first 66 milliseconds of the fall is

$$y = \frac{1}{2} \cdot 981 \cdot (0.066)^2 \text{ cm} = 2.1 \text{ cm},$$

which is less than one inch!

These numbers are for a fall in a vacuum, but a fall through air would have little impact on them so early in the fall. That is, we have the (surprising, I think) result that, except for an extremely brief time interval at the very start of a fall, the fall is in the realm of $Re > 10^3$ and that means it is the quadratic drag force law that is the force law in play over virtually the entire duration of the fall. An implication of this conclusion is that to be in the realm of $Re < 1$, where Stokes's linear drag force law is in play, we must be dealing with a very, very slow fall, e.g., a grain of sand descending through cold, thick honey. A practical rule of thumb is that if you can actually see an object falling (it is large enough to be seen, and you can see it moving), then it is surely already in the quadratic drag force realm.

Now, let me show you two fascinating examples of a quadratic drag force fall. I'll start with the oft-quoted statement in physics textbooks and classroom lectures that Galileo proved, with his famous Leaning Tower of Pisa experiments, that two objects (one heavy and one not so heavy) fall at the same rate. (We'll exclude such complicated objects as feathers.) That is, contrary to Aristotle's supposed claim that the heavier object would fall faster (that is, reach the ground first) than the lighter object, Galileo is said to have shown that both objects would, if dropped simultaneously from the same height, reach the ground at the same instant. There is today, among historians of science and physicists with an interest in history, an understanding that this story is riddled with inaccuracy and misstatement. I'll not go into the historical debate[3] here but rather show you by direct calculation what the mathematical physics says.

Suppose our object is a sphere of radius R and density ρ_o, that is, the mass of our sphere is

$$m = \frac{4}{3}\pi R^3 \rho_o.$$

Starting its fall from rest at time $t = 0$, let's write $v(t)$ as the sphere's speed at any time $t \geq 0$, and so $v(0) = 0$. Now, the downward gravitational acceleration is g, reduced by the effect of air drag. The air drag force law is kv^2 (where $k = 0.2\rho_a \pi R^2$, as discussed in note 2 of Discussion 4, where ρ_a is the air density). The reduction in acceleration is, by Newton's second law of motion, the drag force divided by the sphere's mass, and so the sphere's acceleration is

$$\frac{dv}{dt} = g - \frac{k}{m}v^2. \tag{5.1}$$

If we write the terminal (falling) speed as v_T, which as discussed before in Discussion 4 is the speed when the falling sphere has reached a constant speed and so $\frac{dv}{dt} = 0$, we see (as before) that

$$\frac{k}{m} = \frac{g}{v_T^2}. \tag{5.2}$$

Thus, (5.1) becomes

$$\frac{dv}{dt} = g - \frac{g}{v_T^2}v^2 = g\left(1 - \frac{v^2}{v_T^2}\right). \tag{5.3}$$

Suppose we measure the distance of the fall with the variable $y(t)$, where $y(t=0)=0$ and y increases positively downward toward the ground. Then, the speed of the fall is

$$v = \frac{dy}{dt}.$$

From the chain rule of calculus we have

$$\frac{dv}{dt} = \frac{dv}{dy} \cdot \frac{dy}{dt} = \frac{dv}{dy}v,$$

and so (5.3) becomes

$$v\frac{dv}{dy} = g\left(1 - \frac{v^2}{v_T^2}\right).$$

This is easily integrated, i.e., writing u as a dummy variable of integration we then have, upon separating variables,

$$\int_0^v \frac{u}{1 - \frac{u^2}{v_T^2}} du = \int_0^y g\, du.$$

Then, either by making a simple change of variable (try $z = 1 - \frac{u^2}{v_T^2}$) or by just looking in integral tables, it is easy to arrive at the sphere's speed as a function of how far it has fallen:

$$v(y) = v_T\sqrt{1 - e^{-2gy/v_T^2}}. \tag{5.4}$$

This immediately tells us that as the sphere falls toward Earth, its speed monotonically increases from zero toward the terminal speed, but never actually reaches v_T.

We can integrate (5.3) is a different way, too, to directly find $v(t)$, that is, the sphere's speed as a function of how long it has fallen. That is, write

$$\frac{dv}{1-\frac{v^2}{v_T^2}} = g\,dt = \frac{v_T^2}{v_T^2 - v^2}\,dv = \frac{v_T^2}{(v_T-v)(v_T+v)}\,dv = \left\{\frac{\frac{1}{2}v_T}{v_T-v} + \frac{\frac{1}{2}v_T}{v_T+v}\right\}dv,$$

and so, integrating with $v(0)=0$, we have

$$gt = -\frac{1}{2}v_T \ln(v_T - v) + \frac{1}{2}v_T \ln(v_T + v) = \frac{1}{2}v_T \ln\left\{\frac{v_T+v}{v_T-v}\right\}.$$

Thus,

$$\ln\left\{\frac{v_T+v}{v_T-v}\right\} = 2gt/v_T,$$

or

$$\frac{v_T+v}{v_T-v} = e^{2gt/v_T} = e^{t/t_T},\ t_T = v_T/2g.$$

Note that another expression for t_T can be found from the formulas for m, v_T, and k. Since $v_T = \sqrt{mg/k}$ and $k = 0.2\rho_a \pi R^2$, we have

$$v_T = \sqrt{\frac{\frac{4}{3}\pi R^3 \rho_o g}{\frac{1}{5}\rho_a \pi R^2}} = \sqrt{\frac{20}{3}\cdot\frac{R\rho_o g}{\rho_a}},\quad t_T = \frac{1}{2g}\sqrt{\frac{20}{3}\cdot\frac{R\rho_o g}{\rho_a}} = \sqrt{\frac{5}{3}\cdot\frac{R\rho_o}{\rho_a g}}. \tag{5.5}$$

In any case, with just a bit of algebra we can solve for v to get

$$v(t) = v_T \frac{e^{t/t_T}-1}{e^{t/t_T}+1} = v_T \frac{e^{t/2t_T}(e^{t/2t_T} - e^{-t/2t_T})}{e^{t/2t_T}(e^{t/2t_T} + e^{-t/2t_T})},$$

or, at last,

$$v(t) = v_T \tanh(t/2t_T). \tag{5.6}$$

From (5.6) we next easily get the distance, $y(t)$, fallen by our sphere; from integral tables we have

$$\int \tanh(Ax)\,dx = \frac{1}{A}\ln\{\cosh(Ax)\},$$

and so, with $A = 1/2t_T$, we have

$$y(t) = \int_0^t v(u)\,du = v_T \int_0^t \tanh(u/2t_T)\,du = v_T \cdot 2t_T \left(\ln\{\cosh(u/2t_T)\}\right)\Big|_0^t,$$

or

$$y(t) = 2v_T t_T \ln\{\cosh(t/2t_T)\}. \tag{5.7}$$

You'll notice that (5.7) says $y(t=0) = 0$, as it should.

Equation (5.7), simple as it may look, has great surprises for most who first encounter it. For example, let's go back to what Galileo had to say about two falling balls:

> Aristotle says that "an iron ball of one hundred pounds falling from a height of one hundred cubits [the cubit was an ancient measure of length based on the length of the current king's forearm, and so varied from king to king, but it was typically about a foot-and-a-half] reaches the ground before a one-pound ball has fallen a single cubit." I say that they arrive at the same time. You find, *on making the experiment* [my emphasis], that the larger [ball] outstrips the smaller by two finger-breadths, that is, when the larger has reached the ground, the other is short of it by two finger-breadths.[4]

Iron has a density of about $\rho_o = 7.8\,\text{g/cm}^3$ and so, since there are 454 grams in a pound, the radius of a one-pound iron sphere (ball 1) is about

$$R_1 = \left\{\frac{3 \cdot 454}{4\pi \cdot 7.8}\right\}^{1/3} \text{cm} = 2.4 \text{ cm}.$$

The radius of a ball 100 times heavier (ball 2) would then be $(100)^{1/3} = 4.64$ times greater, i.e., $R_2 = 11.1$ cm. In Figure 5.1 I've plotted $y_2(t) - y_1(t)$, using (5.7), for these two iron balls, and, it is clear, the heavier ball *does* fall faster than the lighter ball.[5] In a vacuum, of course, the two balls would indeed, as Galileo claimed, fall at identical rates, but in air with its quadratic drag they do not—and Galileo was simply wrong in his assertion that an actual experiment would show him correct. It would not. In fact, when the 100-pound ball hits the ground when dropped from a height of 100 cubits (that is, from a height of 150 feet, equal to 4,572 cm) we see that one-pound ball still has over 50 cm to

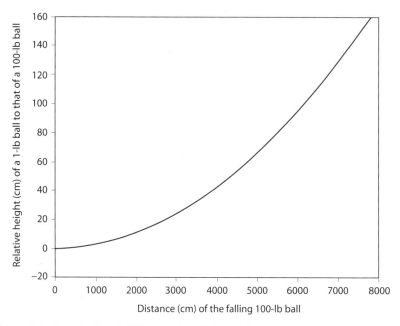

Figure 5.1. Iron balls of different sizes fall at *different* rates!

go, substantially greater than Galileo's "two finger breadths." This gap could, I think, be seen by an observer on the ground.[6] So, I ask the natural question: *Did Galileo really do his historically famous experiment at the Leaning Tower of Pisa?*

For our next surprise, suppose we now take two balls of different material, one with density 11.3 g/cm^3 (ball 1 is a lead ball) and $R_1 = 1$ cm, and the other with density 0.75 g/cm^3 (ball 2 is a wooden ball) and $R_2 = 3$ cm. As before, ball 2 is the heavier ball, as its much lower density is more than compensated for by its much larger volume. In Figure 5.2 I've again plotted $y_2(t) - y_1(t)$, but now it is the less massive (lead) ball that reaches the ground first (that's what $y_2(t) - y_1(t) < 0$ means, of course). With a 150-foot drop, the plot shows that the lighter ball hits the ground when the heavier ball still has about 400 cm to go, a huge gap that a ground observer should easily be able to see. If you tell me *that* doesn't surprise you, well,—I don't believe you! The reason for this is, of course, the greater air drag experienced by the ball with the greater cross-sectional area, irrespective of the fact that it is the heavier ball.

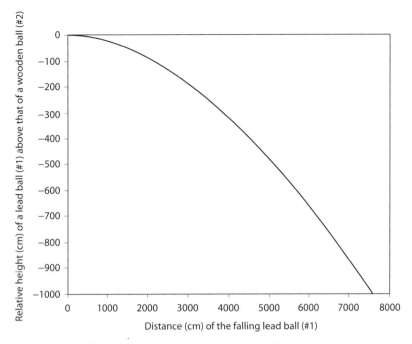

Figure 5.2. The lighter (lead) ball hits the ground first!

Our analyses so far, for a body falling through the atmosphere starting from rest, speeding up, and approaching a terminal speed from below, does motivate an intriguing question: what happens if the body *starts* its fall with an initial speed greater than v_T? For example, suppose a meteor plows into the upper atmosphere at a typical speed for such objects of 20 miles per second (which is far larger than its terminal speed for a fall from rest through the atmosphere)? The answer is that the meteor will slow down and approach v_T from above. This slowing down is achieved via a fantastically large deceleration. For a quadratic drag force law we can write the net (upward) force on a meteor with mass m as

$$F = kv^2 - mg,$$

where I'll assume that the acceleration of gravity at the top of Earth's atmosphere is the same as at the ground (I'll address this assumption and show it is okay in just a bit). So, the deceleration of the meteor is, from Newton's second law,

$$a = F/m = (k/m)v^2 - g.$$

When $v = v_T$ we have $a = 0$ by definition, and so

$$(k/m) = g/v_T^2.$$

Thus,

$$a = \frac{g}{v_T^2}v^2 - g = g\left[\left(\frac{v}{v_T}\right)^2 - 1\right].$$

If, for example, we take $v_T = 100$ m/s and $v = 30{,}000$ m/s at atmospheric entry, we see that $a = 90{,}000\ g$. Most meteors are most likely utterly destroyed by such an enormous deceleration long before they slow down to a speed even approaching v_T.

5.2 Long Falls through A Real Atmosphere

Up to now I have implicitly limited the physics to falls through the atmosphere that are near the surface of Earth. But what if we have a really long fall that starts very high up in the sky? Such a fall might be intentional, for example a military HALO parachute jump,[8] or unintentional, for example the fall from 65,000 feet to the ocean surface below of the intact crew cabin during the *Challenger* space shuttle disaster of 1986.[9] Are such long falls somehow different from the near-Earth falls we have been discussing so far? What we have to do, to answer that question, is look at the assumptions inherent in those 'short' fall analyses. There are two. If you look back at what we did, you'll see we assumed that both the acceleration of gravity, and the density of air, remain constant over the entire duration of the fall. The first assumption continues to hold even for a pretty long fall, but the second does not.

So, first, let me show you now why we can still retain the assumption of a constant value of the acceleration of gravity equal to its value at the surface of Earth (g), even when our falling object is far above the surface of the planet. To see this, suppose the radius of Earth is R_e, and that the falling object is distance z above the surface, i.e., it is distance $R_e + z$ from Earth's center. We then have the gravitational force on the object, from Newton's famous inverse square law of gravity, as

$$G \frac{Mm}{(R_e + z)^2},$$

where G, M, and m are the universal gravitational constant, the mass of Earth, and the mass of the falling object, respectively. Since this force is also equal (from Newton's famous second law of motion) to ma, where a is the instantaneous acceleration of gravity the object experiences, we have

$$a = G \frac{M}{(R_e + z)^2}.$$

When $z = 0$, we know that $a = g$ (by definition), and so $GM = gR_e^2$. Thus,

$$a = \frac{gR_e^2}{(R_e + z)^2} = \frac{g}{(1 + \frac{z}{R_e})^2}.$$

From this you should be able to see that

$$z > R_e \left(\frac{1}{\sqrt{0.99}} - 1 \right) = 0.00504 R_e$$

before a is more than 1% different from g. Since $R_e = 3,960$ miles, this means that z must be greater than 19.95 miles (105,300 feet) before we make even as much as a 1% error in taking the acceleration of gravity as everywhere equal to g during even a very long (i.e., high-altitude) fall through the atmosphere.

So, even at an altitude much higher than that from which *Challenger* fell, we don't have to worry about the variation of the acceleration of gravity with height. It is 1 g all the way down. What we do have to worry about is the variation of air density, which is much less at the start of the fall than it is later as the fall approaches the ground. The mathematics of this complication seems to be regularly rediscovered,[10] but I believe it was first worked out by the German-born American aerodynamicist Max Munk (1890–1986). Munk's analysis[11] is a bit terse (in my opinion), so I'll take you through it in a much more leisurely fashion.

The analysis begins with the fact that the density of the atmosphere decreases exponentially with increasing elevation.[12] If as before we measure the distance of a fall by $y(t)$, where $y(t=0) = 0$ and y increases positively downward, then the speed of the fall is

$$v(t) = \frac{dy}{dt},$$

and the air density is

$$\rho_a(y) = \rho_a(0) e^{\beta y},$$

where $\rho_a(0)$ is the air density at the start of the fall and β is some positive constant with units of m^{-1} if we measure y in units of meters. That is, the air density increases with increasing y, which simply says that the air density increases exponentially as the fall approaches the surface of the planet. (The case of $\beta = 0$ is the special, far less realistic case of a constant air density atmosphere.) Now we simply repeat the argument from the first part of this discussion: the air drag force is $0.2 \rho_a \pi R^2 v^2$, and so the reduction in the gravitational acceleration (from g) of our falling sphere is this force divided by the mass of the sphere (as before, I'll take the sphere to have radius R and density ρ_o), which gives

$$\frac{dv}{dt} = g - \frac{\frac{1}{5}\rho_a(y) \pi R^2 v^2}{\frac{4}{3}\pi R^3 \rho_o} = g - \frac{3}{20} \cdot \frac{1}{R\rho_o} \cdot \rho_a(y) v^2 = g - \frac{3}{20} \cdot \frac{\rho_a(0)}{R\rho_o} e^{\beta y} v^2,$$

or

$$\frac{dv}{dt} = \frac{d^2y}{dt^2} = g - k e^{\beta y} \left(\frac{dy}{dt}\right)^2, \quad k = \frac{3}{20} \cdot \frac{\rho_a(0)}{R\rho_o}, \tag{5.8}$$

where k is a constant (with the dimension cm^{-1} if R is measured in cm).

Now, a brief time-out. Note carefully that the above k is *not* the k of our earlier analysis, the k of either (5.1) or (5.2). There, that earlier k played a central role in determining the terminal falling speed v_T. But notice that now, with variable air density, there is no such speed. That is, for any $\beta > 0$ there is no *constant* solution for $v = dy/dt$ when we set $dv/dt = 0$ in (5.8). This may seem a bit disconcerting, because it was v_T that we used as a unit reference speed, against which we measured all other speeds. But another "natural" speed will emerge just in time to play the role of a reference speed for us. Okay, back to our analysis.

The differential equation of (5.8) looks pretty nasty, but in fact we *can* make some useful progress in its solution. Since

$$v = \frac{dy}{dt},$$

then, again by the chain rule of calculus, we can write

$$\frac{d^2y}{dt^2} = \frac{dv}{dt} = \frac{dv}{dy} \cdot \frac{dy}{dt} = v\frac{dv}{dy},$$

and so (5.8) becomes

$$v\frac{dv}{dy} = g - ke^{\beta y} v^2.$$

Since

$$v\frac{dv}{dy} = \frac{1}{2}\frac{d(v^2)}{dy},$$

then

$$\frac{d(v^2)}{dy} = 2g - 2ke^{\beta y} v^2.$$

If we next define the (dimensionless) variable

$$x = e^{\beta y},$$

and since, again by the chain rule,

$$\frac{d(v^2)}{dy} = \frac{d(v^2)}{dx} \cdot \frac{dx}{dy} = \frac{d(v^2)}{dx} \cdot \beta e^{\beta y} = \frac{d(v^2)}{dx}\beta x,$$

we have

$$\frac{d(v^2)}{dx}\beta x = 2g - 2kxv^2,$$

and so we arrive at a first-order linear (which means we can solve it!) differential equation for v^2:

$$\frac{d(v^2)}{dx} + \frac{2k}{\beta}v^2 = \frac{2g}{\beta x}. \tag{5.9}$$

We can solve (5.9) by the classic technique of using the appropriate integrating factor (see any good book on differential equations), which

in this case is $e^{P(x)}$, where
$$P(x) = \int \frac{2k}{\beta} dx = \frac{2k}{\beta} x.$$

Multiplying through (5.9) by the integrating factor, we have
$$\frac{d(v^2)}{dx} e^{\frac{2k}{\beta} x} + \frac{2k}{\beta} v^2 e^{\frac{2k}{\beta} x} = \frac{2g}{\beta x} e^{\frac{2k}{\beta} x},$$

or
$$\frac{d}{dx} \left\{ v^2 e^{\frac{2k}{\beta} x} \right\} = \frac{2g}{\beta x} e^{\frac{2k}{\beta} x}.$$

This integrates by inspection to give
$$v^2 e^{\frac{2k}{\beta} x} = \frac{2g}{\beta} \int_1^x \frac{e^{\frac{2k}{\beta} u}}{u} du,$$

where the limits on the integral have been chosen so that when $x = 1$ (its initial value, i.e., since $x = e^{\beta y}$, then $x = 1$ when $y = 0$), we will have $v = 0$ (the fall begins from rest). So, at last,
$$v^2 = \frac{2g}{\beta} e^{-\frac{2k}{\beta} x} \int_1^x \frac{e^{\frac{2k}{\beta} u}}{u} du.$$

Or *almost* at last. To put this expression for v^2 into the most convenient form, let's now write the dimensionless constant α as
$$\alpha = \frac{2k}{\beta} \quad (5.10)$$

and then, returning to y (the distance of the fall) as our independent variable, we arrive at
$$v^2 = \frac{2g}{\beta} e^{-\alpha x} \int_1^x \frac{e^{\alpha u}}{u} du = \frac{2g}{\beta} e^{-\alpha e^{\beta y}} \int_1^{e^{\beta y}} \frac{e^{\alpha u}}{u} du.$$

If we next change variable to $z = \alpha u$ ($dz = \alpha du$), we have
$$v^2 = \frac{2g}{\beta} e^{-\alpha e^{\beta y}} \int_\alpha^{\alpha e^{\beta y}} \frac{e^z}{z/\alpha} \cdot \frac{1}{\alpha} dz,$$

or

$$v^2 = \frac{2g}{\beta} e^{-\alpha e^{\beta y}} \int_\alpha^{\alpha e^{\beta y}} \frac{e^z}{z} dz. \tag{5.11}$$

Notice that $\frac{g}{\beta}$ has the units of $\frac{m/s^2}{1/m} = m^2/s^2$, that is, the units of a speed *squared*. So, let's define our unit speed as

$$\text{unit speed} = \sqrt{\frac{2g}{\beta}}. \tag{5.12}$$

Notice, too, that (5.12) is undefined in the case of $\beta = 0$ —the case of a uniformly dense atmosphere—but in that special case, as I discussed earlier, we have v_T, the terminal speed, available to serve as our reference speed. Further, let's define our unit distance as

$$\text{unit distance} = \frac{1}{\beta}. \tag{5.13}$$

These definitions reduce (5.11) to

$$v^2 = e^{-\alpha e^y} \int_\alpha^{\alpha e^y} \frac{e^z}{z} dz.$$

This integral may perhaps look a little unusual to many readers: as Munk (see note 11) put it, the integral is "a mathematical function known as the exponential integral. It is a little known transcendental function [Munk was writing in 1944, when that statement was probably true] which cannot be reduced to the elementary functions." But in fact it is a well-tabulated function in handbooks and is available as a built-in function (just as cos, cosh, sinh, etc., are) in any good modern scientific programming language, such as MATLAB. It is generally defined as follows:

$$\text{Ei}(x) = \int_{-\infty}^{x} \frac{e^u}{u} du, \; x > 0,$$

and so our final result becomes

$$v = \sqrt{e^{-\alpha e^y} \{\text{Ei}(\alpha e^y) - \text{Ei}(\alpha)\}}. \tag{5.14}$$

All we need to do now is to determine the value of β, which sets the values of both the unit distance and the unit speed, and the value of α, and then we can plot v as a function of y once the size of our falling sphere is given. From Mohazzabi and Shea (see note 10) we learn the value of β for Earth's atmosphere is $0.134 \text{ km}^{-1} = 0.000134 \text{ m}^{-1}$. This says, using (5.12) and (5.13), that

$$\text{unit distance} = \frac{1}{\beta} = \frac{1}{0.000134 m^{-1}} = \frac{1000}{0.134} m = 7{,}463 \text{ meters},$$

and

$$\text{unit speed} = \sqrt{\frac{1}{\beta}} \cdot \sqrt{2g} = \sqrt{7{,}463 m} \cdot \sqrt{2 \cdot 9.81 m/s^2} = 383 \text{ meters/second}.$$

Now, what is α? From (5.8) and (5.10) we have

$$\alpha = \frac{2k}{\beta} = \frac{2}{\beta} \cdot \frac{3}{20} \cdot \frac{\rho_a(0)}{R\rho_o} = 0.3 \cdot \frac{1}{\beta} \cdot \frac{1}{R} \cdot \frac{\rho_a(0)}{\rho_o},$$

where, to restate, R is the radius of the falling sphere, ρ_o is the sphere's density, and $\rho_a(0)$ is the density of air at the *start* of the fall. Now, let me make one last definition. If we say the fall *starts* at height h, and if ρ_a is the density of air *at the surface* of the planet, then we have

$$\rho_a(0) = \rho_a e^{-\beta h}.$$

So,

$$\alpha = 0.3 \cdot \frac{1}{\beta} \cdot \frac{1}{R} \cdot \frac{\rho_a}{\rho_o} \cdot e^{-\beta h}.$$

If we measure R in units of centimeters, h in meters, and the density of air in g/cm^3, then with $\rho_a = 1.3 \times 10^{-3} \text{ g/cm}^3$ we have

$$\alpha = 0.3 \cdot 7{,}463 \times 10^2 \cdot \frac{1}{R} \cdot \frac{1.3 \times 10^{-3}}{\rho_o} \cdot e^{-h/7{,}463}$$

or,

$$\alpha = 291 \frac{e^{-h/7{,}463}}{R\rho_o}. \tag{5.15}$$

With (5.13), (5.14), and (5.15) we can now calculate the speed of a falling sphere once we have specified (1) h, the height (in meters) at which the fall starts, (2) R, the radius of the falling sphere

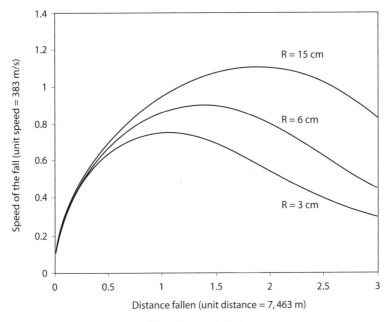

Figure 5.3. Three iron balls falling from 22,389 meters.

(in centimeters), and (3) ρ_o, the density of the falling sphere (in g/cm^3). To give you an idea of the magnitude of the numbers we are talking about for spheres falling to Earth's surface from great heights, Figure 5.3 shows the speeds of three iron spheres ($\rho_o = 7.8$ g/cm^3) of radii 3 cm, 6 cm, and 15 cm (this last sphere weighs a little over 240 pounds, equal to the weight of a large man), each falling from a height of $h = 22{,}389$ meters (about 73,500 feet). The horizontal axis is the distance fallen, and so has a maximum value of 3 ($= 22{,}389/7{,}463$). The vertical axis is the falling speed (in units of 383 m/s).

The speed versus distance fallen curves show that all three balls fall with increasing speed until a maximum speed is reached (the heavier the ball, the larger is that maximum speed and the further the ball falls before reaching it), and then the speed begins to decrease. This makes physical sense, too, if you think about it, as the balls encounter a continually denser atmosphere as they approach the surface of Earth. This is very much different from the monotonic *increasing* nature of the falling speed through a uniform, constant-density atmosphere. The following table shows, for each ball, its maximum speed $V_{\max}$, in m/s, the distance it has fallen when it reaches its maximum speed D_{max}, in m, and the speed with which it impacts the ground V_{impact}, in m/s. Also, just for comparison, I've listed (using (5.6)), the terminal speed v_T (assuming an atmosphere with air density equal everywhere to that at the surface) for each of the spheres, also in m/s. Notice that both the maximum speed and the impact speed for each sphere, when falling through a realistic exponential atmosphere, exceed the terminal speed of the less realistic constant-density atmosphere.

radius	V_{max}	D_{max}	V_{impact}	v_T
3 cm	288	7,829	114	109
6 cm	346	10,284	173	153
15 cm	424	14,060	320	243

CP. P5.1:

Show, by directly differentiating the equation derived immediately after (5.13), $v^2 = e^{-\alpha e^y} \int_{\alpha}^{\alpha e^y} \frac{e^z}{z} dz$, that the v vs. y curves are of the nature shown in Figure 5.3. That is, show that there does indeed exist a speed V_{max} such that $\frac{d(v^2)}{dy} > 0$ for the first portion of the fall and $\frac{d(v^2)}{dy} < 0$ for the remaining portion of the fall. *Hint:* You'll need to know how to differentiate an integral to do this, and if you don't know how to do that, here's *Leibniz's formula:*[13] if $\Psi(y) = \int_{w(y)}^{u(y)} f(x, y) dx$, then

$$\frac{d\Psi}{dy} = \int_{w(y)}^{u(y)} \frac{\partial f}{\partial y} dx + f[u(y), y] \cdot \frac{du}{dy} - f[w(y), y] \cdot \frac{dw}{dy}.$$

Notes and References

1. Craig F. Bohren, "Dimensional Analysis, Falling Bodies, and the Fine Art of *Not* Solving Differential Equations" (*American Journal of Physics*, April 2004, pp. 534–537).

2. I introduced the Reynolds number in Discussion 4 (see its note 2), and you can find much more on it in the beautiful paper by E. M. Purcell, "Life at Low Reynolds Number" (*American Journal of Physics*, January 1977, pp. 3–11). There Purcell wrote of Reynolds, "That was a very great man. He was a professor of engineering, actually. He was the one who not only invented Reynolds number, but he was also the one who showed what turbulence amounts to and that there is instability in flow, and all that. He is also the one who solved the problem of how you lubricate a bearing, which is a very subtle problem that I recommend to anyone who hasn't looked into it."

3. See, for example, Barry M. Casper, "Galileo and the Fall of Aristotle: A Case of Historical Injustice?" (*American Journal of Physics*, April 1977, pp. 325–330), and Carl G. Adler and Byron L. Coulter, "Galileo and the Tower of Pisa Experiment" (*American Journal of Physics*, March 1978, pp. 199–201).

4. *Dialogues Concerning Two New Sciences* (New York: Macmillan, 1914, pp. 64–65) (see note 4 in Discussion 4).

5. This calculation was inspired by a fascinating paper written by Gerald Feinberg, "Fall of Bodies Near the Earth" (*American Journal of Physics*, June 1965, pp. 501–502). There are some minor typographical errors in Feinberg's mathematics, which I've corrected.

6. I should qualify this statement with the proviso that the two balls must be dropped at the same instant. Falling at a speed on the order of 100 ft/s when they hit the ground, a 50-cm gap (i.e., about a foot-and-a-half gap) is equivalent to about 16 milliseconds. The dropping of the two balls must be synchronized to even less than that to avoid masking the gap. I'll leave it to you to think about how that might have been accomplished using only technology that would have been available to Galileo.

7. There is one additional complication that can be added to our analysis of falling balls—the upward buoyant force due to the air displaced by a ball. According to Archimedes' principle, that force is equal to the weight of the air displaced by the ball (which of course is the weight of a volume of air equal to the ball's volume). You can read more on how to incorporate buoyant forces into the analysis of falling balls in Byron L. Coulter and Carl G. Adler, "Can a Body Pass a Body Falling Through the Air?" (*American Journal of Physics*, October 1979, pp. 841–846), and Peter Timmerman and Jacobus P. van der Weel, "On the Rise and Fall of a Ball with Linear or Quadratic Drag" (*American Journal of Physics*, June 1989, pp. 538–546).

8. HALO (high-altitude, low-opening) parachute jumps are used by airborne soldiers attempting to get on the ground fast to avoid "floating" high up in the sky as targets to ground fire. This is done by delaying the opening

of the parachute until literally the last possible moment, consistent with not blasting a hole in the ground with one's body. It is an extremely dangerous maneuver. (You can experience that interesting event by playing the video game *Medal of Honor Airborne*.) You can find an exciting (if fanciful) description of a fictional HALO jump (in a snowstorm!) in Stephen Hunter's novel, *Time to Hunt* (New York: Doubleday, 1998, pp. 480–482 and 493–494). The highest delayed-opening parachute jump to date was made August 16, 1960, by a U.S. Air Force officer from a balloon gondola. The free-fall jump started at a height of 102,800 feet, and at 90,000 feet (thirty seconds into the fall) he reached his maximum speed, an astounding 614 miles per hour, which is nine-tenths the speed of sound at that altitude. The main parachute opened automatically at a height of 17,500 feet (a small, flat-spin stabilization canopy chute had opened at 96,000 feet). See Capt. Joseph W. Kittinger, Jr., "The Long, Lonely Leap" (*National Geographic* December 1960, pp. 854–873), which includes an automatically taken photograph of Kittinger the instant after he stepped over a sign at the foot of the gondola door, reading "Highest step in the world," with the understandable prayer "Lord, take care of me now" on his lips.

9. The orbiter explosion occurred at an altitude of 50,800 feet, but the crew cabin continued on upward to a peak altitude of 65,000 feet, where it began its free fall to the ocean below. The fall took two minutes and forty-five seconds, during which it is generally thought the astronauts were alive. A heartbreaking reconstruction of how the flight crew might have struggled during the fall to regain control of their fatally wounded craft is told by the astronaut Mike Mullane in his terrific book, *Riding Rockets* (New York: Scribner 2006, pp. 249–250).

10. See, for example, Pirooz Mohazzabi and James H. Shea, "High-Altitude Free Fall" (*American Journal of Physics*, October 1996, pp. 1242–1246). See also, all in the *American Journal of Physics*, Ralph Hoyt Bacon, "Motion of a Particle Through a Resisting Medium of Variable Density" (January 1951, p. 64); George Luchak, "The Fall of a Particle Through the Atmosphere" (October 1951, p. 426); William Squire, "Motion of a Particle Through a Resisting Medium of Variable Density" (October 1951, pp. 426–427); and W. A. Bower, "Note on a Body Falling Through a Resisting Medium of Variable Density" (December 1951, pp. 562–564).

11. Max M. Munk, "Mathematical Analysis of the Vertical Dive" (*Aero Digest*, February 15, 1944, pp. 114 and 213).

12. The decrease in air density with increasing height I take to be obvious. The fact that the decrease is exponential is not quite so obvious, but is easily derived (it is a common example in college freshman physics texts, and so any

of a number of good, modern texts will have a mathematical derivation). For us, we'll take it as simply an experimentally verified fact.

13. You can find a freshman calculus derivation of this important formula in my book, *The Science of Radio*, 2nd ed., (New York: Springer, 2001, pp. 416–418).

Really Long Falls

> Hesiod's Heaven is just 1.57 Earth radii from [Hell], near enough for relatively easy comings and goings.
> — B. G. Dick, "Hesiod's Universe" (1983)

6.1 Falling into the Sun

What if, instead of a long fall through Earth's atmosphere, we had the atmosphere itself, the oceans, the continents—*everything, the entire Earth*—fall into the Sun? How long would that take? If we allow our imaginations even more freedom, how long would it take to fall from Heaven to Hell? Outrageous as that question no doubt seems, it is raised and answered in both ancient Greek theology and in more recent English poetry. So, as an inverted form of the Earth-into-the-sun question, we might ask how far it is from Earth to Heaven. Mathematical physics can answer both these questions if we assume Newton's inverse square law of gravity holds in Heaven and Hell, as well as on Earth and in the rest of the universe.

Let's do the more realistic Earth-into-the-Sun analysis first. I'll take Earth as a point object, which is probably not a bad assumption, considering the relative sizes of Earth and the Sun. To set the problem up analytically, I'll make the following definitions:

r_s = radius of the (assumed spherical) Sun ($7 \cdot 10^8$ meters);
r_o = radius of Earth's (assumed circular) orbit ($1.5 \cdot 10^{11}$ meters);
m_s = mass of the Sun ($2 \cdot 10^{30}$ kg);

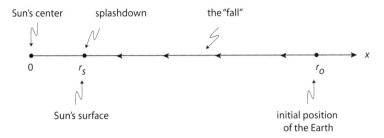

Figure 6.1. Falling into the Sun.

G = universal gravitational constant ($6.67 \cdot 10^{-11} \frac{\text{m}^3}{\text{kg} \cdot \text{s}^2}$);
T = duration of the fall;
$x(t)$ = distance from the sun's center to Earth, $0 \leq t \leq T$.

What I'll now calculate is the time T for the Earth to travel, from an initial position of rest with respect to the Sun (the Earth is somehow brought to a dead stop in its orbit—now *there's* a great scenario for a Hollywood disaster movie!), all the way to the Sun's surface. T, then, is the time interval from the start of the fall to the instant the Earth just starts to splash into the Sun. I will be assuming that the Earth hasn't been vaporized[1] before splashdown. That is, I'll assume that the mass of Earth is unchanged over the entire duration of the fall. Figure 6.1 shows the geometry of the fall.

Conservation of energy says that the sum of the kinetic and the gravitational potential energies of Earth, at any distance x from the Sun's center ($r_s \leq x \leq r_o$), will equal the initial ($x = r_o$) gravitational potential energy. There is, of course, zero initial kinetic energy, because the fall starts from rest. To find the gravitational potential energy of Earth at distance x from the Sun's center requires that we calculate the energy needed to move Earth in from infinity to a position of rest at distance x, which will of course be a negative quantity, because the Sun's attractive gravity is doing the work as Earth moves closer to the Sun. That energy is, with m_e as the mass of Earth,

$$\int_{\infty}^{x} G \frac{m_s m_e}{u^2} du = G m_s m_e \left(-\frac{1}{u}\right)\Big|_{\infty}^{x} = -G \frac{m_s m_e}{x}.$$

So, from conservation of energy (the right-hand side of (6.1) is the initial gravitational potential energy, that is, the energy at $x = r_o$), we have

$$\frac{1}{2}m_e \left(\frac{dx}{dt}\right)^2 - G\frac{m_s m_e}{x} = -G\frac{m_s m_e}{r_o}. \tag{6.1}$$

Notice that m_e is in every term, and so cancels away. (Remember, we are assuming $m_e \neq m_e(t)$.) Thus,

$$\left(\frac{dx}{dt}\right)^2 = 2Gm_s \left(\frac{1}{x} - \frac{1}{r_o}\right) = 2Gm_s \frac{r_o - x}{r_o x}.$$

Since we know Earth is falling toward the Sun, we know that x is decreasing, and so we use the *negative* square root to write

$$\frac{dx}{dt} = -\sqrt{\frac{2Gm_s}{r_o}} \sqrt{\frac{r_o - x}{x}}, \quad x \leq r_o.$$

Notice that $\frac{dx}{dt} = 0$ when $x = r_o$, just as it should for a fall starting from rest. If you plug numbers into this result you can verify that, at splashdown ($x = r_s$), Earth is moving at a speed of 383 miles per second (in its orbit around the Sun, the average speed of Earth is about 18 miles per second).

Our result for $\frac{dx}{dt}$ says that

$$dt = -\sqrt{\frac{r_o}{2Gm_s}} \sqrt{\frac{x}{r_o - x}} dx,$$

or, integrating t from 0 to T as x varies from r_o to r_s,

$$\int_0^T dt = T = -\sqrt{\frac{r_o}{2Gm_s}} \int_{r_o}^{r_s} \sqrt{\frac{x}{r_o - x}} dx. \tag{6.2}$$

To do the integral, change the variable to

$$z^2 = \frac{x}{r_o - x},$$

and so

$$r_o z^2 - x z^2 = x,$$

or
$$x = \frac{r_o z^2}{1+z^2}.$$

Thus,
$$\frac{dx}{dz} = \frac{2r_o z(1+z^2) - 2zr_o z^2}{(1+z^2)^2} = \frac{2r_o z}{(1+z^2)^2}.$$

So,
$$\int_{r_o}^{r_s} \sqrt{\frac{x}{r_o - x}}\, dx = \int_{\infty}^{\sqrt{\frac{r_s}{r_o - r_s}}} z \frac{2r_o z}{(1+z^2)^2} dz = 2r_o \int_{\infty}^{\sqrt{\frac{r_s}{r_o - r_s}}} \frac{z^2}{(1+z^2)^2} dz.$$

Now we have an integral easy to find in standard math tables. In particular,
$$\int \frac{u^2}{(a^2 + u^2)^2} du = \frac{-u}{2(a^2 + u^2)} + \frac{1}{2a}\tan^{-1}\left(\frac{u}{a}\right)$$

and so, with $a = 1$,
$$\int_{r_o}^{r_s} \sqrt{\frac{x}{r_o - x}}\, dx = 2r_o \left[\frac{-\sqrt{\frac{r_s}{r_o - r_s}}}{2(\frac{r_s}{r_o - r_s} + 1)} + \frac{1}{2}\tan^{-1}\left(\sqrt{\frac{r_s}{r_o - r_s}}\right) - \frac{\pi}{4} \right]$$

$$= 2r_o \left[\frac{-\sqrt{\frac{r_s}{r_o - r_s}}}{2(\frac{r_o}{r_o - r_s})} + \frac{1}{2}\cos^{-1}\left(\sqrt{\frac{r_o - r_s}{r_o}}\right) - \frac{\pi}{4} \right],$$

where you can see how we got from the $\tan^{-1}$ to the $\cos^{-1}$ simply by sketching a triangle and then using the Pythagorean theorem; continuing,

$$= r_o \left[-\sqrt{\frac{r_s}{r_o - r_s}} \left(\frac{r_o - r_s}{r_o}\right) + \cos^{-1}\left(\sqrt{\frac{r_o - r_s}{r_o}}\right) - \frac{\pi}{2} \right]$$

$$= r_o \left[-\sqrt{\frac{r_s(r_o - r_s)}{r_o^2}} + \cos^{-1}\left(\sqrt{\frac{r_o - r_s}{r_o}}\right) - \frac{\pi}{2} \right]$$

$$= r_o \left[\cos^{-1}\left(\sqrt{\frac{r_o - r_s}{r_o}}\right) - \frac{\pi}{2} - \sqrt{\frac{r_s}{r_o}\left(1 - \frac{r_s}{r_o}\right)} \right].$$

Thus, putting this result for the integral in (6.2) back into (6.2), we have

$$T = r_o \sqrt{\frac{r_o}{2Gm_s}} \left[\sqrt{\frac{r_s}{r_o}\left(1 - \frac{r_s}{r_o}\right)} + \frac{\pi}{2} - \cos^{-1}\left(\sqrt{\frac{r_o - r_s}{r_o}}\right) \right].$$

Now,

$$\frac{\pi}{2} - \cos^{-1}\left(\sqrt{\frac{r_o - r_s}{r_o}}\right) = \cos^{-1}\left(\sqrt{\frac{r_s}{r_o}}\right)$$

—again, sketch a triangle and use the Pythagorean theorem to see this—and so, at last,

$$T = r_o \sqrt{\frac{r_o}{2Gm_s}} \left[\sqrt{\frac{r_s}{r_o}\left(1 - \frac{r_s}{r_o}\right)} + \cos^{-1}\left(\sqrt{\frac{r_s}{r_o}}\right) \right]. \qquad (6.3)$$

Remember, since m_e cancelled out in (6.1), T is the time for any (non-vaporizing) mass to fall into the Sun, starting from rest at a distance of Earth's orbital radius. (This is the basis for one imaginative proposal for disposing of radioactive waste: somehow reduce the orbital speed of the waste to zero and then let it fall into the Sun.) Inserting numbers, (6.3) tells us that it would take just slightly more than 64.6 days for Earth to fall into the Sun. *More* than enough time, I think, for a movie screen science-hero type guy or gal to save the planet!

6.2 Falling from Heaven to Hell

Let's now take a look at the Heaven-to-Hell travel question. Centuries before Christ, in his *Theogony*, the Greek poet Hesiod (who lived about the time of Homer, 700 BC), wrote "a brazen anvil falling down from Heaven nine nights and days would reach the Earth on the tenth: and again, a brazen anvil falling from Earth nine nights and days would reach Tartarus upon the tenth."[2] Tartarus was a terrible place, believed to be below even Hell, but let's suppose it was Hell that Hesiod had in mind for the anvil's final destination (we are, after all, mathematical physicists, not poets or theologians, and so we won't worry over such piffles). And further, let's go along with the common idea that Hell is at the center of the Earth.[3] From this information we can now solve for the distance that separates Earth from Heaven, and this is the inverse of the Earth-falling-into-the-Sun question, where we knew the distance to fall and wanted to learn the duration of the fall. For the anvil-in-Heaven-falling-to-Earth question we know the duration and want to calculate the distance.

The very same issue is raised in Milton's epic 1667 poem *Paradise Lost*, where we read of Satan being driven by the command of God (along with his crew of fallen angels) out of Heaven into Hell: "Into the wastful Deep; the monstrous sight Strook them with horror backward, but far worse Urg'd them behind; headlong themselves they threw Down from the verge of Heav'n, Eternal wrauth Burnt after them to the bottomless pit.... Nine days they fell...." Milton was quite clearly influenced by Hesiod,[4] although he did change Hesiod's anvil into the Devil and made the duration of the fall from Heaven to Hell a total of just nine days.

Some readers may wonder at the usefulness of calculating anything having to do with what is simply a fanciful tale (and I think that to be a fair description of a literal interpretation of the endless burning of screaming sinners in boiling lakes of molten sulfur), but the extreme thermal nature of Hell has provided physicists with lots of amusing jokes. My favorite is the "proof" that whatever the temperature of Hell may be, the place is certainly at a perfectly uniform temperature. After all, if that isn't the case, then some bad-boy physicist who ended up there (surely, over all of historical time, there is at least one rotten, dead physicist who headed in Hell's direction!) could use his knowledge of thermodynamics to harness the resulting thermal gradient to make a refrigerator—which would go a long way to defeating the entire purpose of being there![5]

To set the analysis up for Hesiod's description, I'll make the following definitions:

r_e = the radius of (a uniform density) Earth;
ρ = constant density of Earth;
G = universal gravitational constant $\left(6.67 \cdot 10^{-11} \dfrac{\mathrm{m}^3}{\mathrm{kg} \cdot \mathrm{s}^2}\right)$;
r_h = distance from Earth's center to Heaven;
T = time to fall either from Heaven to Earth's surface, or from Earth's surface to Hell (Hesiod says the two times are the same);
m = mass of the anvil.

Hesiod is not entirely clear on the point, but let's assume that each of the two parts of the fall, Heaven to Earth and Earth to Hell, starts from rest. Let's do the Earth-to-Hell analysis first. We have the anvil starting from rest at distance $x = r_e$ from Earth's center (which is at $x = 0$). When the anvil is at some intermediate point along its fall into Hell, the gravitational force on it is due only to the mass of the Earth that remains "below" it, that is, the mass of a sphere of density ρ and radius x. (We'll prove this, a result due to Isaac Newton, in Discussion 10.) Thus, the gravitational force on the anvil is given by

$$F = -G\frac{(m)(\frac{4}{3}\pi x^3 \rho)}{x^2} = -\frac{4}{3}\pi G \rho x m,$$

where the minus sign indicates the force is in the direction opposite that of increasing x. Now, from Newton's second law of motion, we have

$$F = m\frac{d^2 x}{dt^2},$$

and so the mass of the anvil cancels away, and we have

$$\frac{d^2 x}{dt^2} = -\frac{4}{3}\pi G \rho x, \quad x(0) = r_e, \quad \left.\frac{dx}{dt}\right|_{t=0} = 0. \tag{6.4}$$

(I am obviously ignoring air drag in this analysis.)

It is well-known that the solution for $x(t)$ is what is called "simple harmonic motion," and I suppose I could just plop the answer down on the page and move on. But it is such a pretty mathematical problem, and one very quick to do, that I'll show it to you here. The trick is to assume the solution is given by

$$x(t) = Ae^{st}, \tag{6.5}$$

where A and s are both constants. Then, under that assumption, we'll see that we can actually find a solution that satisfies both the differential equation and the initial conditions of (6.4), and so (mathematicians tell us) we must have *the* solution. From (6.5) we have

$$\frac{d^2 x}{dt^2} = As^2 e^{st},$$

and so, from (6.4),

$$As^2 e^{st} = -\frac{4}{3}\pi G \rho A e^{st},$$

which says
$$s^2 = -\frac{4}{3}\pi G\rho.$$

That is, there are *two* possible constant values for s, which are given by
$$s_1 = i\sqrt{\frac{4}{3}\pi G\rho} = i\alpha, \quad s_2 = -i\sqrt{\frac{4}{3}\pi G\rho} = -i\alpha,$$

where
$$\alpha = \sqrt{\frac{4}{3}\pi G\rho}, \quad i = \sqrt{-1}.$$

The most general solution to (6.4) is the sum of these two specific solutions, because (6.4) is what mathematicians call a linear differential equation, and so I'll write
$$x(t) = A_1 e^{i\alpha t} + A_2 e^{-i\alpha t}, \tag{6.6}$$

where, to be most general, I have not assumed that the same constant A goes with each of the two different values of s. To find the values of A_1 and A_2 we can use the two initial conditions in (6.4). Since $x(0) = r_e$, we have
$$r_e = A_1 + A_2,$$

and since
$$\frac{dx}{dt} = i\alpha A_1 e^{i\alpha t} - i\alpha A_2 e^{-i\alpha t},$$

then the second initial condition says (as $i\alpha \neq 0$),
$$0 = A_1 - A_2.$$

These two results say that $A_1 = A_2 = \frac{1}{2}r_e$, and so
$$x(t) = \frac{1}{2}r_e e^{i\alpha t} + \frac{1}{2}r_e e^{-i\alpha t},$$

or, using Euler's fabulous formula $e^{i\theta} = \cos(\theta) + i\sin(\theta)$, we have the amazingly simple result of
$$x(t) = r_e \cos(\alpha t), \quad \alpha = \sqrt{\frac{4}{3}\pi G\rho}. \tag{6.7}$$

The anvil reaches Hell when $x = 0$, and that occurs when $\alpha t = \frac{1}{2}\pi$. Since $t = T$ at that fearsomely hot moment, we have

$$T\sqrt{\frac{4}{3}\pi G \rho} = \frac{1}{2}\pi,$$

or

$$T = \frac{1}{2}\sqrt{\frac{3\pi}{4G\rho}}. \tag{6.8}$$

Before continuing with the central goal of this calculation, that of determining the distance between Earth and Heaven, we can use (6.8) to show that while Hesiod was an interesting writer, he certainly wasn't much of a physicist. We can rewrite (6.8) to solve for ρ;

$$\rho = \frac{3\pi}{16GT^2},$$

which, if we use $T = 9$ days $= 777{,}600$ seconds, gives $\rho = 1.46 \cdot 10^{-5}$ g/cm^3, which is far, far too low, many times less than even the density of air, and certainly much less than that of the Earth![6]

Now let's do the Heaven-to-Earth part of the anvil's fall. Looking back at (6.3) for inspiration, we see that we can write

$$T = r_h \sqrt{\frac{r_h}{2Gm_e}} \left[\sqrt{\frac{r_e}{r_h}\left(1 - \frac{r_e}{r_h}\right)} + \cos^{-1}\left(\sqrt{\frac{r_e}{r_h}}\right) \right], \tag{6.9}$$

where I've written r_h for r_o, m_e for m_s, and r_e for r_s. Since Hesiod tells us that the Heaven-to-Earth journey takes the same time as the Earth-to-Hell journey, then we can equate (6.8) and (6.9) to write

$$\frac{1}{2}\sqrt{\frac{3\pi}{4G\rho}} = r_h \sqrt{\frac{r_h}{2Gm_e}} \left[\sqrt{\frac{r_e}{r_h}\left(1 - \frac{r_e}{r_h}\right)} + \cos^{-1}\left(\sqrt{\frac{r_e}{r_h}}\right) \right].$$

Writing $m_e = \frac{4}{3}\pi r_e^3 \rho$, this becomes (with the obvious cancellation of G and ρ on both sides),

$$\frac{1}{2}\sqrt{\frac{3\pi}{4}} = r_h \sqrt{\frac{r_h}{2\frac{4}{3}\pi r_e^3}} \left[\sqrt{\frac{r_e}{r_h}\left(1 - \frac{r_e}{r_h}\right)} + \cos^{-1}\left(\sqrt{\frac{r_e}{r_h}}\right) \right],$$

or, writing $u = r_e/r_h$, and doing a little algebra, we get the transcendental equation

$$u^{3/2}\frac{\pi}{\sqrt{2}} = \sqrt{u(1-u)} + \cos^{-1}(\sqrt{u}), \quad u = r_e/r_h.$$

This equation can be numerically solved using the Newton-Raphson method (see note 5 in Discussion 4 for details), with the result being

$$u = \frac{r_e}{r_h} = 0.636522446\cdots$$

Thus,

$$r_h = \frac{r_e}{0.6365\cdots} = 1.57 r_e,$$

which means Heaven is (as the opening quotation says) just 1.57 Earth radii from Earth's center, that is, from Hell. In other words, Heaven is just 0.57 Earth radii above the surface of the Earth, or 2,257 miles. Astronauts on their way to the moon must have passed right through Heaven!

You should, of course, take all this talk of Heaven and Hell with a grain of salt, since you'll recall that Hesiod's poem didn't even come close to getting the correct value for Earth's density. If Hesiod was equally wrong in locating Heaven, then Heaven is more likely much farther out in space, perhaps a million (billion, trillion?) miles or more beyond the moon. Oh, well, what does it matter where heaven is? How many of us are going there, anyway?[7]

CP. P6.1:

How fast would Hesiod's anvil be moving when it arrived in Hell, assuming a constant-density Earth with Earth's actual average density? Give your answer as a formula in terms only of r_e (Earth's radius) and g (acceleration of gravity at the Earth's surface). For $g = 32.2$ ft/s^2 and $r_e = 3,960$ miles, evaluate your formula in ft/s. *Hint*: Use (6.7).

Notes and References

1. Because of the inverse square law, the rate at which the sun's energy falls on Earth quadruples every time their separation is halved. When still nearly three million miles from the sun's center, the center of the falling Earth will have halved its distance from the sun's center five times, meaning the sun's radiation intensity on Earth will have increased by a factor of $4^5 > 1,000$ times! The atmosphere, the oceans, and the polar ice caps would be long gone, and probably a good chunk of Earth proper would at that point have been vaporized and blown away into space, as well.

2. See *Hesiod, the Homeric Hymns, and Homerica*, edited and translated by H. G. Evelyn-White (Cambridge, Mass.: Harvard University Press, 1964).

3. In Western culture it is the Bible that we might turn to as an "authoritative" source on how people, or at least how Christians, have viewed the location of Hell. In Acts 2:31 we are told that when Jesus died, His soul descended into Hell, while in Matthew 12:40 we learn that "the Son of man [Jesus] be three days and three nights in the heart of the Earth." The Bible greatly influenced all of Western literature that came after; for example, in Dante Alighieri's *The Divine Comedy* (ca. 1320), Hell is in the downward direction. Of course, the casting of Satan out of Heaven is supposed to have occurred when all in the universe was still in chaos, long before the Earth (and its center) was made, and so one could reasonably argue that Hell must be "somewhere else." We'll probably never really know, of course, at least not while we are still here on Earth.

4. Lawrence Babb, *The Moral Cosmos of "Paradise Lost"* (Lansing: Michigan State University Press, 1970, p. 113).

5. All engineering students at Stanford, at least when I was there, are required to take a course in thermodynamics. Usually this is a class offered by the Mechanical Engineering Department, but after looking at the assigned text—stuffed with engineering steam tables!—I decided I would rather have the flu all term than suffer through such stuff. I therefore petitioned the Engineering School to let me take thermo in the Physics Department, and so autumn quarter of 1961 I was lucky enough to be in Physics 173 with the famous Felix Bloch (Nobel Prize in Physics, 1952). That was a memorable class. It was in Professor Bloch's class that I first heard the temperature-of-Hell joke, and it was at the beginning of one lecture that Bloch burst into the room and announced, "No class today. Instead, everybody over to the Physics Department office for cookies and milk—Professor [Robert] Hofstader

[also a professor of physics at Stanford] has just won the Nobel Prize!" Cookies and milk? How times have changed!

6. The paper that served as the primary inspiration for this part of the discussion is B. G. Dick's "Hesiod's Universe" (*American Journal of Physics*, July 1983, pp. 666–667). Nobody, as far as I know, ever commented on the curious value of the starting page number of Dick's paper.

7. Additional interesting calculations of the same nature as the one done in this discussion can be found in Andrew J. Simosen, *Hesiod's Anvil: Falling and Spinning through Heaven and Earth* (Washington, D.C.: The Mathematical Association of America, 2007).

The Zeta Function— and Physics

> Let us now pursue an apparently tangential path. We wish to consider one of the most fascinating and glamorous functions of analysis, the Riemann zeta function.
> — Richard Bellman, *A Brief Introduction to Theta Functions* (1961)

7.1 A Curious Double Integral

Following Bellman's lead, I'm going to take a break from falling objects with this discussion (don't worry, we'll get right back to that topic in the next discussion). This will be a mostly mathematical break, but at the end I will make a (speculative) connection with physics. So, consider the almost surely "too simple-looking to be important" double integral

$$\int_0^1 \int_0^1 \frac{1}{1-xy} dx\, dy. \tag{7.1}$$

The integrand of (7.1), integrated over the unit square, is well behaved everywhere except at the single point of the upper right corner of the square, where the denominator vanishes. Looks can be deceiving, of course, and in fact the above integral is equal to the solution of one of the most famous problems in the history of mathematics. To see that, all we need to notice is the expansion

$$\frac{1}{1-xy} = 1 + xy + (xy)^2 + (xy)^3 + (xy)^4 + \cdots, \quad -1 < xy \leq 1,$$

which is easily confirmed by cross multiplication. Thus, our double integral is equal to

$$\int_0^1 \int_0^1 \{1 + xy + x^2 y^2 + x^3 y^3 + x^4 y^4 + \cdots\} \, dx \, dy$$

$$= \int_0^1 \left\{ x + \frac{1}{2} x^2 y + \frac{1}{3} x^3 y^2 + \frac{1}{4} x^4 y^3 + \frac{1}{5} x^5 y^4 + \cdots \right\} \Big|_0^1 dy$$

$$= \int_0^1 \left\{ 1 + \frac{1}{2} y + \frac{1}{3} y^2 + \frac{1}{4} y^3 + \frac{1}{5} y^4 + \cdots \right\} dy$$

$$= \left\{ y + \frac{1}{2} \cdot \frac{1}{2} y^2 + \frac{1}{3} \cdot \frac{1}{3} y^3 + \frac{1}{4} \cdot \frac{1}{4} y^4 + \frac{1}{5} \cdot \frac{1}{5} y^5 + \cdots \right\} \Big|_0^1$$

and so

$$\int_0^1 \int_0^1 \frac{1}{1 - xy} dx \, dy = 1 + \frac{1}{2^2} + \frac{1}{3^2} + \frac{1}{4^2} + \frac{1}{5^2} + \cdots = \sum_{n=1}^{\infty} \frac{1}{n^2} = 1.64493 \cdots.$$

The evaluation of this sum, of the reciprocals squared of the positive integers, *as a formula* utterly defeated all the mathematicians of the world until the Swiss-born genius Leonhard Euler found, in 1734, that it is $\frac{\pi^2}{6}$. Euler's original solution is diabolically clever, but today there are several "elementary" derivations,[1] one of which is based on a direct evaluation of our double integral in (7.1). What I'll show you next is the simplest and most direct solution, based on Fourier series.

7.2 Fourier Series and the Zeta Function

Let me start by reminding you of the basic result from the theory of Fourier series.[2] Suppose $f(x)$ is a function defined on the interval $-T \leq x \leq T$. Then, over that interval (and, actually, over the entire infinite x-axis as a periodic function with period $2T$) we can write $f(x)$ in the form of an infinite sum of sine and cosine functions; that is, as

$$f(x) = \frac{1}{2} a_o + \sum_{n=1}^{\infty} \left\{ a_n \cos\left(\frac{n \pi x}{T}\right) + b_n \sin\left(\frac{n \pi x}{T}\right) \right\}, \quad (7.2)$$

where the a_n and b_n coefficients are given by

$$a_n = \frac{1}{T}\int_{-T}^{T} f(x)\cos\left(\frac{n\pi x}{T}\right)dx, \quad n = 0, 1, 2, 3, \cdots$$

and

$$b_n = \frac{1}{T}\int_{-T}^{T} f(x)\sin\left(\frac{n\pi x}{T}\right)dx, \quad n = 1, 2, 3, \cdots.$$

Now, let $f(x) = x^2$ and $T = \pi$. Then (either from integral tables or integration by parts)

$$a_o = \frac{1}{\pi}\int_{-\pi}^{\pi} x^2 dx = \frac{1}{\pi}\left(\frac{1}{3}x^3\right)\bigg|_{-\pi}^{\pi} = \frac{2}{3}\pi^2,$$

and

$$a_n = \frac{1}{\pi}\int_{-\pi}^{\pi} x^2 \cos(nx)\,dx = \frac{1}{\pi}\left\{x^2\frac{\sin(nx)}{n} + 2x\frac{\cos(nx)}{n^2} - 2\frac{\sin(nx)}{n^3}\right\}\bigg|_{-\pi}^{\pi}$$

$$= \frac{4}{n^2}\cos(n\pi), \quad n = 1, 2, 3, \cdots.$$

We don't have to actually carry out the b_n integration because we know by inspection that the result is zero; we know that because $f(x) = x^2$ is an even function and so the integrand of the b_n integral is an odd function ($\sin(nx)$ is odd, and even times odd is odd) and the integral of an odd function over an interval symmetrical about zero vanishes.

Inserting these coefficients into (7.2), we have

$$x^2 = \frac{1}{3}\pi^2 + 4\sum_{n=1}^{\infty}\frac{\cos(n\pi)\cos(nx)}{n^2}, \quad -\pi \leq x \leq \pi.$$

In particular, for $x = \pi$ we have

$$\pi^2 = \frac{1}{3}\pi^2 + 4\sum_{n=1}^{\infty}\frac{\cos^2(n\pi)}{n^2} = \frac{1}{3}\pi^2 + 4\sum_{n=1}^{\infty}\frac{1}{n^2}.$$

Thus,

$$\sum_{n=1}^{\infty}\frac{1}{n^2} = \frac{\pi^2 - \frac{1}{3}\pi^2}{4} = \frac{\frac{2}{3}\pi^2}{4} = \frac{\pi^2}{6},$$

a result that made Euler famous—and rightfully so![3]

Euler actually did much more than calculate $\sum_{n=1}^{\infty} \frac{1}{n^2}$, however. If we define the function (now called the *Riemann zeta function*)

$$\zeta(s) = \sum_{n=1}^{\infty} \frac{1}{n^s}, \tag{7.3}$$

then Euler found a general formula for all s that are the positive *even* integers. (In general, s is complex-valued with a real part greater than one, but I'll limit our discussion here to real values.) What he didn't do, and what no mathematician since Euler has been able to do, is find a formula for $\zeta(s)$ when s is equal to a positive *odd* integer (when $s = 1$ the sum in (7.3) diverges). This has become one of the most famous open problems of modern mathematics. Mathematicians can of course calculate the *numerical* value of $\zeta(s)$, for any s as accurately as desired—for example, $\zeta(3) = 1.20205\cdots$ has been computed to one million decimal places!—but it is a *formula* that is wanted. The determination of $\zeta(s)$ for even *one* odd integer value of s would absolutely rock the mathematical world with incredible excitement.

What makes this such a fascinating quest is the apparent ease with which expressions like the double integral in (7.1) can be written. For example, you should now be able to show, using precisely the same approach as above, that

$$\int_0^1 \int_0^1 \int_0^1 \frac{dx\,dy\,dz}{1-xyz} = \sum_{n=1}^{\infty} \frac{1}{n^3} = \zeta(3).$$

But nobody yet has been able to evaluate this triple integral over the unit cube.[4] It has defeated all of history's greatest mathematicians, as well as all their many other attempts to calculate $\zeta(3)$ over the nearly three hundred years since Euler calculated $\zeta(2)$.

Actually, we can write $\zeta(3)$ not as a triple integral but as just a double integral, just as we did $\zeta(2)$. Here's how. Start with

$$I(a) = \int_0^1 \int_0^1 \frac{x^a y^a}{1-xy}\,dx\,dy,$$

where a is a parameter, e.g., $I(0) = \zeta(2)$. Then, using the same approach as before,

$$I(a) = \int_0^1 \int_0^1 x^a y^a \left\{ 1 + xy + x^2 y^2 + x^3 y^3 + x^4 y^4 + \cdots \right\} dx\, dy$$

$$= \int_0^1 x^a \left\{ \int_0^1 \left\{ y^a + xy^{a+1} + x^2 y^{a+2} + x^3 y^{a+3} + x^4 y^{a+4} + \cdots \right\} dy \right\} dx$$

$$= \int_0^1 x^a \left\{ \frac{y^{a+1}}{a+1} + x\frac{y^{a+2}}{a+2} + x^2 \frac{y^{a+3}}{a+3} + x^3 \frac{y^{a+4}}{a+4} + x^4 \frac{y^{a+5}}{a+5} + \cdots \right\}\bigg|_0^1 dx$$

$$= \int_0^1 \left\{ \frac{x^a}{a+1} + \frac{x^{a+1}}{a+2} + \frac{x^{a+2}}{a+3} + \frac{x^{a+3}}{a+4} + \frac{x^{a+4}}{a+5} + \cdots \right\} dx$$

$$= \left\{ \frac{x^{a+1}}{(a+1)^2} + \frac{x^{a+2}}{(a+2)^2} + \frac{x^{a+3}}{(a+3)^2} + \frac{x^{a+4}}{(a+4)^2} + \frac{x^{a+5}}{(a+5)^2} + \cdots \right\}\bigg|_0^1$$

$$= \frac{1}{(a+1)^2} + \frac{1}{(a+2)^2} + \frac{1}{(a+3)^2} + \frac{1}{(a+4)^2} + \frac{1}{(a+5)^2} + \cdots.$$

Thus,

$$\int_0^1 \int_0^1 \frac{x^a y^a}{1 - xy} dx\, dy = \sum_{n=1}^{\infty} \frac{1}{(n+a)^2}. \tag{7.4}$$

Now, let's differentiate this last result with respect to a. From the right-hand side we have

$$\frac{d}{da} \sum_{n=1}^{\infty} \frac{1}{(n+a)^2} = \sum_{n=1}^{\infty} \frac{-2(n+a)}{(n+a)^4} = -2 \sum_{n=1}^{\infty} \frac{1}{(n+a)^3}.$$

To differentiate the left-hand side of (7.4), i.e., to differentiate the double integral, notice that

$$x^a y^a = (xy)^a = e^{\ln\{(xy)^a\}} = e^{a \ln(xy)}.$$

Thus,

$$\frac{d}{da}\int_0^1\int_0^1 \frac{x^a y^a}{1-xy}dx\,dy = \int_0^1\int_0^1 \frac{\ln(xy)e^{a\ln(xy)}}{1-xy}dx\,dy,$$

and so, unwrapping the exponential, we have

$$\int_0^1\int_0^1 \frac{x^a y^a \ln(xy)}{1-xy}dx\,dy = -2\sum_{n=1}^{\infty}\frac{1}{(n+a)^3}.$$

For $a=0$, this reduces to

$$\int_0^1\int_0^1 \frac{\ln(xy)}{1-xy}dx\,dy = -2\sum_{n=1}^{\infty}\frac{1}{n^3} = -2\zeta(3).$$

Since $\ln(xy) = \ln(x) + \ln(y)$, we have

$$\int_0^1\int_0^1 \frac{\ln(xy)}{1-xy}dx\,dy = \int_0^1\int_0^1 \frac{\ln(x)}{1-xy}dx\,dy + \int_0^1\int_0^1 \frac{\ln(y)}{1-xy}dx\,dy,$$

and, since the two integrals on the right are equal, we get the amazingly simple-looking result,

$$\sum_{n=1}^{\infty}\frac{1}{n^3} = \zeta(3) = -\int_0^1\int_0^1 \frac{\ln(x)}{1-xy}dx\,dy. \quad (7.5)$$

Well, amazingly simple-looking it may be, but that double integral on the right of (7.5) still waits for some brilliant, not-yet-thought-of idea to crack it open. Indeed, if Euler couldn't calculate $\zeta(3)$, then it will almost surely take an idea of *nova* brilliance!

And just to show how perverse the numbers gods can be, if one is asking for the sum of the *odd* reciprocal integers cubed with *alternating signs*, that is, for

$$\sum_{n=1}^{\infty}\frac{(-1)^{n+1}}{(2n-1)^3} = 1 - \frac{1}{3^3} + \frac{1}{5^3} - \frac{1}{7^3} + \frac{1}{9^3} - \cdots,$$

then it is not hard to show that the answer is $\frac{\pi^3}{32}$.[5] Why the problem goes from "not hard" to "stupendously hard" just by making what appears to be such minor changes remains an immensely deep mystery.

7.3 The Zeta Function in Physics

Okay, very pretty, you may say, but what does any of this have to do with physics? The fact is, despite appearing to be a purely mathematical construct, the zeta function often appears in physical problems when least expected. Even though the zeta function doesn't have an obvious geometrical origin as do, say, the trigonometric functions, nevertheless the zeta function appears, in a natural (although perhaps surprising) way, in numerous analyses that have a physical basis. I'll show you just one to finish this discussion. A definite integral that occurs in a natural way in both solid-state physics and the theory of what physicists call *blackbody radiation* is the perhaps nasty-looking

$$\int_0^\infty \frac{x^3}{e^x - 1} dx.$$

What I'm going to do is first show you that its value is related to $\zeta(4)$, and then we'll calculate the value of $\zeta(4)$ (and, thus, the value of the integral). We begin by writing the integral in slightly altered form:

$$\int_0^\infty \frac{x^3}{e^x - 1} dx = \int_0^\infty \frac{e^{-x} x^3}{1 - e^{-x}} dx.$$

Now,

$$\frac{1}{1-y} = 1 + y + y^2 + y^3 + y^4 + \cdots,$$

and so, with $y = e^{-x}$, we have

$$\int_0^\infty \frac{x^3}{e^x - 1} dx = \int_0^\infty e^{-x} x^3 (1 + e^{-x} + e^{-2x} + e^{-3x} + \cdots) dx$$

$$= \int_0^\infty e^{-x} x^3 dx + \int_0^\infty e^{-2x} x^3 dx + \int_0^\infty e^{-3x} x^3 dx + \int_0^\infty e^{-4x} x^3 dx + \cdots.$$

From integral tables (or integration by parts),

$$\int_0^\infty e^{ax} x^3 dx = \frac{e^{ax}}{a} \left\{ x^3 - \frac{3x^2}{a} + \frac{6x}{a^2} - \frac{6}{a^3} \right\} \Bigg|_0^\infty.$$

Since $a < 0$ ($a = -n$, where $n = 1, 2, 3, \cdots$), we have

$$\int_0^\infty e^{ax} x^3 dx = \frac{6}{a^4}.$$

Thus, with $a = -n$, we arrive at

$$\int_0^\infty \frac{x^3}{e^x - 1} dx = 6 \sum_{n=1}^\infty \frac{1}{n^4} = 6\zeta(4). \quad (7.6)$$

So, all we have left to do is the calculation of $\zeta(4)$.

Fourier series did the trick for us earlier, with $\zeta(2)$, and it will again for $\zeta(4)$. This time let's define $f(x) = x^4$, $-\pi \leq x \leq \pi$, and so the Fourier coefficients are

$$a_o = \frac{1}{\pi} \int_{-\pi}^\pi x^4 dx = \frac{1}{\pi} \left(\frac{1}{5} x^5 \right) \Big|_{-\pi}^\pi = \frac{2}{5} \pi^4,$$

and (I used a *Mathematica* integrator[6] on the Web for this)

$$a_n = \frac{1}{\pi} \int_{-\pi}^\pi x^4 \cos(nx) dx$$

$$= \frac{1}{\pi} \left\{ \frac{4x^3 \cos(nx)}{n^2} - \frac{24x \cos(nx)}{n^4} + \frac{x^4 \sin(nx)}{n} \right.$$

$$\left. - \frac{12x^2 \sin(nx)}{n^3} + \frac{24 \sin(nx)}{n^5} \right\} \Big|_{-\pi}^\pi$$

$$= \frac{1}{\pi} \left\{ \frac{8\pi^3 \cos(n\pi)}{n^2} - \frac{48\pi \cos(n\pi)}{n^4} \right\} = \left\{ \frac{8\pi^2}{n^2} - \frac{48}{n^4} \right\} \cos(n\pi).$$

As before, $b_n = 0$, since $f(x) = x^4$ is an even function. Putting these coefficients into (7.2), we have

$$x^4 = \frac{1}{5} \pi^4 + \sum_{n=1}^\infty \left\{ \frac{8\pi^2}{n^2} - \frac{48}{n^4} \right\} \cos(n\pi) \cos(nx), \; -\pi \leq x \leq \pi.$$

In particular, for $x = \pi$, this becomes

$$\pi^4 = \frac{1}{5} \pi^4 + 8\pi^2 \sum_{n=1}^\infty \frac{\cos^2(n\pi)}{n^2} - 48 \sum_{n=1}^\infty \frac{\cos^2(n\pi)}{n^4},$$

or

$$\frac{4}{5}\pi^4 = 8\pi^2 \sum_{n=1}^{\infty} \frac{1}{n^2} - 48 \sum_{n=1}^{\infty} \frac{1}{n^4} = 8\pi^2 \zeta(2) - 48\zeta(4),$$

and so

$$\zeta(4) = \frac{8\pi^2 \zeta(2) - \frac{4}{5}\pi^4}{48} = \frac{8\pi^2 \cdot \frac{\pi^2}{6} - \frac{4}{5}\pi^4}{48} = \frac{\frac{40\pi^4}{30} - \frac{24\pi^4}{30}}{48}$$

$$= \frac{16\pi^4}{30 \cdot 48} = \frac{\pi^4}{90}.$$

Thus, from (7.6), we have our result:[7]

$$\int_0^{\infty} \frac{x^3}{e^x - 1} dx = \frac{\pi^4}{15},$$

a wonderful result for solid state physicists[8] as a fabulous gift from the still vastly mysterious zeta function.

I won't say much here about another possible connection of the zeta function with physics because it is still mostly inspired speculation, but perhaps it is worth a brief mention. As I mentioned earlier, the zeta function argument s is generally complex-valued and, for an infinity of particular complex values of s, it true that $\zeta(s) = 0$. These values of s are called the *zeros* of the zeta function.[9] The *complex* zeros (by definition) all lie above and below the real axis in the complex plane; Riemann conjectured (in 1859) that every single one of the complex zeros has the same real part of precisely $\frac{1}{2}$. This conjecture is called the *Riemann hypothesis*, and its proof (or disproof) is the mathematical world's greatest open problem today. (In a list of the top ten open problems I am not sure where the calculation of $\zeta(s)$ for s an odd integer would rank, but I would guess it's in the top five.) Hundreds of billions—maybe even more than a trillion by now–of complex zeros have been calculated, and every last one of them does indeed have a real part of $\frac{1}{2}$; but of course, that sort of computer testing proves nothing. All it would take is the discovery of just one complex zero with a real part $\neq \frac{1}{2}$ and the game would be over. (It is almost surely true, however, that every mathematician alive would be absolutely stunned if the conjecture should be *disproved*.) Now, what has struck many people—who may in fact have too much time on their hands—is that if one looks at how the known complex zeros are distributed along the vertical line $z = \frac{1}{2} + ib$, b any real number, it "appears" that the distribution "seems" to "look a lot like" the energy-level distribution in heavy nuclei atoms (for example, uranium 238). Why is that? I haven't the foggiest idea! I don't think anybody else does, either. So, this is where I'll end on this speculation, but if you're interested you can find more, at a popular level, in the easily accessible literature.[10]

To finish this discussion, let me tell you a charming zeta function story from pure math. In late January 1913 the eminent English mathematician G. H. Hardy (1877–1947) received a mysterious letter. Written by an unknown, self-taught mathematician then working as a lowly clerk in Madras, India, it was a plea for Hardy to look at some results from his labors. Much of it was perplexing, but none as much as the line in a second letter that declared $1 + 2 + 3 + 4 + \cdots = -\frac{1}{12}$. It looked at first like something only a lunatic would write. To Hardy's credit, however, sense was soon read into it, and the mathematical genius Srinivasa Ramanujan (1887–1920) was "discovered." What Ramanujan's sum actually was can be understood by writing $\frac{1}{n}$ as n^{-1}, with n any positive integer. So, what Ramanujan meant (but expressed badly) was that

$$1 + 2 + 3 + 4 + \cdots = \frac{1}{\frac{1}{1}} + \frac{1}{\frac{1}{2}} + \frac{1}{\frac{1}{3}} + \frac{1}{\frac{1}{4}} + \cdots$$
$$= \frac{1}{1^{-1}} + \frac{1}{2^{-1}} + \frac{1}{3^{-1}} + \frac{1}{4^{-1}} + \cdots = \zeta(-1)$$

and, indeed, $\zeta(-1) = -\frac{1}{12}$ (if the extended definition of $\zeta(s)$ is used—see note 9 again). Ramanujan had discovered a special case of the zeta function before he had ever heard of Riemann!

CP P7.1:

Calculate the values of $\int_0^1 \int_0^1 \frac{dx\,dy}{1-x^2 y^2}$ and $\int_0^1 \int_0^1 \frac{dx\,dy}{1+xy}$.

CP P7.2:

We calculated $\zeta(2)$ and $\zeta(4)$ by finding the Fourier series expansions of $f(x) = x^2$ and $f(x) = x^4$, respectively, over the interval $-\pi \leq x \leq \pi$. Explain why we *cannot* calculate $\zeta(3)$ with the Fourier series expansion of $f(x) = x^3$ over the interval $-\pi \leq x \leq \pi$.

Notes and References

1. Dan Kalman, "Six Ways to Sum a Series" (*College Mathematics Journal*, November 1993, pp. 402–421; in particular, see pp. 410–412) The opening double integral, and the other double and triple integrals for $\zeta(2)$ and $\zeta(3)$ in this discussion, are often called *Beukers's integrals*; Professor Frits Beukers, a mathematician at Utrecht University in the Netherlands, has long been in the hunt for $\zeta(3)$.

2. You can find much more on Fourier series and the historical development of the subject in my book, *Dr. Euler's Fabulous Formula* (Princeton N.J.: Princeton University Press, 2006, pp. 114–187).

3. At the end of the paper in note 1 there appears the following very funny little tune, where line 6 up from the bottom refers to (7.1), and line 3 up from the bottom refers to a Fourier series evaluation of $\zeta(2)$ equivalent to the one in this discussion. The other lines will make sense if you read Kalman's entire paper.

Six Ways to Sum a Series
(To the tune of *Fifty Ways to Leave Your Lover*)

This sum converges to a limit, I can see.
The terms decrease in size so very rapidly.
Isn't there some way to tell what the sum turns out to be?
There's got to be at least six ways to sum this series.

I added up one hundred terms it took all night.
I added fifty more, but still it was not right.
"Though adding terms this way won't work,"
I said, "some other method might."
I'll bet someone could find six ways to sum this series.

Tally up the inverse roots, Toots!
Add some bounds that use cotan, Stan!
Double integrate a square, Cher!
Just give it a try.

Calculate a residue, Stu!
Analyze a Fourier, Ray!
Use an imaginary real, Neal!
It's easy as π!

4. A triple integral over a cube that bears at least a passing similarity to our triple integral over the unit cube was evaluated by the English mathematician G. N. Watson (1886–1965) in a 1939 paper in the *Quarterly Journal of Mathematics* ("Three Triple Integrals," pp. 266–276). There he showed that

$$\int_0^\pi \int_0^\pi \int_0^\pi \frac{du\,dv\,dw}{1-\cos(u)\cos(v)\cos(w)} = \frac{1}{4}\Gamma^4\left(\frac{1}{4}\right) = 43.198\cdots,$$

where

$$\Gamma(s) = \int_0^\infty e^{-x} x^{s-1} dx, \quad s > 0,$$

is the *Euler gamma function*. Watson's triple integral first caught the interest of mathematicians when it appeared in a 1938 paper on the physics of ferromagnetism.

5. See *Dr. Euler*, p. 149.

6. Available at http://integrals.wolfram.com/index.jsp.

7. The result in (7.6) is actually a special case of a far more general formula:

$$\Gamma(s)\zeta(s) = \int_0^\infty \frac{x^{s-1}}{e^x - 1} dx,$$

where $\Gamma(s)$ is, as in note 4, the gamma function; for s a positive integer, $\Gamma(s) = (s-1)!$. For example, if $s = 4$, the general formula says

$$3! \cdot \zeta(4) = \int_0^\infty \frac{x^3}{e^x - 1} dx = 6\zeta(4),$$

which is (7.6). For a derivation of this beautiful general result, see my book, (*An Imaginary Tale: The Story of $\sqrt{-1}$* (Princeton N.J.: Princeton University Press, 2007 [corrected ed.], pp. 182–185).

8. Charles Kittel, *Introduction to Solid State Physics*, 6th ed. (New York: John Wiley & Sons, 1986, pp. 106–108). I used an earlier edition of this book when I took Physics 176 during the winter quarter of 1961 at Stanford. To be completely honest about that experience, I must admit I remember only two things from the course. First, it was entertainingly taught by Arthur Schawlow, who, two decades later, won the 1981 Nobel Prize in Physics, and second, (7.6). That's all. Solid-state physics just was not my thing! You can find a discussion of how (7.6) appears in blackbody radiation theory in volume 1 of Richard P. Feynman, *Lectures on Physics* (Reading, Mass.: Addison-Wesley, 1963, p. 45–48).

9. And $\zeta(s) = 0$, too, when s is equal to an even, negative integer; these are the so-called *trivial zeros* of $\zeta(s)$, because they are so easily calculated. (We can

extend the definition of $\zeta(s)$ to $s < 1$ using a technical process called *analytic continuation*.) For more on this, see *An Imaginary Tale*, p. 153 and pp. 185–186.

10. See, for example, Brian Hayes, "The Spectrum of Riemannium" (*American Scientist*, July–August 2003, pp. 296–300). If this is a bit too speculative for your taste, you can find a more "real physics" appearance of $\zeta(3)$ in the paper by Daniel R. Stump, Gerald L. Pollack, and Jerzy Borysowicz, "Magnets at the Corners of Polygons" (*American Journal of Physics*, September 1997, pp. 892–897). See also note 10 in Discussion 15.

Ballistics—
With No Air Drag (Yet)

> The theory of the parabolic trajectory was not asserted to be an entirely accurate representation of the phenomena nor a merely hypothetical picture, but a very close approximation based upon the most satisfactory laws of mechanics with indisputable mathematical logic and capable, therefore, of application in practice. There was no inherent improbability in this. As it was known that a piece of wood in falling could attain an apparently enormous velocity without revealing any retardation by the air, it was reasonable to suppose that such a small heavy body as a musket ball could suffer none. Even in the sixteenth century, from looking at the flight of fire bombs, it had been guessed that they followed a symmetrical curve.
> — A. R. Hall, *Ballistics in the Seventeenth Century* (1952)

8.1 Shooting a Cannon in a Vacuum

After that "relaxing breather" on the pure mathematics of the Riemann zeta function in the previous discussion, let's get back to tossing stuff. In the earlier discussions on air drag, I limited the physics to strictly up-and-down motions. The more general (and, I think, vastly more intricate) case is when the moving object starts on the journey along its flight trajectory with an initial angle θ_0 different

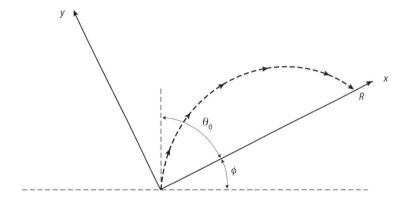

Figure 8.1. For what θ_0 is R maximized?

from the straight-up (or straight-down) case of 90 degrees (measured with respect to the spherical Earth's "local" surface, which serves as our horizontal reference). Two obvious examples of this are a baseball leaving the bat of a hitter (I'm ignoring pop-ups) and an artillery shell leaving the muzzle of a gun. For the purpose of being specific, as well as motivating one little additional twist, I'll focus for now on the shell-gun example. For that problem, a historically important question (one still important today) is, for what initial angle of fire θ_0 does the shell achieve its maximum range R? The twist I'll add to this is to suppose we are not necessarily firing over a horizontal plane (the typical physics textbook case) but rather over a local surface that is itself inclined at angle ϕ to the Earth's surface. If $\phi > 0$, then the gun is firing uphill, while if $\phi < 0$, the gun is firing downhill. (This problem was first studied by the Italian Evangelista Torricelli (1608–1647) in the early 1640s, who found a beautiful geometric solution.)[1] Both cases are of great interest to an artillery gunner, in the first case facing an opposing force that holds the high ground, or in the second case defending the high ground from an attacking force that wishes to capture it for itself. Firing over a horizontal plane is the special, typical textbook case of $\phi = 0$. Modern mobile tank warfare, in which high-speed, rapidly maneuvering tanks bouncing over rough terrain are firing on the run, has kept the mathematics of artillery gunfire as alive today as it was centuries ago. The geometry of the problem is shown in Figure 8.1.

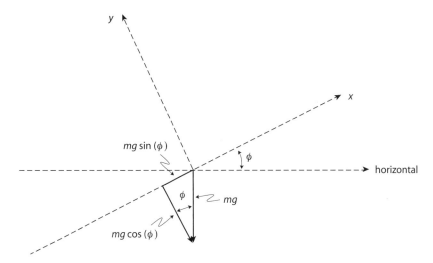

Figure 8.2. Resolving the downward gravitational force on a shell into components along tilted coordinate axes.

If, for a start, we assume that the only force acting on the shell once it leaves the gun muzzle is gravity, that is, if we ignore air resistance, then the analysis is easy. I'll add air drag to the analysis in the next discussion. I'll take the muzzle speed as v_0, the initial firing angle as θ_0, R as the range of the shell, and our *tilted* coordinate system as shown in Figure 8.1. The initial speed components of the shell are, in the x and y directions, respectively, $v_0 \cos(\theta_0)$ and $v_0 \sin(\theta_0)$. Now, the only force acting on the shell during its flight is that of gravity, straight down, and we can resolve that force into its x and y components, as shown in Figure 8.2.

So, from Newton's second law, we have

$$m\frac{d^2x}{dt^2} = -mg\sin(\phi)$$

and

$$m\frac{d^2y}{dt^2} = -mg\cos(\phi),$$

which integrate to

$$\frac{dx}{dt} = -g\sin(\phi)t + C_1$$

and

$$\frac{dy}{dt} = -g\cos(\phi)t + C_2,$$

where C_1 and C_2 are indefinite constants of integration. Using our initial ($t = 0$) speed components, we have

$$\frac{dx}{dt} = -g\sin(\phi)t + v_0\cos(\theta_0)$$

and

$$\frac{dy}{dt} = -g\cos(\phi)t + v_0\sin(\theta_0).$$

These last two equations, in turn, integrate to

$$x = -\frac{1}{2}gt^2\sin(\phi) + v_0 t\cos(\theta_0) + C_3 \qquad (8.1)$$

and

$$y = -\frac{1}{2}gt^2\cos(\phi) + v_0 t\sin(\theta_0) + C_4, \qquad (8.2)$$

where C_3 and C_4 are indefinite integration constants. Since $x(t=0) = y(t=0) = 0$, we have $C_3 = C_4 = 0$ and so

$$x(t) = v_0 t\cos(\theta_0) - \frac{1}{2}gt^2\sin(\phi) \qquad (8.3)$$

and

$$y(t) = v_0 t\sin(\theta_0) - \frac{1}{2}gt^2\cos(\phi). \qquad (8.4)$$

We know $y = 0$ when the gun is fired at $t = 0$, but $y = 0$, too, when $x = R$. Using (8.2) to solve for the non-zero t that gives $y = 0$, we find that solution (the time of impact) is given by

$$t = T = \frac{2v_0\sin(\theta_0)}{g\cos(\phi)}.$$

Now, at $t = T$ we have, immediately, $x(T) = R$, the range. That is, from (8.1),

$$R = v_0 \frac{2v_0 \sin(\theta_0)}{g \cos(\phi)} \cos(\theta_0) - \frac{1}{2} g \frac{4v_0^2 \sin^2(\theta_0)}{g^2 \cos^2(\phi)} \sin(\phi)$$

$$= \frac{2v_0^2 \sin(\theta_0) \cos(\theta_0)}{g \cos(\phi)} - \frac{2v_0^2 \sin^2(\theta_0) \sin(\phi)}{g \cos^2(\phi)}$$

$$= \frac{2v_0^2 \sin(\theta_0)}{g \cos^2(\phi)} \{\cos(\theta_0) \cos(\phi) - \sin(\theta_0) \sin(\phi)\}$$

$$= \frac{2v_0^2 \sin(\theta_0) \cos(\theta_0 + \phi)}{g \cos^2(\phi)}.$$

Remembering the trigonometric identity

$$2 \sin(A) \cos(B) = \sin(A + B) + \sin(A - B),$$

this last expression for R becomes, as a function of the firing angle θ_0,

$$R = \frac{v_0^2 \{\sin(2\theta_0 + \phi) - \sin(\phi)\}}{g \cos^2(\phi)}. \tag{8.5}$$

We obviously maximize R by maximizing $\sin(2\theta_0 + \phi)$, as that is the only term under our control (by varying θ_0), and as it gets bigger, so does R. That is, for maximum R we require $\sin(2\theta_0 + \phi) = 1$, which means

$$2\theta_0 + \phi = \frac{\pi}{2},$$

or, at last, we have the remarkably simple result that to maximize R, we should make

$$\theta_0 = \frac{\pi}{4} - \frac{1}{2}\phi = \frac{1}{2}\left(\frac{1}{2}\pi - \phi\right). \tag{8.6}$$

In words, the maximum range of a gun is achieved when the gun fires at the angle that bisects the angle formed by the vertical axis and the ground, *whatever the slope of the ground may be.* (Notice that nowhere did I set a derivative to zero to find this maximum.) This elegant result was first stated by Isaac Newton's friend Edmond Halley (1656–1742),

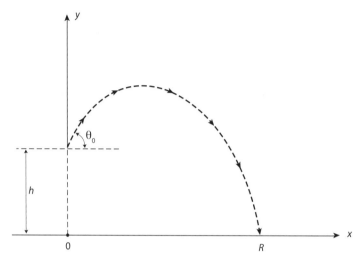

Figure 8.3. The geometry of the shot put.

of comet fame, in a paper in the *Philosophical Transactions of the Royal Society* of 1695.[2] Using the θ_0 of (8.6) in (8.5) gives us the maximum range:

$$R_m(\phi) = \frac{v_0^2\{1 - \sin(\phi)\}}{g\cos^2(\phi)} = \frac{v_0^2\{1 - \sin(\phi)\}}{g\{1 - \sin^2(\phi)\}},$$

or

$$R_m(\phi) = \frac{v_0^2}{g\{1 + \sin(\phi)\}}. \tag{8.7}$$

Notice that when firing over a horizontal plane ($\phi = 0$) we see from (8.6) that a firing angle of $\theta_0 = 45$ degrees gives the maximum range—from (8.7)—of $\frac{v_0^2}{g}$, a result known to Galileo, and even earlier.

8.2 What Makes a Champion Shot-Putter?

The analysis so far assumes that the shell is fired from the coordinate origin, that is, from (0,0), and so starts its flight at distance zero from the ground. What if, instead, the trajectory starts at distance $h > 0$ above the origin, as shown in Figure 8.3? To pick an example from sports, suppose we think of a shot-putter launching his shot from $(0,h)$ at an initial speed of v_0. What now is the launch angle that maximizes the range? Let's assume that $\phi = 0$, that is, our shot-putter is on a level track field. There is, then, a very pretty, quite elementary general solution showing that now the optimal launch angle is *less* than 45 degrees for $h > 0$. Of course, the general result should reduce to 45 degrees for $h = 0$. As before, we can also solve this optimization problem without calculating a single derivative.

Returning to (8.1) and (8.2), setting $\phi = 0$, and recalculating C_4 under the new initial condition of $y(t=0) = h$, we have

$$x(t) = v_0 t \cos(\theta_0) \tag{8.8}$$

and

$$y(t) = h - \frac{1}{2}gt^2 + v_0 t \sin(\theta_0). \tag{8.9}$$

Suppose that $t = T$ is the time at which the shot hits the ground, distance R from directly below the launch point., i.e., at $x = R$ and $y = 0$. Then,

$$R = v_0 T \cos(\theta_0) \tag{8.10}$$

and

$$0 = h - \frac{1}{2}gT^2 + v_0 T \sin(\theta_0),$$

or

$$v_0 T \sin(\theta_0) = \frac{1}{2}gT^2 - h. \tag{8.11}$$

Squaring both (8.10) and (8.11) and then adding, we arrive at

$$v_0^2 T^2 \sin^2(\theta_0) + v_0^2 T^2 \cos^2(\theta_0) = R^2 + \left(\frac{1}{2}gT^2 - h\right)^2 = v_0^2 T^2,$$

or

$$R^2 = v_0^2 T^2 - \left(\frac{1}{2}gT^2 - h\right)^2. \tag{8.12}$$

Expanding (8.12) and remembering the high school algebra technique of completing the square used to solve the general quadratic equation,

$$R^2 = v_0^2 T^2 - \frac{1}{4}g^2 T^4 + g T^2 h - h^2 = (v_0^2 + gh)T^2 - \frac{1}{4}g^2 T^4 - h^2$$

$$= -\frac{1}{4}g^2 \left\{ T^4 - \frac{4(v_0^2 + gh)}{g^2} T^2 \right\} - h^2$$

$$= -\frac{1}{4}g^2 \left\{ T^4 - \frac{4(v_0^2 + gh)}{g^2} T^2 + \frac{4(v_0^2 + gh)^2}{g^4} - \frac{4(v_0^2 + gh)^2}{g^4} \right\} - h^2$$

$$= -\frac{1}{4}g^2 \left\{ T^4 - \frac{4(v_0^2 + gh)}{g^2} T^2 + \frac{4(v_0^2 + gh)^2}{g^4} \right\} + \frac{(v_0^2 + gh)^2}{g^2} - h^2$$

$$= -\frac{1}{4}g^2 \left\{ T^2 - \frac{2(v_0^2 + gh)}{g^2} \right\}^2 + \frac{(v_0^2 + gh)^2}{g^2} - h^2,$$

or, more transparently,

$$R^2 = \frac{(v_0^2 + gh)^2}{g^2} - h^2 - \frac{1}{4}g^2 \left\{ T^2 - \frac{2(v_0^2 + gh)}{g^2} \right\}^2. \tag{8.13}$$

The first two terms on the right of (8.13) are constants, and the third term (excluding the minus sign) is, *at its least*, zero, as anything squared cannot be negative. Clearly, since we are subtracting that third term in (8.13), we maximize R^2 by making the third term equal to zero, which says, for maximum range, we want

$$T^2 = \frac{2(v_0^2 + gh)}{g^2}. \tag{8.14}$$

The maximum range is then given by

$$R_m^2 = \frac{(v_0^2 + gh)^2}{g^2} - h^2 = \frac{v_0^4 + 2ghv_0^2 + g^2h^2 - g^2h^2}{g^2},$$

or

$$R_m = \frac{v_0 \sqrt{v_0^2 + 2gh}}{g}. \tag{8.15}$$

From (8.10) and (8.11) we have

$$\frac{v_0 T \sin(\theta_0)}{v_0 T \cos(\theta_0)} = \frac{\frac{1}{2}gT^2 - h}{R} = \tan(\theta_0),$$

and so, inserting (8.14) and (8.15) for T^2 and R_m for R, respectively, we have the launch angle θ_0 that gives us the maximum range:

$$\tan(\theta_{0\,\text{max}}) = \frac{\frac{1}{2}g\frac{2(v_0^2+gh)}{g^2} - h}{\frac{v_0\sqrt{v_0^2+2gh}}{g}} = \frac{v_0^2}{v_0\sqrt{v_0^2+2gh}},$$

or

$$\tan(\theta_{0\,\text{max}}) = \frac{v_0}{\sqrt{v_0^2+2gh}}. \tag{8.16}$$

Notice that if $h = 0$, then $\tan(\theta_{0\,\text{max}}) = 1$, and so $\theta_{0\,\text{max}} = 45$ degrees, as derived earlier. But, if $h > 0$, then $\tan(\theta_{0\,\text{max}}) < 1$, and so $\theta_{0\,\text{max}} < 45$ degrees. Now, just to see how the numbers go, the world outdoor record for the shot-put is just short of 76 feet. If we assume the shot is launched from a shoulder height of $h = 6$ feet at the optimal angle, then (8.15) says (where I'm using English units and so, in particular, v_0 will be in ft/s)

$$76 = \frac{v_0\sqrt{v_o^2 + 2 \cdot 32.2 \cdot 6}}{32.2},$$

or, after a little quadratic equation algebra, we get

$$v_0 = 47.6\,\text{ft/s}.$$

If instead the shot was launched with this speed at a 45 degrees angle, then (8.14) says

$$T = \frac{\sqrt{2(v_0^2+gh)}}{g} = \frac{\sqrt{2(47.6^2+32.2\cdot 6)}}{32.2} = 2.18\,\text{seconds},$$

and so (8.10) gives a range of

$$R = 47.6 \cdot 2.18 \cdot \cos(45°) = \frac{103.8}{\sqrt{2}} = 73.4\,\text{feet},$$

considerably less than the record. The optimal launch angle is not 45 degrees, of course, but rather, from (8.16),

$$\theta_{0\,\text{max}} = \tan^{-1}\left(\frac{47.6}{\sqrt{47.6^2 + 2 \cdot 32.2 \cdot 6}}\right) = \tan^{-1}(0.924) = 42.7 \text{ degrees}.$$

I suspect that what makes the difference between a gold and a silver medal at the Olympics (or between a silver and a bronze) is less how brawny a competitor is—*all* Olympic-caliber shot-putters look pretty strong to me!—but rather how well each comes to achieving a launch angle that is as close as possible to *their* $\theta_{0\,\text{max}}$ (which in turn depends on their individual values of v_0 and h). The value of v_0 *is*, of course, strength dependent.

8.3 Another Cannon Question

To complete the present discussion, let me show you a result following directly from our analyses that surprises most people when they first see it. Our work here has ignored air drag and so what follows is for a gun shooting in a vacuum—on the moon, say—but the result is such that even in a vacuum, the conclusion catches attention. If you walk up to most people and say, "Quick, true or false: if I fire a cannon, then the shell is always traveling away from the muzzle," most people will at first reply, "of course." And then, after a few seconds of thought, most will add, "unless, of course, you fire it straight up—then the shell will first travel away from the muzzle as it rises up to its maximum height, but then it will fall back down on your head!"

That's *mostly* correct. But the scenario does motivate a curious question. Suppose that, instead of firing exactly straight up (that is, at an angle of $\theta_0 = 90$ degrees with respect to the horizontal), we fire the cannon at an angle of $\theta_0 = 89$ degrees. I think it clear that while the shell now won't fall on your head in this case, it will still return to Earth fairly close to the cannon. That means that for the latter part of its trajectory, the shell's distance from the cannon was indeed *decreasing*. What about a firing angle of 88 degrees? Of 87 degrees? In those cases, too, there will be intervals of time when the shell is traveling *back down toward the cannon.* So, here's the question. As we continue to reduce the firing angle, eventually we'll reach a critical angle θ_c such that, if $\theta_0 < \theta_c$, only then will the shell *always* be moving away from the cannon muzzle; what is θ_c? I think you'll be surprised by how small θ_c turns out to be. In the analysis to follow I'll assume both that $\phi = 0$ (the cannon is firing over a level field) and that $h = 0$. And now I *will* calculate a derivative.

From (8.3) and (8.4) we can write the instantaneous distance squared of the shell from the cannon as

$$l^2 = x^2(t) + y^2(t) = [v_0 t \cos(\theta_0)]^2 + [v_0 t \sin(\theta_0) - \frac{1}{2}gt^2]^2,$$

or, with very little algebra,

$$l^2 = v_0^2 t^2 - g v_0 t^3 \sin(\theta_0) + \frac{1}{4} g^2 t^4.$$

(I'll be working with l^2 rather than l because the two quantities increase and decrease in step, and l^2 avoids the complication of the square-root operation that comes with l.) From the last result we can write the instantaneous rate of change of l^2 as

$$\frac{d(l^2)}{dt} = 2v_0^2 t - 3g v_0 t^2 \sin(\theta_0) + g^2 t^3,$$

or

$$\frac{d(l^2)}{dt} = g^2 t \left[t^2 - \frac{3v_0}{g} \sin(\theta_0) t + \frac{2v_0^2}{g^2} \right]. \quad (8.17)$$

Since $g^2 t \geq 0$ for all $t \geq 0$ (that is, over the entire duration of the shell's trajectory), we see that whether or not the shell is travelling away from or toward the cannon is completely determined by the algebraic sign of the quantity in brackets in (8.17). That is, when $f(t) = t^2 - \frac{3v_0}{g} \sin(\theta_0) t + \frac{2v_0^2}{g^2} > 0 \Longrightarrow$ the shell is traveling *away* from the cannon, and when $f(t) = t^2 - \frac{3v_0}{g} \sin(\theta_0) t + \frac{2v_0^2}{g^2} < 0 \Longrightarrow$ the shell is traveling *toward* the the cannon.

The function $f(t)$ describes a *derivative parabola*, as shown in Figure 8.4. Whether or not that parabola lies—as shown—completely above the t-axis ($f(t) > 0$ so that the shell is traveling away from the cannon) or instead cuts through the t-axis ($f(t) < 0$ between the two values

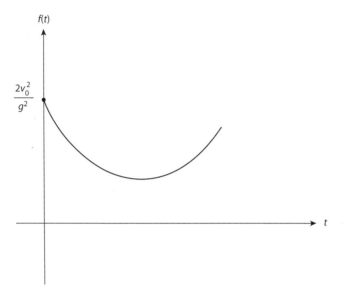

Figure 8.4. Derivative parabola.

of t, where $f(t) = 0$ which thus defines an interval of time during which the shell is traveling back toward the cannon) is determined by the value of $\sin(\theta_0)$, that is, by θ_0. In fact, if we solve the quadratic equation $f(t) = 0$, we get

$$t = \frac{1.5 v_0}{g} \sin(\theta_0) \pm \frac{0.5 v_0}{g} \sqrt{9 \sin^2(\theta_0) - 8}. \tag{8.18}$$

For the straight-up case of $\theta_0 = 90$ degrees, (8.18) says that $f(t) = 0$ has two solutions, $t = \frac{v_0}{g}$ and $t = \frac{2v_0}{g}$. These two times—the first is the time at which the shell reaches its maximum height and then begins to fall back to Earth, and the second is the time of impact (on your head!)—correctly define the interval over which the shell is moving back toward the cannon.

If $9 \sin^2(\theta_0) - 8 < 0$, then the solutions of (8.18) are complex (there are no real solutions), which geometrically means the derivative parabola lies completely above the t-axis and the shell is always traveling away from the cannon. Clearly, the value of θ_c is that value of θ_0 such that $9 \sin^2(\theta_c) - 8 = 0$, which means the derivative parabola just touches the t-axis. Thus, there will be times when the shell is traveling back toward the cannon *unless* $\theta_0 < \theta_c$, where

$$\theta_c = \sin^{-1}\left(\frac{\sqrt{8}}{3}\right) = \cos^{-1}\left(\frac{1}{3}\right) = 70.53 \text{ degrees},$$

a value that is, I think, remarkably less than 90 degrees.

CP. P8.1:

Suppose a gun, with its breech (where the shell sits) height h above a level field, has a maximum range of R_m. If the gun barrel is then leveled to be parallel with the ground, in preparation for what is commonly called a point-blank shot, then what is the point-blank range R_{pb} in terms only of h and R_m? *Hint*: Use (8.15) and, when useful, the approximation $R_m >> h$. If $R_m = 20{,}000$ feet and $h = 4$ feet, calculate R_{pb}. Are you surprised by the size of the result?

Notes and References

1. For the $\phi > 0$ case, a geometrical construction is given in A. R. Hall, *Ballistics in the Seventeenth Century* (Cambridge: Cambridge University Press, 1952, p. 94).

2. You can find much more on Halley's observation in my book, *When Least Is Best* (Princeton N.J.: Princeton University Press, 2004 [corrected ed. 2007], pp. 165–171).

Ballistics—With Air Drag

> The history of the mathematical investigation of the path described by a projectile in a resisting medium... is somewhat curious.
> — Francis Bashworth, *A Mathematical Treatise on the Motion of Projectiles* (1873)

9.1 Thin Air *Cannot* Be Ignored!

In this discussion I'll essentially repeat the previous one, but now with the realistic complication of air drag. You'll see, when we compare the numerical predictions of the two analyses, that to ignore air drag in ballistic calculations is to accept potentially huge errors. The problem of ballistics with air drag is an old one that can be traced back to a very nasty conflict between two mathematicians in 1718. One, the Scottish-born John Keill (1671–1721), formulated the air drag ballistic problem as a challenge to the other, the Swiss Johann Bernoulli (1667–1748). The two had tangled before in a bitter conflict,[1] and Keill's challenge problem was clearly an attempt to embarrass Bernoulli. Keill intended his problem to stump Bernoulli and thus humiliate him in the eyes of other mathematicians. It was, for Keill, payback time.

In his monumental *Principia* (1687), Newton had treated the ballistic problem with air drag for the unrealistic case of a linear drag force. For Newton this was a purely mathematical exercise, however, and he well knew the lack of any real-life application of his result; as he wrote in

Principia, the "resistance of bodies in the ratio of the velocity is more a mathematical hypothesis than a physical one. In mediums void of all tenacity, the resistances made to bodies are as the square of the velocities." But he did not present a solution to that more realistic problem. Why not? Probably because it is a much tougher problem. But that doesn't mean he couldn't have solved it. As you'll soon see, it *is* a treatable problem.

Since Newton had not published a solution to the realistic case of a square law drag force, Keill no doubt thought it must be a very, very hard problem, and so believed it would be the perfect one with which to challenge (and embarrass) Bernoulli. In February 1718 he challenged Bernoulli to find the trajectory of a projectile shot through a medium of uniform density and a uniform gravitational field, under the assumption of a square law air drag force. "This will put that bloody Swiss fool in his place," Keill may have chuckled to himself. If so, alas for Keill, because Bernoulli, who was a far better mathematician than Keill, quickly found a solution for a drag force that varied as *any* power of velocity, and then he turned the tables on Keill and counterchallenged him to produce a solution to his own problem. *That* the outfoxed Keill could not do (important lesson: never ask a challenge question for which you don't already know the answer!), and Bernoulli had the last laugh; he finally published his solution in 1721.

It was not an easy solution, however. As one later writer put it, "The problem of the motion of a projectile in a medium resisting as any power of velocity was first solved by [Johann] Bernoulli in 1718, but he expressed the coordinates of any point of the trajectory in a form where a final integration is required to determine them numerically; and, as far as the general problem is concerned, no later mathematician has improved on his results."[2] With the easy access today to powerful desktop computers running sophisticated scientific programming applications like MATLAB, the numerical number-crunching obstacles that seemed so formidable in the eighteenth, nineteenth, and most of the twentieth century have been completely removed. What I'll show you next is how a modern analysis solves Keill's challenge problem.[3] The mathematics is quite clever.

In Figure 9.1 I've shown our projectile of mass m launched at an initial speed and angle (measured with respect to the horizontal) of

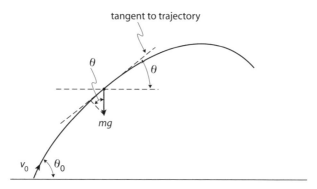

Figure 9.1. The geometry of a projectile's trajectory in a resistive medium.

v_0 and θ_0, respectively. At an arbitrary point (speed v and angle θ) on its trajectory there are two forces acting on the projectile: the gravitational force directed, *always*, straight down, and and the air drag force directed in the instantaneously *opposite* direction to that of the projectile's motion (the air resistance is said to be *antiparallel* to the projectile's velocity). If we write the air drag as kv^2, then the component of the gravitational force *along* the path of the projectile is $mg\sin(\theta)$, and so we immediately have, from Newton's second law,

$$m\frac{dv}{dt} = -mg\sin(\theta) - kv^2. \tag{9.1}$$

Now, what of the gravitational force component *normal* to the projectile's path? Here we use the concept of the instantaneous radius of curvature, ρ, due to Newton: ρ is the radius of the circular path that, instantaneously, at any given point, best matches the actual motion of the projectile (by *best*, it is meant that circular path that has its first and second derivatives equal to those of the actual trajectory at the given point[4]). Clearly, $\rho = \rho(t)$, as that best circular path is generally different for every point along the trajectory. In any case, we can write the instantaneous centripetal force on the projectile as

$$\frac{mv^2}{\rho} = mg\cos(\theta). \tag{9.2}$$

Now, since by definition (see note 4 again),

$$\rho = -\frac{ds}{d\theta}, \tag{9.3}$$

where $s(t)$ is the arc-length of the trajectory traversed at time t (notice that $v = \frac{ds}{dt}$), then by the chain rule of calculus we have

$$\rho = -\frac{ds}{dt} \cdot \frac{dt}{d\theta} = -v\frac{dt}{d\theta}.$$

Inserting this into (9.2) we have

$$\frac{v^2}{-v\frac{dt}{d\theta}} = g\cos(\theta) = -\frac{v}{\frac{dt}{d\theta}} = -v\frac{d\theta}{dt},$$

which reduces to

$$dt = -\frac{v}{g\cos(\theta)}d\theta. \tag{9.4}$$

Putting this into (9.1),

$$\frac{m\,dv}{-\frac{v}{g\cos(\theta)}d\theta} = -mg\sin(\theta) - kv^2 = -\frac{mg\cos(\theta)}{v} \cdot \frac{dv}{d\theta},$$

or

$$\frac{dv}{d\theta} = v\frac{\sin(\theta)}{\cos(\theta)} + \frac{k}{mg} \cdot \frac{v^3}{\cos(\theta)},$$

or

$$dv\cos(\theta) - v\sin(\theta)\,d\theta = \frac{k}{mg}v^3\,d\theta. \tag{9.5}$$

The reason for writing (9.5) in the form shown is because the differential of $v\cos(\theta)$, that is, $d[v\cos(\theta)]$, is given by the product rule for differentials as

$$d[v\cos(\theta)] = dv\cos(\theta) + v\,d[\cos(\theta)] = dv\cos(\theta) - v\sin(\theta)\,d\theta,$$

which is precisely the left-hand side of (9.5). Thus,

$$d[v\cos(\theta)] = \frac{k}{mg}v^3\,d\theta,$$

and so

$$\frac{d[v\cos(\theta)]}{v^3\cos^3(\theta)} = \frac{\frac{k}{mg}v^3\,d\theta}{v^3\cos^3(\theta)},$$

or
$$\frac{d[v\cos(\theta)]}{[v\cos(\theta)]^3} = \frac{k}{mg} \cdot \frac{d\theta}{\cos^3(\theta)}. \tag{9.6}$$

Despite what may appear to be a nasty-looking expression in (9.6), we have actually made great progress, because both sides of (9.6) are immediately integrable. For the left-hand side, just think of $\int \frac{dz}{z^3} = -\frac{1}{2z^2}$, with $z = v\cos(\theta)$. From integral tables we have, with $-\frac{1}{2}C$ as a constant of indefinite integration,

$$\frac{-1/2}{[v\cos(\theta)]^2} = \frac{k}{mg}\left[\frac{1/2\sin(\theta)}{\cos^2(\theta)} + 1/2\ln\left\{\tan\left(\frac{\pi}{4} + \frac{\theta}{2}\right)\right\}\right] - 1/2C.$$

Or

$$\frac{1}{[v\cos(\theta)]^2} = -\frac{k}{mg}\left[\frac{\tan(\theta)}{\cos(\theta)} + \ln\left\{\tan\left(\frac{\frac{\pi}{2}+\theta}{2}\right)\right\}\right] + C. \tag{9.7}$$

Remembering the half-angle formula for the tangent, that is

$$\tan\left(\frac{\alpha}{2}\right) = \frac{1-\cos(\alpha)}{\sin(\alpha)},$$

we have, with $\alpha = \frac{\pi}{2} + \theta$,

$$\tan\left(\frac{\frac{\pi}{2}+\theta}{2}\right) = \frac{1-\cos(\frac{\pi}{2}+\theta)}{\sin(\frac{\pi}{2}+\theta)}$$

which becomes, using the well-known formulas for the sine and cosine of the sum of two angles,

$$\tan\left(\frac{\frac{\pi}{2}+\theta}{2}\right) = \frac{1-[\cos(\frac{\pi}{2})\cos(\theta) - \sin(\frac{\pi}{2})\sin(\theta)]}{\sin(\frac{\pi}{2})\cos(\theta) + \cos(\frac{\pi}{2})\sin(\theta)}$$

$$= \frac{1+\sin(\theta)}{\cos(\theta)} = \sec(\theta) + \tan(\theta).$$

Thus, (9.7) becomes

$$\frac{1}{[v\cos(\theta)]^2} = -\frac{k}{mg}[\tan(\theta)\sec(\theta) + \ln\{\sec(\theta) + \tan(\theta)\}] + C. \tag{9.8}$$

Finally, since $v = v_0$ when $\theta = \theta_0$, the constant of integration is

$$C = \frac{1}{v_0^2\cos^2(\theta_0)} + \frac{k}{mg}[\tan(\theta_0)\sec(\theta_0) + \ln\{\sec(\theta_0) + \tan(\theta_0)\}].$$

Putting this result for C back into (9.8), and then doing just a bit of easy algebra to solve for v (as a function of θ), we have

$$v(\theta) = \frac{\sec(\theta)}{\sqrt{\left[\frac{\sec(\theta_0)}{v_0}\right]^2 + \frac{k}{mg}\left[\tan(\theta_0)\sec(\theta_0) - \tan(\theta)\sec(\theta) + \ln\left\{\frac{\sec(\theta_0)+\tan(\theta_0)}{\sec(\theta)+\tan(\theta)}\right\}\right]}}.$$

(9.9)

Now, with $v(\theta)$ in hand from (9.9), we can easily find the projectile's trajectory, either in the form $y = y(x)$ or as $y = y(t)$, as follows. From the chain rule we have

$$\frac{dx}{d\theta} = \frac{dx}{ds} \cdot \frac{ds}{d\theta},$$

or, using (9.3) and the geometry shown in Figure 9.1 that relates the differentials ds and dx, we have

$$\frac{dx}{d\theta} = -\rho \frac{dx}{ds} = -\rho \cos(\theta),$$

and this in turn, using (9.2), becomes

$$\frac{dx}{d\theta} = -\frac{v^2}{g}. \quad (9.10)$$

Also,

$$\frac{dy}{d\theta} = \frac{dy}{ds} \cdot \frac{ds}{d\theta} = -\rho \frac{dy}{ds} = -\rho \sin(\theta) = -\rho \cos(\theta)\tan(\theta),$$

or

$$\frac{dy}{d\theta} = -\frac{v^2}{g}\tan(\theta). \quad (9.11)$$

And finally,

$$\frac{dt}{d\theta} = \frac{dt}{ds} \cdot \frac{ds}{d\theta} = -\rho \frac{dt}{ds} = -\rho \frac{1}{v} = -\frac{\rho \cos(\theta)}{v \cos(\theta)} = -\frac{\frac{v^2}{g}}{v \cos(\theta)},$$

or

$$\frac{dt}{d\theta} = -\frac{v}{g}\sec(\theta). \quad (9.12)$$

And now the formal solution to the problem is obvious. We find $x(\theta)$, $y(\theta)$, and $t(\theta)$ by simply integrating (9.10), (9.11), and (9.12), respectively, that is,

$$x(\theta) = -\frac{1}{g} \int_{\theta_0}^{\theta} v^2(u)\, du \qquad (9.13)$$

and

$$y(\theta) = -\frac{1}{g} \int_{\theta_0}^{\theta} v^2(u) \tan(u)\, du \qquad (9.14)$$

and

$$t(\theta) = -\frac{1}{g} \int_{\theta_0}^{\theta} v(u) \sec(u)\, du. \qquad (9.15)$$

From (9.13) and (9.14) we have $y = y(x)$, from (9.13) and (9.15) we have $x = x(t)$, and from (9.14) and (9.15) we have $y = y(t)$. Also, in the same way, we can calculate the arc-length of the trajectory from (9.2). That is,

$$\rho = \frac{v^2}{g \cos(\theta)},$$

and so, since

$$\rho = -\frac{ds}{d\theta},$$

then

$$s = -\int_{\theta_0}^{\theta} \rho\, d\theta = -\frac{1}{g} \int_{\theta_0}^{\theta} \frac{v^2(u)}{\cos(u)}\, du,$$

or, finally,

$$s(\theta) = -\frac{1}{g} \int_{\theta_0}^{\theta} v^2(u) \sec(u)\, du. \qquad (9.16)$$

9.2 Air Drag and Baseball

It was of course the "doing" of these integrations that so bedeviled early analysts—particularly so, given the extraordinarily complicated nature of $v(\theta)$ in (9.9)—but today there is no problem. Indeed, let me next show you a practical application of these equations and how they are evaluated on a modern desktop computer. Imagine a baseball player at bat on a level playing field who gets a good hit on a pitch. The ball leaves his bat with an initial speed of $v_0 = 100$ miles per hour (= 146.7 ft/s) at an initial angle of $\theta_0 = 60$ degrees. What is the trajectory of the ball? (Let's agree to say that the trajectory starts at the instant the ball leaves the bat and ends when the ball is again at its initial height. Alternatively, we'll imagine our batter is very short!) More specifically, we could ask detailed questions, such as how high will the ball go? how far will the ball go? how long is the ball in the air? with what angle does the ball hit the ground? how fast is the ball moving when it hits the ground? and so on. In a vacuum the answers are easy to calculate. If you look back at the previous discussion, you'll find that, for a level playing field ($\phi = 0$), the horizontal distance traveled by the ball is, from (8.5),

$$R = \frac{v_0^2}{g} \sin(2\theta_0) = \frac{(146.7)^2}{32.2} \sin(120°) = 578.8 \text{ feet},$$

and that the flight time of the ball is

$$T = \frac{2v_0}{g} \sin(\theta_0) = \frac{2 \cdot 146.7}{32.2} \sin(60°) = 7.89 \text{ seconds}.$$

By symmetry (in a vacuum) the ball reaches its maximum height at time $t = \frac{T}{2}$, and so, from (8.4), the maximum height is

$$H = v_0 \frac{T}{2} \sin(\theta_0) - \frac{1}{2} g \frac{T^2}{4} = \frac{v_0^2}{g} \sin^2(\theta_0) - \frac{v_0^2}{2g} \sin^2(\theta_0)$$

$$= \frac{v_0^2}{2g} \sin^2(\theta_0) = \frac{(146.7)^2}{2 \cdot 32.2} \sin^2(60°) = 250.6 \text{ feet}.$$

Again by symmetry, we expect the ball traveling through a vacuum to return to the ground with a speed equal to the initial speed, at an angle equal to the negative of the launch angle. How much are these numbers changed by taking into account air drag? As you'll see next, by a lot.

The numerical values for k and mg that appear in (9.9), for a major-league baseball, are (in English units, which means all speeds are in feet per second and all distances are in feet),[6] $k = 1.63 \cdot 10^{-5}$ lb·s²/ft² and $mg = 0.32$ lb (this is the weight of a regulation baseball). So, for

$\theta_0 = 60$ degrees and for $v_0 = 146.7$ ft/s, we have the numerical values

$\tan(\theta_0) = \sqrt{3} = 1.732$,
$\sec(\theta_0) = 2$,
$\tan(\theta_0)\sec(\theta_0) = 3.464$,
$\sec(\theta_0) + \tan(\theta_0) = 3.732$,
$\left[\frac{\sec(\theta_0)}{v_0}\right]^2 = 1.86 \cdot 10^{-4}$,
$\frac{k}{mg} = 5.1 \cdot 10^{-5}$.

Thus, from (9.9) we have, for our baseball problem,

$$v(\theta) = \frac{\sec(\theta)}{\sqrt{1.86 \cdot 10^{-4} + 5.1 \cdot 10^{-5}\left[3.464 - \tan(\theta)\sec(\theta) + \ln\left\{\frac{3.732}{\sec(\theta) + \tan(\theta)}\right\}\right]}}.$$

(9.17)

It is this expression that is used in (9.13) and (9.14) to find $x(\theta)$ and $y(\theta)$ and then, by simply plotting y vs. x, to obtain the trajectories $y = y(x)$ that are shown in Figure 9.2. The two curves in that figure are labeled "no drag" and "drag" for the cases of a trajectory in a vacuum and a trajectory in air, respectively. As you can see, there is a big difference between the two trajectories.

Like the other computer-generated art in this book, I used MATLAB to easily generate Figure 9.2. What was once a huge obstacle to the analysts of yesteryear is now reduced to a freshman physics, math, and computer programming assignment. As part of that computer code, I also used (9.15) to find $t(\theta)$ and (9.17) to find $v(\theta)$, and this allowed the construction of the following table of comparison values for R, T, and H, between the "no drag" and "drag" cases. The "angle of return" and the "speed of the return," both at the instant of ground impact, were also determined for the two cases. The first two columns of the table provide a partial check on how well the code works—and the numbers do check.

	no drag (theoretical)	no drag (computer)	drag (computer)
R	578.8 feet	578.4 feet	318 feet
T	7.89 seconds	7.88 seconds	6.53 seconds
H	250.6 feet	250.5 feet	172.3 feet

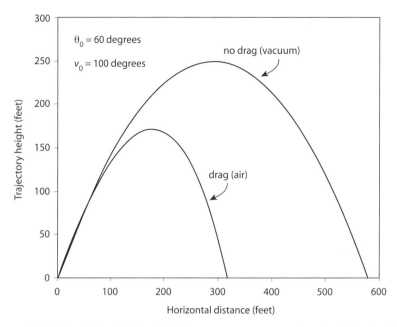

Figure 9.2. The flight of a baseball in a vacuum compared to its flight in air.

The angle of descent, for the no drag case, is indeed -60 degrees, but in the drag case the descent angle at impact is steeper: -70.1 degrees. For the no drag case the ball speed at impact is, as argued earlier, equal to the launch speed, but it is considerably slower in the case of air drag; 96.2 ft/s = 65.6 mph. And finally, in the case of no drag, the ball has its minimum speed, as you might expect, at the top of its trajectory (73.3 ft/s = 50 mph), but in the case of air drag the minimum speed (47.9 ft/s = 32.7 mph) occurs shortly after the top of the trajectory is passed, on the decending portion of the trajectory.

To conclude this discussion, let me show you one last comparison between the vacuum and air trajectories of a baseball. This particular comparison is especially interesting because, unlike the earlier comparisons, we can do the calculations *exactly*, even for the air drag case, with no need to turn to a computer (although, of course, we can use the computer to confirm the theoretical results). Specifically, I'll derive an expression for the arc length S from trajectory start ($\theta = \theta_0$) to trajectory top ($\theta = 0$ degrees, the value of θ at the top of *any* trajectory, for *any* value of k); I'll start with (9.9) for $v(\theta)$ and put it into (9.16) with $\theta = 0$ (and $S = s(0)$) to get

$$S = -\frac{1}{g} \int_{\theta_0}^{0} v^2(u) \sec(u) \, du$$

$$= -\frac{1}{g} \int_{\theta_0}^{0} \frac{\sec^3(\theta)}{\left[\frac{\sec(\theta_0)}{v_0}\right]^2 + \frac{k}{mg}\left[\tan(\theta_0)\sec(\theta_0) - \tan(\theta)\sec(\theta) + \ln\left\{\frac{\sec(\theta_0)+\tan(\theta_0)}{\sec(\theta)+\tan(\theta)}\right\}\right]} d\theta,$$

which, with a little easy algebra and trigonometric rewriting (which you should do) becomes

$$S = \int_{\theta_0}^{0} \frac{-\frac{v_0^2 \cos^2(\theta_0)}{g \cos^3(\theta)}}{1 - \frac{k v_0^2 \cos^2(\theta_0)}{mg}\left[\ln\left\{\frac{\sec(\theta)+\tan(\theta)}{\sec(\theta_0)+\tan(\theta_0)}\right\} + \frac{\sin(\theta)}{\cos^2(\theta)} - \frac{\sin(\theta_0)}{\cos^2(\theta_0)}\right]} d\theta. \quad (9.18)$$

This may look like an impossible mess, but (I think astonishingly) the integral of (9.18) can be evaluated exactly.

To show you this, let's define the function $f(\theta)$ as follows:

$$f(\theta) = 1 - \frac{k v_0^2 \cos^2(\theta_0)}{mg}\left[\ln\left\{\frac{\sec(\theta)+\tan(\theta)}{\sec(\theta_0)+\tan(\theta_0)}\right\} + \frac{\sin(\theta)}{\cos^2(\theta)} - \frac{\sin(\theta_0)}{\cos^2(\theta_0)}\right],$$

that is, as the denominator of the integrand of the integral in (9.18). Then, if we calculate $\frac{df(\theta)}{d\theta}$ we find, again after a bit of easy algebra (that you should confirm) that

$$\frac{df(\theta)}{d\theta} = -\frac{2k v_0^2 \cos^2(\theta_0)}{mg \cos^3(\theta)},$$

or

$$-\frac{v_0^2 \cos^2(\theta_0)}{g \cos^3(\theta)} d\theta = \frac{m}{2k} df(\theta).$$

Thus, the initially horrible-looking (9.18) collapses to the wonderfully easy

$$S = \frac{m}{2k} \int_{\theta=\theta_0}^{\theta=0} \frac{df(\theta)}{f(\theta)} = \frac{m}{2k} \left(\ln\{f(\theta)\}\right)|_{\theta=\theta_0}^{\theta=0} = \frac{m}{2k} \left[\ln\{f(0) - \ln f(\theta_0)\}\right],$$

or, as $f(\theta_0) = 1$ and so $\ln f(\theta_0) = 0$, we have

$$S = \frac{m}{2k} \ln \{f(0)\} = \frac{m}{2k} \ln \left\{ 1 - \frac{kv_0^2 \cos^2(\theta_0)}{mg} \right.$$
$$\left. \times \left[\ln \left\{ \frac{1}{\sec(\theta_0) + \tan(\theta_0)} \right\} - \frac{\sin(\theta_0)}{\cos^2(\theta_0)} \right] \right\}.$$

Since

$\sec(60°) = 2,$
$\tan(60°) = \sqrt{3},$
$\sin(60°) = \sqrt{3}/2,$
$\cos(60°) = 1/2,$

then

$$S = \frac{m}{2k} \ln \left\{ 1 - \frac{kv_0^2}{4mg} \left[\ln \left\{ \frac{1}{2+\sqrt{3}} \right\} - 2\sqrt{3} \right] \right\}. \quad (9.19)$$

Putting numbers into (9.19) for our baseball (notice that $m = mg/g$), we have the length of the ascending portion of the ball's trajectory as

$$S = \frac{0.32}{2 \cdot 32.2 \cdot 1.63 \cdot 10^{-5}} \ln \left\{ 1 - \frac{1.63 \cdot 10^{-5} (146.7)^2}{4 \cdot 0.32} [\ln(0.268) - 3.464] \right\},$$

or

$$S = 255.2 \text{ feet.}$$

As a check, the same computer code that I used earlier gave a value for S of 255 feet (and for the entire trajectory, from start to ground impact, the code gave a length of 490.9 feet). In comparison, when the code was run for a vacuum trajectory ($k = 0$), the value of S was declared to be the much longer 399.1 feet (and 798.3 feet for the entire trajectory length; this doubling is just what we would expect for the symmetrical vacuum trajectory). There are many, many more such questions we could explore with computer code, but I think this is enough on air drag.[7] So, I'll stop here and simply ask you to try and imagine what Johann Bernoulli would have given for a modern desktop computer.

CP. P9.1:

Take the limit $k \to 0$ in (9.19) and use the result to compute S for the vacuum trajectory. How well does the answer agree with the computer code's estimate of 399.1 feet? *Hint*: Remember that $\ln(1-x) \approx -x$ as $x \to 0$.

Notes and References

1. You can read more about this conflict in a paper by Niccolò Guicciardini, "Johann Bernoulli, John Keill and the Inverse Problem of Central Forces" (*Annals of Science*, November 1995, pp. 537–575). Keill and Bernoulli had fought even before the central force debate; as Guicciardini writes, "Keill was one of the most loyal and assiduous among the defenders of Newton in the priority dispute with Leibniz over the invention of the calculus. Keill's bad temper and offensive attitude poisoned the debate at a level that embarrassed even Newton." Bernoulli was a vigorous supporter of Leibniz over Newton, a position that enraged the sycophantic Keill, who in 1708 had accused Leibniz of plagiarizing Newton's calculus (a claim rejected by modern historians of science as absurd). For more on the calculus debate and Keill's role in it, see A. Rupert Hall, *Philosophers at War* (Cambridge: Cambridge University Press, 1980). I should tell you that not everyone agrees with the biographical interpretation of my story of Keill versus Bernoulli. See, for example, D. T. Whiteside, "The Mathematical Principles Underlying Newton's *Principia Mathematica*" (*Journal for the History of Astronomy* 1, 1970, pp. 116–138 [in particular, see Whiteside's note 39 on p. 135]).

2. Frank Gilman, "The Ballistic Problem" (*Annals of Mathematics*, April 1905, pp. 127–137).

3. In this discussion I am following the technical presentation in the paper by A. Tan, C. H. Frick, and O. Castillo, "The Fly Ball Trajectory: An Older Approach Revisited" (*American Journal of Physics*, January 1987, pp. 37–40).

4. The *curvature* of our ballistic trajectory is a measure of how fast the trajectory is bending at each point. More technically, it is the rate of change of the angle θ of the tangent line to the trajectory with respect to the arc-length s traversed by the projectile, from its initial firing point to the tangent point. That is, using the traditional κ (kappa) to denote the curvature, we have

$$\kappa = \frac{d\theta}{ds}.$$

As Figure 9.1 indicates, $\theta = \theta_0$ at the start of the trajectory and thereafter continually decreases (reaching, for example, $\theta = 0$ at the top of the trajectory and then going negative on the falling portion of the trajectory). On the other hand, s starts with the value zero and thereafter continually increases. Therefore, the curvature of the ballistic trajectory is everywhere negative. Now, from Figure 9.1 we have

$$\tan(\theta) = \frac{dy}{dx},$$

while from the Pythagorean theorem we have

$$ds = \sqrt{(dx)^2 + (dy)^2} = dx\sqrt{1 + \left(\frac{dy}{dx}\right)^2}.$$

So

$$\frac{ds}{dx} = \sqrt{1 + \left(\frac{dy}{dx}\right)^2}.$$

Since

$$\theta = \tan^{-1}\left(\frac{dy}{dx}\right),$$

then

$$\frac{d\theta}{dx} = \frac{d}{dx}\tan^{-1}\left(\frac{dy}{dx}\right) = \frac{\frac{d^2y}{dx^2}}{1 + \left(\frac{dy}{dx}\right)^2}.$$

This gives us

$$\kappa = \frac{d\theta}{ds} = \frac{\frac{d\theta}{dx}}{\frac{ds}{dx}} = \frac{\frac{\frac{d^2y}{dx^2}}{1+\left(\frac{dy}{dx}\right)^2}}{\sqrt{1 + \left(\frac{dy}{dx}\right)^2}} = \frac{\frac{d^2y}{dx^2}}{\left\{1 + \left(\frac{dy}{dx}\right)^2\right\}^{3/2}}.$$

Let's write ρ as the radius of the circle that is tangent to the trajectory at a given point and which has the same curvature as does the trajectory itself at that point. This circle, called by Leibniz the *osculating* (or "kissing") circle, is the circle that has the most intimate contact with the trajectory at that point among all possible tangent circles. We can find an expression for ρ, called the *radius of curvature*, as follows. Let the center of the osculating circle be at (a,b), that is, the equation of the circle is

$$(x-a)^2 + (y-b)^2 = \rho^2. \tag{1}$$

Differentiating with respect to x, we get

$$2(x-a) + 2(y-b)\frac{dy}{dx} = 0, \qquad (2)$$

or

$$\frac{dy}{dx} = -\frac{x-a}{y-b}. \qquad (3)$$

Differentiating (2) with respect to x gives

$$2 + 2\left(\frac{dy}{dx}\right)^2 + 2(y-b)\frac{d^2y}{dx^2} = 0,$$

or

$$\frac{d^2y}{dx^2} = -\frac{2 + 2\left(\frac{dy}{dx}\right)^2}{2(y-b)} = -\frac{1 + \left(\frac{dy}{dx}\right)^2}{(y-b)}. \qquad (4)$$

Using (3) in (4), we have

$$\frac{d^2y}{dx^2} = -\frac{1 + \frac{(x-a)^2}{(y-b)^2}}{(y-b)} = -\frac{(y-b)^2 + (x-a)^2}{(y-b)^3},$$

or, using (1),

$$\frac{d^2y}{dx^2} = -\frac{\rho^2}{(y-b)^3}$$

or

$$\rho^2 = -(y-b)^3 \frac{d^2y}{dx^2}. \qquad (5)$$

From (4) we have

$$(y-b)^3 = -\left\{\frac{1 + \left(\frac{dy}{dx}\right)^2}{\frac{d^2y}{dx^2}}\right\}^3,$$

and so (5) becomes

$$\rho^2 = \left\{\frac{1 + \left(\frac{dy}{dx}\right)^2}{\frac{d^2y}{dx^2}}\right\}^3 \frac{d^2y}{dx^2} = \frac{\left\{1 + \left(\frac{dy}{dx}\right)^2\right\}^3}{\left\{\frac{d^2y}{dx^2}\right\}^2},$$

or

$$\rho = \pm \frac{\left\{1 + \left(\frac{dy}{dx}\right)^2\right\}^{3/2}}{\frac{d^2y}{dx^2}} = \pm\frac{1}{\kappa}.$$

Remembering that $\kappa < 0$ everywhere on the ballistic trajectory, we pick the minus sign so that $\rho > 0$ (as a good circular *radius* should be). That is,

$$\boxed{\rho = -\frac{ds}{d\theta}.}$$

Newton certainly had this formula in hand for the radius of curvature by 1671 (probably earlier, in fact); you can read about why Newton found such analyses essential in the essay by J. Bruce Brackenridge, "The Critical Role of Curvature in Newton's Developing Dynamics," in *The Investigation of Difficult Things*, ed. P. M. Harman and Alan E. Shapiro), (Cambridge: Cambridge University Press, 1992, pp. 231–260). Isn't that a wonderful title? It sounds like one of Harry Potter's assigned textbooks at Hogwarts School of Witchcraft and Wizardry! As the editors explain, however, it is actually from Newton's own writings.

5. Peter J. Brancazio, "Trajectory of a Fly Ball" (*The Physics Teacher*, January 1985, pp. 20–23).

6. If you haven't had enough yet of air drag, look up the fascinating paper by Richard H. Price and Joseph D. Romano, "Aim High and Go Far: Optimal Projectile Launch Angles Greater Than 45°" (*American Journal of Physics*, February 1998, pp. 109–113).

Gravity and Newton

> There will never be a Newton of the grass-blade.
> — The German philosopher Emmanuel Kant, who declared that, unlike physics, there could not possibly be a mathematical theory of biology

> Orbits are not difficult to comprehend. It is gravity which stirs the depths of insomnia.
> — Norman Mailer, *Of a Fire on the Moon* (1970)

> [T]his strange spirit, who was tempted by the Devil to believe at the time when within these walls he was solving so much, that he could reach *all* the secrets of God and Nature by the pure power of mind—Copernicus and Faustus in one.
> — Concluding words of Lord Keynes in a talk on Newton given at Trinity College, Cambridge (reprinted in the *Newton Tercentenary Celebration*, 1947)

10.1 The Beginnings of Modern Gravity

With two exceptions, all of the previous discussions have used Newton's analytical formulation of gravity, and in this discussion I want to continue that thread, but now in a much broader context than that of simply working with falling objects: here I'll add a bit more historical commentary, along with some additional interesting calculations. Before we are through, we will blow up planets and shrink stars! In the crudest sense, it has never been hard to "discover" gravity. After all, everybody who has ever lived did that the first time they fell down. But why does everything on Earth, without exception, experience a downward (toward the center of the Earth) directed force? Before Newton, only philosophers thought the question worthy of consideration. The common man of the day would have been inclined to think anyone who might ask such a question as loony: "What else *could* happen? Are you daft, man?"

Among philosophers, however, the nature of gravity was a mysterious, befuddling puzzle of the first rank, and all sorts of mechanical explanations were offered up by some pretty smart people. Renè Descartes (1596–1650), for example, a man whose arrogance and self-esteem were exceeded only by his genius, put forth his so-called *vortex theory*, in which gravity is the result of cosmic swirls of—*something*—that sweep the planets around the sun in the orbits. (How such a theory explains why, when you trip over the family cat in the dark, you always fall down the stairs and never up, escapes me.) The vortex theory, described by one historian of science as "a pictorial fancy," and all the other erroneous philosophical theories of gravity before and after, had one great, fatal common flaw: they were nothing but grand talk, and with them one could calculate and predict *nothing*. It was Newton who took the giant intellectual step of refusing to be sidetracked into inventing a mechanical mechanism for gravity but simply put forth an *axiomatic mathematical* starting point from which all else followed on purely analytical grounds. As Newton himself famously declared, "Hypotheses non fingo" ("I do not feign hypotheses").[1]

Pre-Newton philosophers felt a need to have a mechanical basis for gravity to avoid what is called "instantaneous action-at-a-distance," that is, a force that appears instantly between any two masses separated by a void of empty space of arbitrary vastness, and so with no visible means for supporting that interaction. Action-at-a-distance theories, such as is Newton's, were thought by many to be occult theories, bordering on the supernatural. Newton didn't like action-at-a-distance either. As he wrote in a famous passage (in a letter dated February 25, 1693, to the English cleric Richard Bentley),

> That gravity should be innate, inherent and essential to matter so that one body may act upon another at a distance through a vacuum without the mediation of anything else by and through which their action or

force may be conveyed from one to another is to me so great an absurdity that I believe no man who has in philosophical matters any competent faculty of thinking can ever fall into it.

Newton, however, didn't let his inability to explain gravity—he was as mystified as anyone else about the underlying origin of gravity[2]— stop him from attempting to mathematically *describe* gravity. At *that* attempt—with his famous inverse square law[3]—he was magnificently successful. The law is simple to state: If two point masses m_1 and m_2 are separated by distance x, then each experiences an attractive force F given by

$$F = G\frac{m_1 m_2}{x^2}, \qquad (10.1)$$

where G is the *universal gravitational constant* that has the same value everywhere in the universe. Newton's gravitational force is directed along the line joining the two masses: the gravitational force is said to be a *central force*.

It is important to realize that, while correctly expressing mathematically how Newton started his theory, (10.1) is a modern formulation; as far as is known, Newton never wrote anything like (10.1). When he did gravity calculations he always worked with ratios, not absolute quantities. For example, in the *Principia* he worked out the relative densities of the Sun, Jupiter, Saturn, and Earth, but he did not know the actual density of any of these bodies. And Newton had no conception of G; that and (10.1) came to physics long after his death.

A mass m at the surface of the Earth experiences the gravitational force mg, which we call the weight of the mass. That is, if R and M are the radius and the mass of a spherical Earth, respectively, then

$$G\frac{Mm}{R^2} = mg,$$

or

$$GM = gR^2.$$

We already know the acceleration of gravity g at the surface from the bouncing rubber ball experiment in Discussion 2, and R has been known with increasing accuracy since ancient times. Inserting numbers, we have (with $R = 3{,}960$ miles)

$$GM = (32.2 \text{ ft/s}^2)(20{,}908{,}800 \text{ ft})^2 = 1.41 \cdot 10^{16} \text{ ft}^3/\text{s}^2$$

in English units. Converting to the physicists' MKS units is easy (remembering that 1 meter equals 39.37 inches $= 39.37/12$ ft):

$$GM = 1.41 \times 10^{16} \times (12/39.37)^3 \text{m}^3/\text{s}^2 = 4 \times 10^{14} \text{ m}^3/\text{s}^2. \quad (10.2)$$

This was all in the realm of the knowable in Newton's time.

Getting the values of G and M separately, however, had to wait until 1798, when the English scientist Henry Cavendish (1731–1810) performed an incredibly delicate and most difficult experiment. Using a torsion pendulum apparatus that was actually designed and built by a friend (the English geologist John Michell [1724–1793], who died before he could use it[4]), Cavendish measured the deflection of small lead balls caused by nearby, much more massive lead balls. His published account in the *Philosophical Transactions of the Royal Society* ("Experiments to Determine the Density of the Earth") is fifty-eight pages long and is not easy reading. What Cavendish actually measured and published, as you might guess from the title of his paper, was not the value of G, and not the value of M. Rather, he determined that the average density of the Earth is

$$\rho = 5.54 \text{ g/cm}^3 = 5.54 \frac{(10^{-3}\text{kg})}{(10^{-2}\text{m})^3} = 5{,}540 \text{ kg/m}^3.$$

The density of water is 1 g/cm^3, and so, on average, the Earth is more than five times as dense as water. That's dense! This result immediately told Cavendish that the density of the interior of the Earth must be even greater—it is now believed that the Earth's inner core is nearly thirteen times as dense as water—and so all the ancient stories of a hollow Earth (a belief held, for example, by Newton's friend Edmond Halley[5]) must be fantasy.

Most college-level physics textbook descriptions of Cavendish's experiment are not particularly detailed[6] and in fact are commonly wrong on what the experiment was even about. They claim Cavendish measured the *force* between two masses and from that calculated the value of G. Not true.[7] In any case, with Cavendish's ρ, and knowledge of the Earth's radius R, the mass M of the Earth is calculated (but not by Cavendish) to be

$$M = \rho \frac{4}{3}\pi R^3 = 5.54 \cdot 10^3 \text{ kg/m}^3 \cdot \frac{4}{3}\pi (3{,}960 \text{ miles} \cdot 1{,}609 \text{ m/mile})^3$$
$$= 6 \cdot 10^{24} \text{ kg}.$$

And from this, finally, we (but not Cavendish) get the value of G from (10.2):

$$G = \frac{4 \cdot 10^{14} m^3/s^2}{6 \cdot 10^{24} kg} = 6.67 \cdot 10^{-11} \frac{m^3}{kg \cdot s^2}.$$

10.2 Newton's Superb Theorems

One of the more wonderful of Newton's inverse square law calculations was his extension of it to nonpoint masses. In particular, in the *Principia* he showed that a uniformly dense sphere of radius R (more generally, a sphere with a radially symmetric mass density) acts just like a point mass *with the mass of the sphere at the sphere's center* when it interacts with other masses at distance $r \geq R$ from the center of the sphere. (What happens when $r < R$, we'll take up in just a bit.) That is, if we have a sphere with constant density ρ, then its mass is

$$m_1 = \frac{4}{3}\pi R^3 \rho,$$

and so, if we have a point mass m_2 distance $r \geq R$ from the center of the sphere, then m_2 experiences an attractive force of

$$G \frac{\frac{4}{3}\pi R^3 \rho m_2}{r^2} = G' \frac{m_2}{r^2}, \; r \geq R,$$

where $G' = G\frac{4}{3}\pi R^3 \rho$ is simply a new constant. This result is taught in undergraduate physics classes today, but it was a very difficult result for Newton to establish, requiring his full genius to finally crack it. It is now believed by historians of science that it was this stupendous intellectual struggle, one that lasted twenty years, that delayed Newton's publication of anything about his gravity researches until the *Principia* appeared.[8] This amazing result is doubly amazing as it is true only for the inverse square law and for one other unphysical law, which I'll mention in just a bit.

An equally stunning result was Newton's discovery that the gravitational force on a point mass inside a hollow spherical shell of matter, no matter where inside the shell the point mass may be located, is *zero*. This second result, when combined with the first one, allows the description of gravity inside a *solid* sphere, that is, for $r < R$.

The proofs of these two results (the two together have been called the "superb theorems" since the end of the nineteenth century) were presented in the *Principia* as tour de force geometric proofs, but today the first is universally proved by calculus (Newton's geometric proof of zero gravity inside a spherical shell is so beautifully simple, however, that it has remained a common textbook proof as well). A calculus proof of both results is based on thinking of a sphere as a layered onion, that is, as made up of an infinity of differentially thin spherical shells. The differential force on a point mass at $r > R$ due to each shell is found, and then all the differential forces are added (integrated) to give the total force. It is not the way Newton actually proved his result in the *Principia*, but it has been suggested by some that it is the first way he did do it for himself; he is imagined to have decided to give a geometric proof in the *Principia* because the integral calculus was still far from universally understood by Newton's contemporaries in the 1680s.[9]

The classic onion-layer proof is clever and even a bit tricky, but it is easily worth the effort to understand.[10] I'll start by establishing a preliminary result. Imagine a thin circular ring of matter with mass μ and constant density, as shown in Figure 10.1. The center of the ring defines the origin of an x,y,z coordinate system, with the ring lying in the yz plane. There is a point mass m located on the x-axis such that the point mass is distance s from any point on the ring. The angle that the line joining the point mass to any point on the ring makes with the x-axis is therefore a constant, θ. The magnitude of the differential force $d\mathbf{F}$ of attraction between the point mass and a differential mass $d\mu$ of the ring (at angle ϕ from the y-axis, as shown in Figure 10.1) is, therefore, from (10.1)

$$dF = |d\mathbf{F}| = G\frac{m\,d\mu}{s^2}.$$

Now, on the directly opposite side of the ring (that is, at angle $\phi + \pi$) is another differential ring mass $d\mu$, resulting in a differential force of

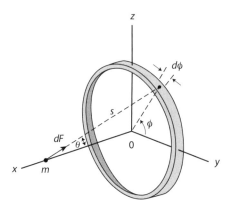

Figure 10.1. Calculating the axial gravitational force due to a thin circular ring of matter.

attraction also of magnitude dF. The components of these two forces that are normal to the x-axis cancel, while the two components along the x-axis *add*. Since the x-axis force components are $dF\cos(\theta)$, then the net attractive differential force experienced by the point mass due to the two $d\mu$ differential masses is along the x-axis, directed toward the ring center, and has magnitude

$$dF_x = 2dF\cos(\theta) = 2G\frac{m\,d\mu}{s^2}\cos(\theta).$$

Since, because of the constant density assumption, we also have

$$d\mu = \mu\frac{d\phi}{2\pi},$$

then

$$dF_x = 2G\frac{m\mu\frac{d\phi}{2\pi}}{s^2}\cos(\theta),$$

or

$$dF_x = G\frac{m\mu}{\pi s^2}\cos(\theta)\,d\phi.$$

To find the total x-axis attractive force F_x, we integrate over the entire ring, that is, we let ϕ run from 0 to π (not 2π, since our symmetry argument of the two $d\mu$ differential masses, π radians apart, has already taken into account both halves of the ring). That is,

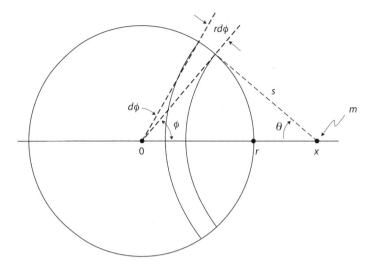

Figure 10.2. Calculating the gravitational force due to a thin spherical shell of matter.

remembering that G, μ, m, s, and θ are all constants, we have

$$F_x = G\frac{m\mu}{\pi s^2}\cos(\theta)\int_0^\pi d\phi,$$

or

$$F_x = G\frac{m\mu}{s^2}\cos(\theta). \tag{10.3}$$

In words, the ring's gravitational force on a point mass lying on the ring axis and distance s from the ring itself (not from the ring's center) is found by multiplying $G\frac{\cos(\theta)}{s^2}$ by the product of the ring mass (μ) and the point mass (m). With this preliminary result established, we can proceed to the main argument.

Imagine that we have a spherical shell with a differentially thin wall of radius r with mass density $\rho(r)$ centered at $x = 0$, where x is measured along a horizontal axis, as shown in Figure 10.2. On the horizontal axis, at distance x from the center of the shell ($x > r$), we imagine there is a point mass m. Now, imagine further that from the point mass we draw a line to the surface of the shell with length s that makes angle θ with the horizontal axis, and then we rotate the line around the horizontal axis to sweep out a circle on the shell.

A line from the center of the shell to that circle makes angle ϕ with the horizontal axis. If we increase ϕ by $d\phi$, then we'll sweep out a slightly larger circle on the shell, and these two circles on the shell define a thin circular ring. From our preliminary result, we know that the gravitational attractive force on the point mass, due to that ring, is directed along the horizontal axis and has magnitude

$$d^2 F_x = G \frac{m \, d^2\mu}{s^2} \cos(\theta),$$

where $d^2\mu$ is the mass of the ring. Notice that I've written both the force and the ring mass as second-order differentials, since at this stage we have the force on the point mass due to a differentially thin ring taken from a differentially thin shell.

The radius of the ring is $r \sin(\phi)$, its width is $r \, d\phi$, and its thickness (the thickness of the shell) is dr. Thus, as the circumference of the ring is $2\pi r \sin(\phi)$, we have

$$d^2\mu = \rho(r) 2\pi r \sin(\phi) r \, d\phi \, dr = 2\pi \rho(r) r^2 \sin(\phi) \, d\phi \, dr,$$

and so

$$d^2 F_x = G \frac{2\pi m \rho(r) r^2 \sin(\phi) \, d\phi \, dr}{s^2} \cos(\theta), \quad (10.4)$$

where you'll notice that the right-hand side of (10.4) has the product of two (first-order) differentials, consistent with the second-order differential $d^2 F_x$ on the left-hand side of (10.4). You'll also notice that I've written $\rho = \rho(r)$, that is, as a function of the shell radius.

Now, from the law of cosines we have

$$s^2 = r^2 + x^2 - 2rx \cos(\phi),$$

and so, remembering that r and x are constants, differentiating with respect to ϕ gives

$$2s \frac{ds}{d\phi} = 2rx \sin(\phi),$$

or

$$\frac{s \, ds}{x} = r \sin(\phi) \, d\phi. \quad (10.5)$$

And again, from the law of cosines,
$$r^2 = s^2 + x^2 - 2xs\cos(\theta),$$
or
$$\cos(\theta) = \frac{s^2 + x^2 - r^2}{2xs}. \tag{10.6}$$

Substituting (10.5) and (10.6) into (10.4), we have

$$\begin{aligned}
d^2F_x &= 2\pi G m \rho(r) \frac{\frac{s\,ds}{x} r\,dr}{s^2} \cdot \frac{s^2 + x^2 - r^2}{2xs} \\
&= 2\pi G m \rho(r) \frac{r\,dr\,ds}{2x^2 s^2}(s^2 + x^2 - r^2) \\
&= \frac{\pi G m \rho(r) r}{x^2}\left(\frac{x^2 - r^2}{s^2} + 1\right) dr\,ds.
\end{aligned}$$

To integrate over the entire shell, we simply integrate over all possible s, that is, we integrate from $s = x - r$ to $s = x + r$ (remember, x is measured from the *center* of the shell). So, the (first-order) differential force on the point mass, due to a differentially thin spherical shell, is

$$dF = \frac{\pi G m \rho(r) r}{x^2} dr \int_{x-r}^{x+r} \left(\frac{x^2 - r^2}{s^2} + 1\right) ds. \tag{10.7}$$

Notice that I have dropped the subscript x from dF_x because of the symmetry of a sphere, i.e., we no longer have to distinguish between different directions from the shell center because it doesn't matter where the point mass is, just that it is distance $x > r$ from the shell center.

The integral is easy to do:

$$\begin{aligned}
\int_{x-r}^{x+r} \left(\frac{x^2 - r^2}{s^2} + 1\right) ds &= (x^2 - r^2)\left[\left(-\frac{1}{s}\right)\Big|_{x-r}^{x+r}\right] + (s)|_{x-r}^{x+r} \\
&= (x^2 - r^2)\left(\frac{1}{x-r} - \frac{1}{x+r}\right) + (x+r) - (x-r) \\
&= (x^2 - r^2)\frac{x+r-x+r}{x^2 - r^2} + 2r = 2r + 2r = 4r.
\end{aligned}$$

So,
$$dF = Gm\frac{4\pi r^2\, dr\, \rho(r)}{x^2}$$
or, as $4\pi r^2\, dr\, \rho(r)$ is the total differential mass of the differentially thin shell—let's call that differential mass dM—we see that
$$dF = G\frac{m\, dM}{x^2}.$$
But this is just (10.1). That is, we can consider *all* of the mass of the spatially extended shell to be concentrated at the center of the shell (at $x = 0$). Now, a solid sphere can be thought of as an onion layering of shells (with each shell generally of a different density), and the mass of each such shell can be treated as concentrated at its center (which is, of course, the same $x = 0$ for all the shells). So, for a solid sphere of total mass M and radius R, we have
$$F = \frac{Gm}{x^2} \int\limits_{\substack{entire\\ sphere}} dM = \frac{GmM}{x^2} \tag{10.8}$$
as the sphere's gravitational force on a point mass m located at distance $x > R$ from the center of the sphere. With $\rho(r)$ as the radially symmetric mass density, the sphere's mass M is given by
$$M = 4\pi \int_0^R r^2 \rho(r)\, dr. \tag{10.9}$$
For $\rho(r) = \rho$, a constant, (10.9) integrates easily in this special case to the obvious
$$M = \frac{4}{3}\pi R^3 \rho.$$

It is a mathematical curiosity that the inverse square law of (10.1), for two point masses, is not the only possibility that has the feature we just derived. That is, if you assume that rather than (10.1) we begin instead with
$$F = Gm_1 m_2 x,$$
that is, with a linearly increasing (mathematical) force law (which of course is physically absurd, as the universe would have collapsed long

ago with such a law), then it is easy to show that the gravitational force a solid sphere of radius R and mass M exerts on a point mass m at distance $x > R$ from the center of the sphere is

$$F = GmMx,$$

where M is again given by (10.9). Newton was aware of this and included a brief analysis of it in the *Principia*. On the other hand, the inverse square law is the only law for which the gravitational force *inside* a spherical shell is zero.[11]

A particularly attractive feature of the calculus derivation is that it is easily extended to give us the second of Newton's superb theorems, namely, that the gravitational force on a point mass anywhere inside a spherical shell is zero. Even though the point mass is now *inside* the spherical shell, none of the preliminary equations leading up to (10.7) change;[12] what does change are the limits on the integral. That is, we now have

$$\int_{r-x}^{r+x} \left(\frac{x^2 - r^2}{s^2} + 1 \right) ds = (x^2 - r^2) \left[\left(-\frac{1}{s} \right) \Big|_{r-x}^{r+x} \right] + (s)|_{r-x}^{r+x}$$

$$= (x^2 - r^2) \left(\frac{1}{r-x} - \frac{1}{r+x} \right) + (r+x) - (r-x)$$

$$= (x^2 - r^2) \frac{r+x-r+x}{r^2 - x^2} + 2x = -2x + 2x = 0,$$

and so the total net force on the point mass anywhere inside the shell, due to the shell, vanishes. What an amazing result—completely surrounded by matter and not even the tiniest pull of gravity on the point mass! (If anybody tells you this is an "obvious" result, well, don't believe them!) It is the word *surrounded* that is key here. Each part of the shell does indeed exert a force on the interior point mass, but when the entire shell is taken into account, all those individual forces exactly cancel. Newton's geometric proof gives this cancellation effect center stage, while the onion-layer calculus proof I've used here pretty much disguises it. But the calculus proof works in both cases, with the point mass inside ($x < r$) *or* outside ($x > r$) the shell, while Newton had to create separate geometric proofs for the two cases.

10.3 The Moon Test and Blowing Up Planets

The first of the two superb theorems—the mass of a spherically symmetric sphere can be thought of as concentrated at the sphere's center as a point mass—was particularly pleasing to Newton because, until he proved it so, he wasn't at all sure it wasn't just simply a good approximation valid at distances much greater than the sphere's radius. He made that approximation during his first test of the inverse square law, which he claimed to have carried out two decades before the *Principia*. This so-called Moon test was the mathematics of *assuming* it is Earth's gravitational pull that keeps the Moon in its orbit (remember, before Newton it was Descartes's mysterious vortices that did the job). Newton essentially reasoned as follows. The radius of the Moon's orbit (taken as circular) is sixty times Earth's radius, that is, $r_{mo} = 60 r_e$. At the Moon, then, Earth's gravity (if inverse square) is 3,600 times smaller than it is at the Earth's surface. Now, if it is Earth's gravity that accounts for the Moon's orbit, then the Moon's orbital centripetal acceleration must be $g/3,600$. If v_m is the Moon's speed along its circular orbit, then the centripetal acceleration is v_m^2/r_{mo}, and so

$$\frac{g}{3,600} = \frac{v_m^2}{r_{mo}}.$$

If T is the time for the Moon to make one complete orbit about the Earth, then

$$v_m = \frac{2\pi r_{mo}}{T},$$

and so

$$g = 3,600 \left(\frac{2\pi r_{mo}}{T}\right)^2 \frac{1}{r_{mo}} = \frac{864,000 \pi^2 r_e}{T^2}.$$

The value of T is 27 days, 8 hours (approximately), or $T = 2,361,600$ seconds. Thus,

$$g = \frac{864,000 \pi^2 r_e}{(2,361,600)^2} \frac{1}{s^2} = 1.529 \cdot 10^{-6} r_e \frac{1}{s^2}.$$

So, to finish this calculation, all Newton needed was Earth's radius. This is where the historical story gets just a bit murky. (Just about all of it is from second-hand versions—not all exactly the same—that

Newton told to different friends at different times when he was an old man.) Supposedly Newton initially (in 1666) used the value of one degree of latitude at the equator to be sixty miles, which gives a grossly undersized circumference of 21,600 miles (a radius of 3,438 miles), and further he took a mile to be 5,000 feet. Thus,

$$g = 1.529 \cdot 10^{-6} \cdot 5{,}000 \cdot 3{,}438 \frac{f}{s^2} = 26.3 \frac{f}{s^2},$$

which is significantly less than the observed value of 32.2 $\frac{f}{s^2}$ (recall your author out in his garage with a bouncy rubber ball?). It was this discrepancy that historians of science long argued was the reason for Newton's delay in publishing anything on gravity (this is *not* the belief today, of course, as I discussed earlier). Only later, so continues this curious tale,[13] did Newton learn that one degree of latitude at the equator is actually 69.5 miles, which gives a circumference of 25,020 miles (a radius of 3,982 miles), with 5,280 feet in a mile, which gives

$$g = 1.529 \cdot 10^{-6} \cdot 5{,}280 \cdot 3{,}982 \frac{f}{s^2} = 32.1 \frac{f}{s^2},$$

which is very close to the observed value.

Whatever the historical details may be (and we will never really know, short of a time machine becoming available), Newton must have been very pleased indeed to learn, with the first[14] of his superb theorems, that reducing the Earth and the Moon to point masses in the Moon test didn't *require* that the two be widely separated. The feared "approximation" was actually mathematically perfect. Newton had made a very great discovery; the inverse square gravity of Earth was the responsible agent controlling the motion of a remote celestial body. This was the first extension of earthly physics into space.

We can now use both of Newton's superb theorems together to determine the answer to a question that remained mysterious even to those who before Newton had guessed the inverse square law for distances at or above the Earth's radius. That question is, how does gravity behave below the surface, inside Earth? The deep interior of our home world is right beneath our feet, and yet it remains, even today, virtually unexplored. Or unexplored in person; we can do a lot of "exploring" with mathematics. And we'll do that in the next two discussions. What I want to do next is show you a neat calculation using just the first of the superb theorems.

At the beginning of this discussion I teased you with the promise that eventually we would blow planets up. How much energy would it take to do that, to completely and utterly obliterate a planet? Or, as it is sometimes less violently put, to *disassemble* a planet? In fiction, at least, this remarkable event has been described in some detail. Superman's home planet Krypton was shown exploding in the 1978 movie, while long before that the planet Altair IV was vaporized in the 1956 film *Forbidden Planet* (your author sat, in a trancelike state, through multiple first-run theater showings of this film when he was fifteen). And, of course, the Death Star of the original *Star Wars* movie (1977) disassembled Alderaan.

No matter the ease with which Hollywood destroys planets, it really would take a lot of energy to do that. Let's imagine that our target planet consists of a spherical cloud of very fine particles held together by just their mutual gravitational attraction. Think of a planet made of talcum powder! Further, let's imagine the planet has a constant density ρ. These are both grossly simplifying assumptions: the first because the matter in real planets is mostly held together by the atomic binding forces in chemical compounds (it's pretty hard to pull a granite rock apart with your bare hands), and the second because a constant density assumption ignores the effect of increasing pressure as we move from the surface to the center of the sphere. Despite these objections, however, we'll at least get a ballpark estimate of the minimum required energy to disassemble a planet.

Now, assuming our planet starts with radius R and a mass M, let's assume we've disassembled it down to a radius of x. Imagine further that we're ready to reduce the radius by an additional thickness of dx. The differential mass involved is $dm = 4\pi x^2 \rho dx$, which we wish to move out to infinity (nobody is ever going to put *this* planet back together again!). The differential energy dW required to do that is the energy required to move dm out to infinity against the attractive force of the remaining part of the planet (which has mass $\frac{4}{3}\pi x^3 \rho$), and so the force on the differential mass we are currently removing, at any distance r ($x \leq r < \infty$), is

$$F = G\frac{(\frac{4}{3}\pi x^3 \rho)(4\pi x^2 \rho dx)}{r^2} = G\frac{16\pi^2 x^5 \rho^2 \, dx}{3r^2}.$$

We find dW as

$$dW = \int_x^\infty F\,dr = G\frac{16\pi^2 x^5 \rho^2 \, dx}{3}\int_x^\infty \frac{dr}{r^2} = G\frac{16\pi^2 x^5 \rho^2 \, dx}{3}\left(-\frac{1}{r}\right)\bigg|_x^\infty$$

$$= G\frac{16\pi^2 x^4 \rho^2 \, dx}{3}.$$

To find the total energy W, we simply integrate over all x, that is,

$$W = \int dW = G\frac{16\pi^2 \rho^2}{3}\int_0^R x^4 \, dx = G\frac{16\pi^2 \rho^2 R^5}{15}.$$

Since

$$M = \frac{4}{3}\pi R^3 \rho,$$

then

$$M^2 = \left(\frac{4}{3}\pi R^3 \rho\right)^2 = \frac{16\pi^2 R^6 \rho^2}{9},$$

and so

$$\frac{W}{M^2} = \frac{G\frac{16\pi^2 \rho^2 R^5}{15}}{\frac{16\pi^2 R^6 \rho^2}{9}} = \frac{9G}{15R} = \frac{3G}{5R},$$

or, at last,

$$W = \left(\frac{3}{5}\right)\frac{GM^2}{R}.$$

This is often called the *gravitational binding energy* of our doomed planet. How big is it?

For a planet the size of Earth, we have $M = 5.98 \cdot 10^{24}$ kg and $R = 6.37 \cdot 10^6$ m and so, in the MKS unit of energy (the joule),

$$W = 0.6\frac{6.67 \cdot 10^{-11}(5.98 \cdot 10^{24})^2}{6.37 \cdot 10^6}J = 2.25 \cdot 10^{32} J.$$

To give this enormous number some connection to the real world, a pound of TNT releases $2 \cdot 10^6$ J when exploded. Thus, two thousand pounds (a ton) of TNT releases $4 \cdot 10^9$ J, and so, finally, a one-megaton

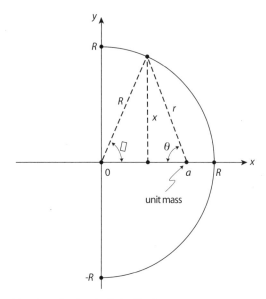

Figure 10.3. A thin, hemispherical shell of matter.

nuclear warhead (enough to pretty much flatten Boston) releases $4 \cdot 10^{15}$ J. To completely disassemble a planet of Earth's size would require the energy of *at least*

$$\frac{2.25 \cdot 10^{32}}{4 \cdot 10^{15}} = 56 \cdot 10^{15}$$

of these one-megaton warheads, that is, 56 million billion of them. Planet Earth (if not its inhabitants) is safe—at least for a while.

10.4 A Surprising Gravity Calculation

Newton's inverse square law sometimes strikes students as too simple to be behind the many various, often exotic gravitational behaviors we actually observe. So, let me show you next a calculation that beautifully illustrates the sort of surprise the inverse square law can present when we least expect it. Imagine, as shown in Figure 10.3, a very thin, uniform-density hemispherical shell of matter (that is, a bowl) with radius R. Our problem[15] is that of calculating the gravitational force due to the shell on a unit point mass that is on the x-axis, anywhere in the interval $0 \leq x < R$. Before we do that, however, let's see how good your intuition is. Obviously, the gravitational force on our unit mass when it is at $x = 0$ will, by symmetry, be directed straight down the positive x-axis, because the entire shell mass is to the right of the unit mass. Then, as we move the unit mass to the right along the x-axis, more and more of the shell mass will be to the left of the unit mass (and so will generate a force on the unit mass that is directed to the left); as the location of the unit mass approaches $x = R$, in fact, almost all the shell mass will be to the left of the unit mass. Do you think that there is a point in the interval $0 \leq x < R$ where the net force on the unit mass is zero? That is, is there a point on the x-axis where the gravitational force due to the shell mass at the left of the unit mass just cancels the gravitational force due to the shell mass at the right of the unit mass? You think about this while we calculate the answer. I think the calculated answer will surprise you.

As shown in Figure 10.3, let the unit mass be at $x = a$. Then, in the notation of the figure, there is a differential ring of matter in the shell at angle θ with radius $x = r \sin(\theta)$, where r is the distance from the point mass to any point in the ring. The angle from $x = 0$ to the ring is α. If the shell has thickness d, if the ring has width $R\, d\alpha$, and if the mass density is ρ, then the differential mass dm of the ring is

$$dm = \rho(R\, d\alpha)[2\pi r \sin(\theta)] d. \tag{10.10}$$

As I've drawn Figure 10.3, with $\theta < 90$ degrees, you can see that the differential gravitational force due to the ring on the unit mass at $x = a$ will be directed to the left, opposite to the direction of increasing x, and so I will write that differential force with an explicit negative sign. From Figure 10.3, then, we have

$$dF_x = -G \frac{dm}{r^2} \cos(\theta). \tag{10.11}$$

Notice that (10.11) is (10.3), except (besides the minus sign) we now have a differential mass dm rather than a mass μ, and so we have a differential force dF_x instead of F_x. When $\theta < 90$ degrees (10.11) gives $dF_x < 0$ (force directed to the left), but when $\theta > 90$ degrees then $\cos(\theta) < 0$, and so $dF_x > 0$ (force directed to the right), just as it should be.

Inserting (10.10) into (10.11),

$$dF_x = -\frac{G\rho R 2\pi r d \sin(\theta)\cos(\theta)d\alpha}{r^2},$$

or

$$dF_x = -\frac{KR}{r}\sin(\theta)\cos(\theta)d\alpha, \qquad (10.12)$$

where $K = G\rho 2\pi d$, a constant. As it stands, (10.12) expresses dF_x in terms of three variables (r, θ, and α), but we can reduce that to a single variable, as follows. From Figure 10.3 and the law of sines,

$$\frac{R}{\sin(\theta)} = \frac{r}{\sin(\alpha)},$$

or

$$\sin(\theta) = \frac{R}{r}\sin(\alpha). \qquad (10.13)$$

Also from the law of cosines,

$$R^2 = a^2 + r^2 - 2ar\cos(\theta),$$

or

$$\cos(\theta) = \frac{a^2 + r^2 - R^2}{2ar}. \qquad (10.14)$$

And finally, again from the law of cosines,

$$r^2 = R^2 + a^2 - 2aR\cos(\alpha),$$

or, differentiating with respect to α,

$$2r\frac{dr}{d\alpha} = 2aR\sin(\alpha),$$

and so

$$\sin(\alpha)\,d\alpha = \frac{r}{aR}\,dr. \qquad (10.15)$$

Inserting (10.13) and (10.14) into (10.12) gives

$$dF_x = -\frac{KR^2}{2a} \cdot \frac{a^2 + r^2 - R^2}{r^3}\sin(\alpha)\,d\alpha,$$

and then using (10.15) gives

$$dF_x = -\frac{KR}{2a^2} \cdot \frac{a^2 + r^2 - R^2}{r^2}\,dr,$$

which expresses dF_x in terms of the single variable r.

From Figure 10.3 you can see that we integrate over the entire hemispherical shell if we let r vary from $R-a$ to $\sqrt{R^2+a^2}$. Thus,

$$F_x = -\frac{KR}{2a^2}\left[\int_{R-a}^{\sqrt{R^2+a^2}} dr + \int_{R-a}^{\sqrt{R^2+a^2}} \frac{a^2-R^2}{r^2}dr\right]$$

$$= -\frac{KR}{2a^2}\left[\sqrt{R^2+a^2}-(R-a)+(a^2-R^2)\left\{-\frac{1}{\sqrt{R^2+a^2}}+\frac{1}{R-a}\right\}\right].$$

This may look a bit messy, but, remarkably, if you do the algebraic simplification (it's not very long), this reduces to the very nice

$$F_x = K\frac{1}{(\frac{a}{R})^2}\left[1-\frac{1}{\sqrt{(\frac{a}{R})^2+1}}\right]. \tag{10.16}$$

Notice that as long as $(\frac{a}{R}) > 0$, then $F_x > 0$, and, while F_x does indeed decrease[16] as $\frac{a}{R}$ increases (as the unit mass moves ever farther along the positive x-axis into the shell), F_x is never zero. It never even gets close to zero. In fact,

$$\lim_{(\frac{a}{R})\to 1} F_x = K\left[1-\frac{1}{\sqrt{2}}\right] = \frac{\sqrt{2}-1}{\sqrt{2}}K.$$

Calculating F_x when the unit mass is at $x = 0$ (that is, $a = 0$) is only a bit more subtle because (10.16) is the indeterminate $\frac{0}{0}$ at $(\frac{a}{R}) = 0$. To do this, simply notice that as $(\frac{a}{R}) \to 0$, we have $\sqrt{(\frac{a}{R})^2+1} \to 1+\frac{1}{2}(\frac{a}{R})^2$, and so

$$\lim_{(\frac{a}{R})\to 0} F_x = \lim_{(\frac{a}{R})\to 0} K\frac{1}{(\frac{a}{R})^2}\left[1-\frac{1}{1+\frac{1}{2}(\frac{a}{R})^2}\right] = \lim_{(\frac{a}{R})\to 0} K\frac{1}{(\frac{a}{R})^2}\left[\frac{\frac{1}{2}(\frac{a}{R})^2}{1+\frac{1}{2}(\frac{a}{R})^2}\right]$$

$$= \lim_{(\frac{a}{R})\to 0} K\frac{\frac{1}{2}}{1+\frac{1}{2}(\frac{a}{R})^2} = \frac{1}{2}K.$$

So,

$$\frac{\lim_{(\frac{a}{R})\to 1} F_x}{\lim_{(\frac{a}{R})\to 0} F_x} = \frac{\frac{\sqrt{2}-1}{\sqrt{2}}K}{\frac{1}{2}K} = 2\frac{\sqrt{2}-1}{\sqrt{2}} = 0.58578\cdots.$$

This says that the gravitational force of the shell on the unit mass, even as $a \to R$, does indeed decrease from what it is at $a = 0$, *but not by that much*; it is still, as $a \to R$, more than 58% of what it is at $a = 0$.

Is that what you thought would be the answer when I asked you this question at the start of the analysis? If so, good for you! (*I*, alas, got it wrong.)

How can we understand this nearly paradoxical result? How can the gravitational force on the unit mass still be directed to the right even when the unit mass is practically at the bottom of the shell and so essentially all the shell mass is to the left of the unit mass? It's almost as if gravity were repulsive instead of attractive! The answer is that even though, as $a \to R$, there is indeed only a small amount of shell mass still to the right of the unit mass, that small amount of mass is very close to the unit mass. Because of the inverse square nature of Newtonian gravity, a small amount of very close shell mass is more than a match for all the shell mass that is at the left—and *far away* (relatively) from the unit mass.

Another dramatic example of how the inverse square law can make an unexpected appearance comes from what astronomers call the *Roche limit*. This is the name given to the closest center-to-center distance that a satellite orbiting a much more massive body can approach that body before being torn apart by gravitational *tidal forces* (so named because it is the same effect that causes the tides on Earth). The massive body exerts a greater gravitational force on the nearer parts of the satellite than it does on the the more remote parts, and the difference in these forces can, under certain circumstances, grow large enough to literally tear the satellite apart. The Roche limit—named after the French astronomer Édouard Roche (1820–1883), who calculated it in 1848—is a common calculation in elementary physics textbooks, so I won't repeat it here.[17] The end result, after some simple algebra, for the case of a satellite held together by just its own self-gravity, is that the satellite will be torn apart if it is closer to the massive body than $R(2\frac{\rho_M}{\rho_s})^{1/3}$, where R is the radius of the massive body and ρ_M and ρ_s are the densities of the massive body and the satellite, respectively. Notice that if $\rho_s > 2\rho_M$, that is, if the satellite is more than twice as dense as the massive body it orbits (for example, a planet made of solid rock orbiting a gas giant star), then the Roche limiting distance is inside (less than R) the massive body, and so the tidal force will not destroy the satellite, no matter how close it gets to the massive body.

The tidal force is the mechanism that is believed to have created the rings of Saturn. The rings are thought to be the debris produced when Saturn's tidal force ripped apart either an ancient moon or a passing mass that ventured too close to the planet. The Roche limit for Saturn is about 150,000 km, and the rings are about 275,000 km in diameter, putting the rings at 135,000 km from Saturn's center, that is, just inside Saturn's Roche limit.[18]

10.5 Gravitational Contraction

One of the great scientific mysteries of nineteenth-century physics was that of finding an explanation of the energy source of the stars, in particular that of our own Sun. (Our Sun, a quite ordinary star, radiates energy at the absolutely prodigious rate of $3.92 \cdot 10^{26}$ watts!)[19] The average man on the street might have had the mental image of a giant lump of coal burning in the sky, but scientists of the day knew that was absurd. "Burning in what?" they would have asked in reply. Certainly the void between the planets and the Sun is just that, a void, and not a vast expanse of oxygen! And, come on, a giant lump of flaming coal about which all the planets spin? Isn't that at about the same childish level of make-believe as elves dancing around a candy-striped pole in front of Santa's house at the North Pole? In 1854 the German physicist Hermann von Helmholtz (1821–1894) put forth a more scientifically plausible suggestion: might the stars, he asked, get their energy from a process called *gravitational contraction*?

The underlying idea behind gravitational contraction is straightforward. Imagine a vast cloud of gas in space, with each molecule in the cloud attracting every other molecule. To keep things simple, suppose the cloud is spherically symmetric, and so each spherical *shell* of molecules "feels" only the inward-directed gravitational force of the molecules located inside the shell (by Newton's second superb theorem), and so each shell moves inward. That is, the gas cloud *contracts* or *self-compresses*. The compression causes the gas cloud to become hotter; after all, the inward-falling molecules are continually increasing in speed (that is, gaining kinetic energy as the shrinking cloud's gravitational potential energy decreases). The ever-increasing speeds of the gas molecules are, in fact, directly related to what we call *temperature*. When the cloud gets hot enough, speculated von Helmholtz, we see a star!

It's an intriguing idea, and it was soon explored by others, too. In particular, in 1862 the Scottish genius William Thomson (1824–1907), better known later as Lord Kelvin, examined all the possible (according to known nineteenth-century physics) mechanisms that might power the Sun; among them was gravitational contraction. He came to the general conclusion that for the Sun to be radiating energy at its present rate it could, with certainty, be at most 500 million years old, and most probably was actually only 100 million years old. Kelvin concluded his essay with the interesting (and prophetic) words,

> As for the future, we may say, with equal certainty, that inhabitants of the earth can not continue to enjoy the light and heat essential to their life for many million years longer *unless sources now unknown to us are prepared in the great storehouse of creation* [my emphasis].

Kelvin's conclusion prompted a huge debate that lasted decades because, while 100 million years perhaps seems a long time, it is only a small fraction of the enormous time span geologists (and the newly emerging science of evolution) required, which was already in line with today's modern estimates for the age of the Earth as 4.5 *billion* years.[20] And certainly the Sun is at least as old as the Earth. This conflict was unsettling, to say the least. As the father of evolution Charles Darwin declared in 1869, "I am greatly troubled at the short duration of the world according to Sir W. Thomson, for I require for my theoretical views a very long period *before* the Cambrian formation." And in 1893, in a paper published in *The American Journal of Science*, the American geologist Clarence King wrote, "The age assigned to the sun by Helmholtz and Kelvin communicated a shock from which geologists have never recovered."

Today we know that the Sun, and indeed all stars, are powered by thermonuclear fusion reactions taking place deep within their interiors. These reactions, converting mass into pure energy, are physical processes that wouldn't have seemed possible to Kelvin even in his wildest fantasies. In the case of our own Sun, over four million *tons* of its mass is totally converted to energy each *second*. In 1967 the American physicist Hans Bethe (1906–2005) received the Nobel Prize for his work in the 1930s that explained how these reactions work, and at last removed the conflict between physics and geology and evolution. Such fusion reactions have powered the Sun for billions of years, and the Sun has sufficient fuel (hydrogen) left to continue shining just as it does today for billions of years into the future.

Still, while not the answer to the stellar energy source question, gravitational contraction remains a valid physical process that provides yet another beautiful example of the reach of Newton's theory of gravity. With some irony, you see, modern theory says that it is gravitational contraction that initially *forms* the balls of hydrogen gas[21] that eventually become stars, and then *ignites* them, that is, starts the fusion reactions by creating via compression the high temperature (15 million degrees) and high density (150 times denser than water) required for hydrogen to "burn" into helium. Once thermonuclear ignition occurs, the enormous outpouring of fusion energy stabilizes the star, and the contraction stops.

In the rest of this section I'll show you the mathematical physics of gravitational contraction,[22] and to give you some historical motivation to the analysis, let me quote one sentence from Kelvin's 1862 paper:[22] "The amount of work performed on a contraction of one-tenth per cent. of the diameter [of the Sun], if the density remained uniform throughout the interior, would be... equal to 20,000 times the mechanical equivalent of the amount of heat... estimated to be radiated from the sun in a year."[23] In his paper, published in a popular magazine of the day and not a scientific journal, Thomson only cites numbers and provides no mathematical analysis of any kind. So, what I'll do now is show you how that "20,000" comes from the assumption of a "one-tenth per cent." contraction.

Our technical problem is thus: a spherical, uniform-density gas cloud with mass M and initial radius R gravitationally contracts *in a uniform way* (more on this soon) to radius nR, where $n < 1$. (For Kelvin's one-tenth of a percent contraction, $n = 0.999$.) At the start of the contraction, all the molecules in the gas cloud are motionless.

At the end of the contraction, our question is, what is the total kinetic energy of the inward-moving gas cloud? This energy is the "mechanical equivalent of the amount of heat" that Kelvin mentions.

I'll write g' as the acceleration of gravity on the surface of the gas cloud just before the start of the contraction. Then, if we focus on a molecule in the cloud at distance r from the cloud's center, the initial inward acceleration for that molecule will be $g'\frac{r}{R}$. As the molecule moves inward during the contraction (from r to nr—and that is what is meant by a uniform contraction), it rides along on the surface of a sphere of ever-decreasing radius but of constant mass (with, of course, increasing density). Let x be the instantaneous distance of the molecule from the cloud's center, where of course $nr \leq x \leq r$. We know that the instantaneous acceleration of the molecule will vary inversely as the square of x, that is, as $\frac{r^2}{x^2}$, and so the inward acceleration at distance x is $(g'\frac{r}{R})(\frac{r^2}{x^2}) = g'\frac{r^3}{Rx^2}$.

If we call the mass of the molecule w then Newton's second law tells us that

$$w\frac{d^2x}{dt^2} = -wg'\frac{r^3}{Rx^2} = w\frac{dv}{dt} = w\frac{dv}{dx}\cdot\frac{dx}{dt} = w\frac{dv}{dx}v,$$

where $v = \frac{dx}{dt}$ is the speed of the molecule during the contraction. Thus,

$$wv\frac{dv}{dx} = -wg'\frac{r^3}{Rx^2},$$

or, finally,

$$v\,dv = -g'\frac{r^3}{R}\cdot\frac{dx}{x^2}. \tag{10.17}$$

Integrating (10.17) over the contraction,

$$\int v\,dv = -g'\frac{r^3}{R}\int_r^{nr}\frac{dx}{x^2} = -g'\frac{r^3}{R}\left(-\frac{1}{x}\right)\Big|_r^{nr},$$

and so, with C some constant, we have

$$\frac{1}{2}v^2 + C = -g'\frac{r^3}{R}\left(\frac{1}{r} - \frac{1}{nr}\right) = g'\frac{r^2}{R}\left(\frac{1}{n} - 1\right),$$

or

$$\frac{1}{2}v^2 + C = g'\frac{r^2}{R}\cdot\frac{1-n}{n}.$$

We know $v = 0$ if $n = 1$ ($n = 1$ means there *isn't* a contraction, and so the initially motionless gas molecules remain motionless), which says $C = 0$. Thus, the speed v of a molecule starting at distance r with zero speed is, when it reaches distance nr, given by

$$\frac{1}{2}v^2 = \frac{g'(1-n)}{nR}r^2. \tag{10.18}$$

Notice that the units on both sides of (10.18) do indeed agree (ft^2/s^2).

Now, consider not just a single molecule but instead a spherical shell of gas with differential thickness dr at initial distance r from the cloud's center. Its differential mass dm is, with ρ as the gas density (which, like Kelvin, I'll take as a constant during the entire contraction), given by

$$dm = 4\pi r^2 \rho \, dr = 4\pi r^2 \frac{M}{\frac{4}{3}\pi R^3} dr = \frac{3r^2 M}{R^3} dr. \tag{10.19}$$

(The assumption of a constant ρ shouldn't be a poor one for the small contraction we are considering.) The differential shell mass of (10.19) remains constant during the entire (uniform) contraction. The shell mass starts contracting with zero speed and ends with speed v. Thus, the differential kinetic energy dW of the contracted differential shell is given by, combining (10.18) and (10.19),

$$dW = \frac{1}{2}v^2 dm = \frac{g'(1-n)}{nR}r^2 \cdot \frac{3r^2 M}{R^3} dr,$$

or

$$dW = \frac{3g'M(1-n)}{nR^4} r^4 \, dr. \tag{10.20}$$

To find the total kinetic energy of the contracted cloud we integrate (10.20) over the entire cloud, and so

$$W = \int_0^R dW = \frac{3g'M(1-n)}{nR^4} \int_0^R r^4 \, dr,$$

or

$$W = \frac{3g'M(1-n)R}{5n}, \tag{10.21}$$

where M is is the mass of the cloud and R is the *initial* radius of the cloud. Since a mass m on the surface of a spherically symmetric sphere

with mass M and radius R experiences a gravitational force of
$$mg' = \frac{GMm}{R^2},$$
then
$$g' = \frac{GM}{R^2},$$
and so from (10.21) the total kinetic energy of our contracted cloud is
$$W = \frac{3}{5} \cdot \frac{GM^2}{R} \cdot \frac{1-n}{n}. \tag{10.22}$$

We can normalize (10.22) with respect to Earth by writing $M = km_e$ and $R = jr_e$, where m_e and r_e are, respectively, the mass and radius of Earth. Then,
$$W = \left(\frac{3}{5} \cdot \frac{Gm_e^2}{r_e}\right) \cdot \frac{k^2}{j} \cdot \frac{1-n}{n}. \tag{10.23}$$

This normalization is convenient because, as you'll recall from Section 10.3, where we blew up the Earth, we've already calculated the factor in parentheses as the Earth's gravitational binding energy. Thus, (10.23) becomes
$$W = 2.25 \cdot 10^{32} \cdot \frac{k^2}{j} \cdot \frac{1-n}{n} \text{J}. \tag{10.24}$$

To finish our calculations, all we need do now is plug in the appropriate values for k, j, and n. For the sun they are $k = 329{,}000$, $j = 109$, and (using Kelvin's assumed contraction, as explained earlier) $n = 0.999$. This gives a total contracted energy of
$$W = 2.25 \cdot 10^{32} \frac{(329{,}000)^2}{109} \cdot \frac{0.001}{0.999} \text{J} = 2.24 \cdot 10^{38} \text{J}.$$

At the sun's radiating rate of $3.92 \cdot 10^{26}$ watts (= J/s) this would require a time duration of
$$\frac{2.24 \cdot 10^{38} \text{ J}}{3.92 \cdot 10^{26} \text{ J/s}} = 5.71 \cdot 10^{11} \text{ seconds},$$
or, converting to years,
$$\frac{5.71 \cdot 10^{11} \text{ s}}{24 \cdot 60 \cdot 60 \cdot 365 \text{ s/y}} = 18{,}106 \text{ years},$$
which is fairly close to Kelvin's value of 20,000 years.

GRAVITY AND NEWTON   **163**

CP. P10.1:

Imagine an immensely large, thin circular ring (such as illustrated in Figure 10.1) of constant density matter in space, of total mass μ and radius R. On the axis of the ring, at distance nR from the center of the ring ($n \geq 0$) at time $t = 0$, a brave astronaut is positioned, initially at rest with respect to the ring. The ring's gravitational attraction will accelerate the astronaut along the ring axis in toward the ring center; at some time $t > 0$ the astronaut will whiz through the ring's center with speed V. What is V, in terms of G, μ, R, and n? Such a ring, of incredibly huge size, is the basis for Larry Niven's highly imaginative 1970 novel *Ringworld*, which won science fiction's two biggest awards that year (the Hugo and the Nebula). Built from a mass equal to that of Jupiter, Niven's ring has a radius of 93 million miles (Earth's orbital radius), is one million miles wide, and is 1,000 meters thick. Spinning around an axis (passing through a central star) normal to its plane at a speed of 770 miles per second, the apparent "gravity" at the inner surface of the ring is very nearly $1\,g$. (As part of this problem, verify this claim.) Beings living on the ring's inner surface would find it quite roomy, since it has an area almost three million times that of Earth's surface area. (Check this claim, too.) Niven flunked out of Caltech as an undergraduate (he later earned a BA degree in math elsewhere), but I suspect he is today the best known of all those with whom he entered the Institute in 1956!

CP. P10.2:

What would gravity be like inside a hollow cavity located somewhere in the Earth? Suppose we model the Earth as a sphere with constant density ρ. If we imagine a spherical cavity totally inside the Earth *but otherwise arbitrarily located*, then show that the gravitational force on a mass anywhere inside the cavity is a constant. As a special case, if the cavity is concentric with Earth's center, then the value of the constant is zero, by Newton's second superb theorem. If the cavity and the Earth are not concentric, however, then the gravitational force will be some non-zero constant. *Double hint*: (1) Remember that force is a vector, and (2) mathematically "create" the cavity (which of course has a density of zero) by superimposing a solid sphere with positive density ρ (Earth) with a (smaller) solid sphere with negative density $-\rho$. That is, the smaller sphere has negative mass, something theoretical physicists can easily do on paper but which experimental physicists can only dream of doing in the lab!

Notes and References

1. From the final General Scholium of the *Principia*, in direct reference to Newton's work on gravity. For an historian's view of how this famous statement of Newton's has long been corrupted, see Alexandre Koyré, *Newtonian Studies* (New York: Chapman & Hall, 1965, pp. 35–36). Newton did put forth a lot of hypotheses in his other work, however, such as on the nature of light in his *Optiks*. I should also tell you that Newton did, probably until 1682, subscribe to Descartes's vortex hypothesis. What is imagined to have occurred that fateful year is the great comet of 1682—Halley's comet—which orbited the Sun in the direction opposite that of the planets. How could a swirling "something" sweep the planets around in one direction and a comet in the reverse direction? See Nicholas Kollerstrom, "The Path of Halley's Comet and Newton's Late Apprehension of the Law of Gravity" (*Annals of Science*, October 1999, pp. 331–356).

2. This mystery was described by the English experimental physicist C. V. Boys, in a lecture he gave at the Royal Institution on the evening of June 8, 1894 (see *Nature*, August 2, 1894, p. 330). As he said that night,

> Unlike any other known physical influence, it [gravity] is independent of medium, it knows no refraction, it cannot cast a shadow. It is a mysterious power, which no man can explain; of its propagation through space, all men are ignorant. It is in no way dependent on the accidental size or shape of the Earth; if the solar system ceased to exist it would remain unchanged. I cannot contemplate this mystery, at which we ignorantly wonder, without thinking of the altar on Mars' hill. When will a St. Paul arise able to declare it unto us? Or is gravitation, like life, a mystery that can never be solved?

Even as Boys uttered these melodramatic words, of course, a fifteen-year-old Albert Einstein was already thinking on matters that would, while Boys was still alive (he died in 1944), result in the general theory of relativity and its explanation of gravity as the purely geometric consequences of something called *warped space-time* (a mathematical concept that revolutionized physics as much as did Newton's dynamics more than two centuries earlier). Unlike Newton's instantaneous action-at-a-distance theory, Einstein's theory of gravity is a local field theory, just as is Maxwell's theory of the electromagnetic field, with gravitational effects propagating at a finite speed (the speed of light).

3. Newton showed that the planetary laws of motion formulated decades earlier by the German astronomer Johannes Kepler (1571–1630) imply a central, inverse square force law. I'll show you how this works in the next discussion. The inverse square law itself long predates Newton, but it was Newton alone who could go beyond simply saying "gravity is inverse square." Newton showed how to calculate the implications of that assertion. For the history of the law, see the two-part paper by Ofer Gal and Raz Chen-Morris, "The Archaeology of the Inverse Square Law" (*History of Science*, December 2005, pp. 391–414, March 2006, pp. 49–67). If one is thinking about how energy flows outward into space from a local source (radiation from a star, for example) then the inverse square law *follows* immediately from conservation of energy and the mathematical result that the area of a sphere in three-dimensional space increases as the square of the radius. The law is not so obvious for gravity, however—after all, what is "flowing" in gravity? Here's one way a modern physicist might argue that a power law, at least, makes sense for gravity (the argument, itself, isn't that new, as you can find it in Maxwell's *Treatise on Electricity and Magnetism* from the middle of the nineteenth century). A physically plausible gravity force law $f(r)$ "should have" the property that the ratio of the gravitational force at distance r_1 to the force at distance r_2 be a function only of the *ratio* of r_1 to r_2. That is, there should exist a function h such that $\frac{f(r_1)}{f(r_2)} = h(\frac{r_1}{r_2})$. This is called a *functional equation*, and the theory of that subject shows that the only solutions are of the form $f(r) = Ar^k$, where A and k are constants; $k = -2$ is the inverse square law.

4. Michell is remembered today by physicists as being the first to speculate about the possible existence of black holes. This was done in a letter (dated May 26, 1783) to Cavendish, in which Michell proposed a way to measure the mass of a star by measuring how much the speed of light was reduced as a result of the assumed "gravitational drag" on particles of light (Newton's corpuscular theory of light) escaping from the star. Michell concluded that for a star with the same average density as the Sun but with a diameter 500 times larger, "a body falling from infinite height towards it, would have acquired at its surface a greater velocity than that of light, and consequently... all light

emitted from such a body would be made to return towards it... by its own proper gravity." Michell's analysis is not a modern one, of course, which would use the general theory of relativity, but we can forgive Michell for this, since Einstein's birth was still almost a century in the future.

5. N. Kollerstrom, "The Hollow World of Edmond Halley" (*Journal for the History of Astronomy* 23, 1992, pp. 185–192). Halley was not a crackpot; his belief was based on an erroneous calculation by Newton (in the *Principia*) of the relative density of the Moon. Newton declared the Moon to be much more dense than the Earth, and Halley, supposing there to be no reason why the material of the Moon should be different from that of the Earth, "explained" a less dense Earth (on average) by having it be hollow. He wasn't alone in finding a hollow Earth reasonable. Halley, who died in 1743, lived long enough to have possibly heard of a 1741 Danish novel by Ludvig Holberg (*Journey of Niels Klim to the World Underground*), but he certainly missed the 1751 "world beneath our feet" novel by Robert Paltock (*The Life and Adventures of Peter Wilkins, A Cornish Man*). I'm only guessing, of course, but I think he would have liked Jules Verne's *A Journey to the Center of the Earth* (1864), made into a wonderfully atmospheric 1959 movie, and perhaps even Edgar Rice Burroughs's *At the Earth's Core* (1922). I'm not at all sure just what he would have made of the 1951 movie *Superman and the Mole-Men* (Lois Lane and Clark Kent visit the site of the world's deepest oil well, only to discover "mole men" using it to access the surface!). And finally, after watching the absolutely absurd 2003 movie *The Core* (winner of the "All-Time Stupid Movie" award from Tom Rogers, author of the fascinating book *Insultingly Stupid Movie Physics* [New York: Sourcebooks, 2007]), I think Halley would have been smart enough to at least wonder about the reasonableness of the film's premise: an earthquake weapon has stopped the Earth's core from rotating, prompting a manned expedition to the core of a *solid* Earth through fearsome pressure and heat.

6. A mathematically detailed account of the workings of Cavendish's torsion pendulum is given in K. E. Bullen, *The Earth's Density* (New York: Chapman and Hall, 1975, pp. 16–18).

7. A nice tutorial of what Cavendish actually published (the average density of the Earth) and how his work was incorrectly warped into the claim that he had measured G (by Boys, in his 1894 lecture—see note 2) is in B. E. Clotfelter, "The Cavendish Experiment as Cavendish Knew It" (*American Journal of Physics*, March 1987, pp. 210–213).

8. You can find an interesting discussion of this delay in Newton's gravity work in Florian Cajori, "Newton's Twenty Years' Delay in Announcing the Law of Gravitation," in *Sir Isaac Newton 1727–1927: A Bicentenary Evaluation of His*

Work (New York: Williams & Wilkins, 1928, pp. 127–188). The suggestion that it was the mathematical difficulty Newton faced in establishing the inverse square law for outside a sphere that delayed Newton was first put forth in 1887, two hundred years after the publication of the *Principia*. You can find an early (obscure) reference to that proposal in Boys's Royal Institution lecture (see note 2). In that lecture Boys said, "Only last night [did] I learn that it was the difficulty of proving [the sphere result] that delayed the publication of Newton's discovery [of his theory of gravity] for so long."

9. See, in particular, the paper by the English mathematician J. E. Littlewood, "Newton and the Attraction of a Sphere" (*The Mathematical Gazette* 32, 1948, pp. 179–181). Littlewood asserts that even Newton's published *geometric* proof "must have left its readers in helpless wonder." Newton's interest in spheres was prompted, of course, by the fact that stars and planets (and their moons) are all pretty nearly spheres.

10. An alternative calculus proof can be found in James C. Rainwater, "Gravity of the Earth by Means of Conical Shells" (*American Journal of Physics*, August 1977, pp. 768–769. This proof is mathematically less demanding than is the classic onion-layer proof I give, but it has the disadvantage of requiring the spherical mass to be of constant density. The onion-layer proof relaxes the density constraint to that of being simply spherically symmetric. Perhaps this is why, even after more thirty years in print, Professor Rainwater's proof is still waiting to be discovered by textbook writers.

11. S. K. Stein, "Inverse Problems for Central Forces" (*Mathematics Magazine*, April 1996, pp. 83–93).

12. This may not be entirely obvious at first glance. There are two possible complications. First, if you imagine that in Figure 10.2 the point mass is moving to the left from outside the spherical shell to inside the shell but is still to the right of the shell's center at $x = 0$, then the geometry is indeed unchanged. If the point mass is further moved to the left of $x = 0$, however, then ϕ is no longer an interior angle in the s, x, r triangle but rather is an *exterior* angle. But we don't have to even consider that case since, by symmetry, if the shell force on the point mass is zero for $0 < x < r$, then it is, too, for $-r < x < 0$. Second, what if the point mass is somewhere inside the shell that is not on the x-axis? Well, then just redraw the figure so that the x-axis *does* pass through the point mass! We can argue that we can always do this because of, again, symmetry.

13. One modern historian of science has called these tales of Newton "bogus history," stories created by Newton to "prove" that he had been in possession of the inverse square law decades before the publication of

the *Principia*. Newton supposedly did this to refute the energetic claims of Robert Hooke that Newton had actually gotten the inverse square idea much more recently from *him* (more on Hooke and this nasty business in the next discussion). See I. Bernard Cohen, *The Newtonian Revolution* (Cambridge: Cambridge University Press, 1980, p. 248).

14. Hanging in London's National Portrait Gallery is visual evidence for Newton's own regard for the first of the superb theorems. That painting shows Newton in his old age (83), just months before his death in 1727, sitting at a desk with the newly published third edition of the *Principia* open before him. The artist has not merely brushed in a meaningless mishmash of 'something that looks like printed pages,' but rather has quite carefully reproduced the pages with such clarity it is easy to see that they are pages 204 and 205. This was clearly not a random choice by the painter but instead must have been that of Newton himself. And just what is on those pages? The first superb theorem!

15. This calculation was motivated by Ronald G. Tabak, "Gravity on a Half Shell" (*American Journal of Physics*, December 1987, pp. 1096–1098). I have filled in a lot of missing steps, no doubt omitted from Professor Tabak's paper because of space limitations in the journal.

16. Can you prove this? It's a routine problem in differentiation. You should do it.

17. A nice discussion can be found on the Web at http://en.wikipedia.org/wiki/Roche_limit.

18. In science fiction, Larry Niven's classic 1966 short story "Neutron Star" is a beautiful (if gruesome) illustration of the gravitational tidal force, and of the very bad things that can happen when a starship pilot gets too close to a neutron star.

19. This huge number can be found either through experiment or by theory. The solar power density incident on the Earth's upper atmosphere is about $1,400\,W/m^2$, and so, using the Earth's orbital radius, the entire power output of the sun is a quick calculation. Alternatively, one could use the Stefan-Boltzman law, which says a black body radiator at absolute temperature T radiates at a power density of σT^4, where the *Stefan-Boltzman constant* $\sigma = 5.67 \cdot 10^{-8}\,W/(m^2 \cdot K^4)$. This law is named after the Austrian mathematical physicists Josef Stefan (1835–1893), who discovered it experimentally in 1879, and his student Ludwig Boltzman (1844–1906), who five years later derived it from the laws of thermodynamics. The sun's surface temperature, T (about 5,800 K), can be estimated from optical observation of its color (for example, blue stars are hotter than red stars).

20. For a superb modern scientific discussion of how the age of the Earth has been determined, with some interesting historical and social history as well, see G. Brent Dalrymple, *The Age of the Earth* (Stanford: Stanford University Press, 1991). More historical discussion is in Chapter 11 ("The Age-of-the-Earth Controversy") of my book *Oliver Heaviside* (Baltimore: Johns Hopkins University Press, 2002, pp. 241–258).

21. Present theory on the origin of the universe is that, in the beginning, just after the Big Bang, the only elements present were hydrogen, helium, and a little bit of lithium. The local variations in the distribution of these elements allowed gravitational contraction to form stars, which all began their fusion lives by "burning" hydrogen into helium. Some of those stars had structures such that, after all their hydrogen had been so transmuted, they entered new phases of fusion that successively created the heavier elements through different fusion reactions (a process called nucleosynthesis). Those stars, at the end of their lives, then exploded and scattered those fusion-created elements throughout space... and then billions of years later planets, dinosaurs, cats, birds, and people were made from those elements. It's all far more spectacular than any myth one could make up sitting around a campfire!

22. My analysis here will use nineteenth-century Newtonian physics, as Kelvin must have. A modern analysis would use relativistic physics, as well as incorporate the details of the so-called equation of state that recognizes how the matter of the Sun actually behaves as it grows ever more dense and hot during the contraction. All of that physics was discovered long after Kelvin's time, of course.

23. William Thomson, "On the Age of the Sun's Heat" (*Macmillan's Magazine*, March 5, 1862, pp. 288–293). Kelvin's fascination with such calculations never faded. In an 1897 address to the Victoria Institute he declared, "The age of the earth as an abode fitted for life is certainly a subject which largely interests mankind in general. For geology it is of vital and fundamental importance—as important as the date of the battle of Hastings is for English history."

Gravity Far Above the Earth

> ... he has done nothing & yet written in such a way as if he knew & had sufficiently hinted all but what remained to be determined by ye drudgery of calculations & observations, excusing himself from that labour by reason of his other business: whereas he should rather have excused himself by reason of his inability. For tis plain by his words he knew not how to go about it. Now is not this very fine? Mathematicians that find out, settle & do all the business must content themselves with being nothing but dry calculators & drudges & another that does nothing but pretend & grasp at all things must carry away all the invention.
> — A very angry Isaac Newton, venting in a letter (June 20, 1686)

11.1 Kepler's Laws of Planetary Motion

In the opening quotation, taken from a letter Newton wrote to his friend Edmond Halley (the man who got the *Principia* into print), Newton's anger was directed at Robert Hooke (1635–1703), the Curator of Experiments at the Royal Society of London. Halley had just written to Newton that Hooke wanted to be mentioned in the *Principia* as at least a co–discoverer of the inverse square law of gravity simply because he had in the past proposed it. Hooke could do nothing beyond proposing, however, and all the wonderful mathematical discoveries in the *Principia* are due to Newton's analyses alone. When an outraged Newton refused to give Hooke the credit he demanded, Hooke publicly accused Newton of plagiarism. From that moment on it is fair to say that Newton, who already greatly disliked Hooke, now absolutely detested the man.

After three centuries of historical scholarship, the precise nature of how Hooke may or may not have influenced Newton has divided historians of science into two camps: Hooke was important versus Hooke was secondary. Most physicists probably could not care less about this issue ("just show me the equations!"), and I am certainly not a historical scholar myself, but I think the case for Hooke has been improving in recent years.[1] Hooke's provocative suggestions to Newton on technical matters almost surely prompted Newton to think more deeply on those matters, or even of new approaches, than he might have otherwise done. But no matter which camp you are in, I can't imagine that anyone would seriously argue that Hooke was Newton's intellectual equal. So, what did Newton do that Hooke could not do?

Decades before Newton's birth it was known, through direct astronomical observation, that the visible planets move through space about the Sun according to certain mathematical laws. Famously known as *Kepler's three laws of planetary motion*, they are:

(1) each planet travels along an elliptical orbit about the Sun, with the Sun at one focus;
(2) the instantaneous orbital speed of each planet varies in such a way that a line joining the Sun and the planet sweeps over equal areas in equal times;
(3) the square of a planet's orbital period (that planet's "year") is proportional to the cube of its mean distance from the Sun (by *mean*, Kepler meant the average of the minimum and the maximum distances of the planet from the Sun—the *perihelion* and the *aphelion* distances, respectively—which equals the length of the semimajor axis of the elliptical orbit); the proportionality constant is the same for all the planets.

As you'll see as we get into this discussion, these statements are strictly true only for a Sun much more massive than its planets, which is a very good approximation in the case for the solar system.

From the first two of these three laws (the first and the second are dated to 1609, and the third to 1619),[2] Newton—and nobody else until Newton (certainly not Hooke)—was able to analytically deduce the inverse square, central force law of gravity. An elementary way for how Newton (and Hooke, too, if only he could have worked through the easy algebra) might have done this is to imagine that we start with an assumed circular orbit (a special, easy-to-work-with case of the general elliptical orbit, and all the planetary orbits in the solar system *are* almost circular, that is, they are elliptical with small eccentricities) and do something like the following.

From Kepler's second law, applied to a circular orbit, it is clear that the planet's orbital speed v is constant. If we write T as the orbital period, r as the orbit radius, and k as the constant in the third law, then

$$T^2 = kr^3, \qquad (11.1)$$

where

$$T = \frac{2\pi r}{v}.$$

Thus,

$$\frac{4\pi^2 r^2}{v^2} = T^2 = kr^3,$$

or

$$v^2 = \frac{4\pi^2 r^2}{kr^3} = \frac{c}{r},$$

where $c(=4\pi^2/k)$ is a constant. Now, the centripetal orbital acceleration of the planet is $a_c = \frac{v^2}{r}$ and so the gravitational force F (due to the Sun) on the planet, with m as the mass of the planet, is

$$F = ma_c = \frac{mv^2}{r} = \frac{m\frac{c}{r}}{r} = \frac{mc}{r^2} = \frac{C}{r^2},$$

where C is the constant mc. And so there you have it—the inverse square law. (The centrality of the force, for this special case, follows from the high degree of symmetry of the *assumed* circular orbit—the gravitational force on the planet is directed toward the Sun at the *center* of the orbit.) Newton's *Principia* actually has a more general derivation, for an assumed elliptical orbit that may not be simply circular.[3]

Let me now show you a neat use of Kepler's third law that quickly solves a problem we considered in an earlier discussion. Recall from

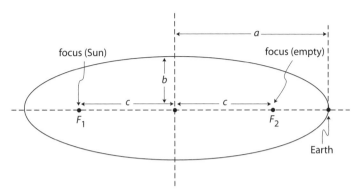

Figure 11.1. The geometry of an elliptical orbit.

Discussion 6 the problem of calculating how long it would take the Earth, if suddenly stopped in its orbit, to fall into the Sun. That analysis was fairly complicated; Kepler's third law, however, makes the analysis almost a one-liner. But first let me review the geometry of the elliptical orbit of the Earth around the sun.

In Figure 11.1 I've drawn a general elliptical orbit with semimajor axis a and semiminor axis b. The two foci, F_1 (the sun) and F_2, are equal distances from the ellipse's center. I'll call the distance from the center to either focus c. Notice that the Earth is shown at the far right, at its maximum distance from the Sun. Now, think back to high school geometry, where the definition of an ellipse is as the path of a point moving in such a way that the sum of the instantaneous distances of that point from two given points (the foci F_1 and F_2) is a constant. In particular, look at the point on the ellipse directly above the center. For that particular point it is clear by inspection that the sum of its distances from F_1 and F_2 is $2\sqrt{b^2+c^2}$.

Now look at the Earth, at the far right in Figure 11.1. Its distance from F_1 is $a+c$ and its distance from F_2 is $a-c$, for a sum of $2a$. Thus, by the very definition of an ellipse, we have

$$2\sqrt{b^2+c^2} = 2a,$$

or

$$b^2 + c^2 = a^2,$$

and so

$$c = \sqrt{a^2 - b^2}.$$

The *eccentricity* of the elliptical orbit is defined as

$$e = \frac{c}{a} = \frac{\sqrt{a^2 - b^2}}{a} = \sqrt{1 - \left(\frac{b}{a}\right)^2}. \quad (11.2)$$

For the circular orbit we considered earlier, F_1 and F_2 are on top of each other, at the center of the orbit, and so $c = 0$. That is, a circular orbit ($a = b$) is an elliptical orbit with zero eccentricity. Now, at the other extreme, imagine the orbit collapses to a very narrow ellipse—in the limit, to an ellipse with $b = 0$ so that the "orbit" degenerates into a straight line segment. This is the "orbit" of our suddenly stopped Earth, which then begins to fall into the Sun. Since $b = 0$, then $e = 1$, which says that $c = a$. That is, our two foci are now *on* the ellipse rather than inside, as shown in Figure 11.1, with F_1 (the sun) at the far left and the Earth initially at F_2 at the far right.

To use Kepler's third law to solve our falling Earth problem (again), let's agree to measure time in units of the original Earth orbital period (one year) and distances in units of the original Earth orbit size (since the Earth's orbit is very nearly circular[4] we have the minimum and maximum Sun-Earth separations nearly equal, with a mean value of one so-called *astronomical unit*, or, 1 A.U., which is about 93,000,000 miles). Since the new straight-line orbit has a maximum separation of 1 A.U. (at the start of the fall) and a minimum separation of zero (the new orbit passes right through the Sun), then the new orbit's mean distance from the sun—the semimajor axis of our "falling elliptical orbit"—is $\frac{1}{2}$ A.U. Kepler's third law says the constant for all planetary orbits around the Sun is

$$k = \frac{T^2}{a^3},$$

where again the semimajor axis of an orbit is a. So, if T is the period of the new "falling" orbit, we have

$$\frac{1^2}{1^3} = \frac{T^2}{\left(\frac{1}{2}\right)^3} = 8T^2,$$

or

$$T = \frac{1}{\sqrt{8}} \text{ years} = \frac{1}{2\sqrt{2}} \text{ years}.$$

The time for the Earth to fall into the Sun is $\frac{1}{2}T$ (the Earth will not, of course, survive to finish the second half of the orbit). That is, the answer to our question is

$$\frac{1}{2} \cdot \frac{1}{2\sqrt{2}} \text{ years} = \frac{1}{4\sqrt{2}} \text{ years} = \frac{365.25}{4\sqrt{2}} \text{ days} = 64.6 \text{ days},$$

just as we calculated—with a great deal more labor—back in Discussion 6.

11.2 Weighing the Planets

Newton was able to apply Kepler's laws in ways that astonished his contemporaries. One such application was the calculation of the relative masses of several of the planets in the solar system. This achievement was rightfully characterized, twenty-one years after Newton's death, by the Scottish mathematician Colin Maclaurin (1698–1746) in his posthumously published *An Account of Sir Isaac Newton's Philosophical Discoveries* (1748), as follows:

> To measure the matter in the Sun and planets was an arduous problem, and, at first sight, seemed above the reach of human art. But the principles of this philosophy afforded a natural and easy solution of it in the most important cases, and Sir Isaac Newton has determined the proportions of the matter that is in the Sun, Jupiter, Saturn, and the Moon, to that in our Earth; that is, he has showed how many earths might form a Sun, a Jupiter, or a Saturn.

To develop Newton's clever way of doing what might seem the impossible, short of making a trip to Jupiter or Saturn or the Moon and repeating the Cavendish experiment there, let me start by showing you a possible alternative approach[5] that, while physically correct, has a serious mathematical shortcoming that Newton's method avoids. This will make you appreciate Newton's genius even more!

In the most elementary conceptualization, most people think of the Moon as revolving around the Earth. But why not think instead of the Earth as revolving around the Moon? Even better, however, is thinking of the Earth and the Moon as revolving not around each other but rather as each individually revolving around their *center of mass* (c.m.). To keep the analysis elementary, let's take both the Earth and the Moon

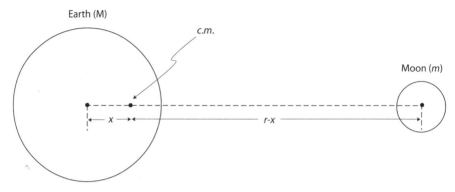

Figure 11.2. The Earth-Moon system and its center of mass (c.m.).

as being in circular orbits about the c.m., with the orbital period of each being T. In Figure 11.2 I've shown the Earth (mass M) and the Moon (mass m) as uniform spheres with their centers distance r apart. The c.m. is distance x from the center of the Earth; I've shown the c.m. as *inside* the Earth, and you'll see how that comes out of the analysis. That is, the Earth's motion is a circular orbit of radius x centered on the c.m., and the Moon's motion is a circular orbit of radius $r - x$ centered on the c.m.

The location of the c.m., that is the value of x, is defined by the equation that says the first moment of the two masses M and m is zero, that is,
$$xM = (r - x)m,$$
which is easily solved for x:
$$x = \frac{rm}{m + M}. \tag{11.3}$$
From (11.3) we find $r - x$ as
$$r - x = \frac{rM}{m + M}. \tag{11.4}$$
Now, if v is the orbital speed of the Moon, and if a_c is the centripetal force of gravity exerted on the moon by the Earth, then
$$\frac{GmM}{r^2} = ma_c, \tag{11.5}$$
where
$$a_c = \frac{v^2}{r - x}. \tag{11.6}$$

Since
$$T = \frac{2\pi(r-x)}{v},$$
then
$$v^2 = \frac{4\pi^2(r-x)^2}{T^2},$$
and this, together with (11.4) and (11.6), gives
$$a_c = \frac{4\pi^2(r-x)}{T^2} = \frac{4\pi^2 \left(\frac{rM}{m+M}\right)}{T^2}. \tag{11.7}$$
From (11.5) we have
$$\frac{GM}{r^2} = a_c,$$
and using this with (11.7) gives
$$\frac{G}{r^3} = \frac{4\pi^2}{(m+M)T^2} = \frac{4\pi^2}{M\left(1+\frac{m}{M}\right)T^2},$$
or
$$T^2 = \frac{4\pi^2}{GM\left(1+\frac{m}{M}\right)} r^3 \tag{11.8}$$
which is, of course, Kepler's third law for the Earth-Moon system *without* the assumption that $M >> m$.

A brief aside: if we reinterpret M and m as the mass of the sun and the mass of any planet in the solar system, respectively, then you see that (11.8) shows that the "constant" in Kepler's third law is not really the same for all the planets;[6] rather, the idea that it is the same constant for all the planets is an approximation that is good only if $M >> m$ (and then the constant k in (11.1) is $4\pi^2/GM$).

Solving (11.8) for m/M, we have
$$\frac{m}{M} = \frac{4\pi^2 r^3}{GMT^2} - 1.$$
Since a mass m' on the surface of the Earth experiences a gravitational force (its weight) of $m'g$, then if r_e is the Earth's radius we have
$$\frac{GMm'}{r_e^2} = m'g.$$

Thus, $GM = gr_e^2$, and so at last we have

$$\frac{m}{M} = \frac{4\pi^2 r^3}{gr_e^2 T^2} - 1. \tag{11.9}$$

All the quantities on the right-hand side of (11.8) are known. In particular,

1. $r = 238{,}850$ miles $= 238{,}850 \cdot 5{,}280$ feet $= 1.261 \cdot 10^9$ feet;
2. $r_e = 3{,}960$ miles $= 3{,}960 \cdot 5{,}280$ feet $= 2.09 \cdot 10^7$ feet;
3. $g = 32.2 \, \text{ft/s}^2$;
4. $T = 27\frac{1}{3}$ days $= 2.3616 \cdot 10^6$ seconds.

Putting these values into (11.8), we get

$$\frac{m}{M} = \frac{4\pi^2(1.261 \cdot 10^9)^3}{32.2(2.09 \cdot 10^7)^2(2.3616 \cdot 10^6)^2} - 1 = 1.0091 - 1 = 0.0091,$$

or

$$\frac{m}{M} = \frac{1}{110}. \tag{11.10}$$

That is, the Earth is 110 times more massive than the Moon. This elementary calculation is, in fact, reasonably close to the results of more detailed analyses, which put the Earth at about 81 times more massive than the Moon. Since the Moon has a diameter about one-fourth that of Earth, the Moon's volume is about 1/64-th that of Earth. To get a value for m/M smaller than 1/64 means the average density of the Moon must be *less* than the average density of the Earth (which Cavendish showed, you'll recall from Discussion 10, to be 5.54 g/cm^3). And indeed, the average density of the Moon is put at about 3.4 g/cm^3 (you'll see where this comes from soon). From (11.3) we have the c.m. of the Earth-Moon system located a distance x from the Earth's center, where

$$x = \frac{r\frac{m}{M}}{\frac{m}{M}+1} = \frac{r\frac{1}{110}}{\frac{1}{110}+1} = \frac{r}{111} = \frac{238{,}850}{111} \text{ miles} = 2{,}151 \text{ miles},$$

well inside the Earth, as shown in Figure 11.2. Using the more accurate value of $m/M = 1/81$ gives $x = 2{,}912$ miles, still over a thousand miles below the surface of the planet.

But let me be honest—these numerical calculations are all in some jeopardy. Our calculation of the relative mass of the Moon has ended up as the difference between two very nearly equal numbers, and the result in (11.10) is very sensitive to the values used on the right-hand side of (11.9). How accurate are the values we used for r, r_e, g, and T? Even slight changes in them can give dramatically different results in (11.10). More specifically, just to see how the numbers work, suppose (contrary to fact) that the actual, true value of m/M is 0.01, and so the actual, true value of $1 + (m/M)$ is 1.01. If we imagine that we make just a one-half of 1% error in calculating $1 + (m/M)$ from (11.9)—let's suppose the error is in the direction of too high—then our calculated value of $1 + (m/M)$ will be 0.00505 too high, that is, it will be 1.01505. Thus, the calculated value of m/M will be 0.01505, compared to the actual, true value of 0.01. That is, m/M will be 0.00505 too high, compared to the actual, true value of 0.01, and this is a 50.5% error!

So, this approach to calculating the Moon's mass isn't a good one. We can, however, use it to calculate the masses of the planets (relative to the Sun) for those planets that have satellites themselves, including Earth. This is what Newton did in the *Principia*. For $M \gg m$ (this assumption is *most* important) we have, from (11.8), that

$$T^2 = \frac{4\pi^2}{GM} r^3,$$

which says that the period of the orbit of a satellite around *any* mass M, at a given distance r, depends *only* on M. Newton used this fact to determine $\frac{M_{planet}}{M_{Sun}}$ in terms of the period of a planet orbiting the sun, and the period of a satellite orbiting the planet, each orbiting object at a known distance. To be clear about this, we need the following four quantities:

1. R_p = orbital radius of the planet around the Sun;
2. R_{sat} = orbital radius of the satellite around the planet;
3. T_{sat} = period of the satellite around the planet;
4. T_p = period of the planet around the Sun.

We'll take $M_{Sun} \gg M_{planet}$ and $M_{planet} \gg M_{sat}$ (the mass of the satellite). So,

$$T_{sat}^2 = \frac{4\pi^2}{GM_{planet}} R_{sat}^3 \qquad (11.11)$$

and
$$T_p^2 = \frac{4\pi^2}{GM_{Sun}} R_p^3. \tag{11.12}$$

Dividing (11.12) by (11.11), we have
$$\frac{T_p^2}{T_{sat}^2} = \frac{R_p^3}{R_{sat}^3} \cdot \frac{M_{planet}}{M_{Sun}},$$

or
$$\frac{M_{planet}}{M_{Sun}} = \left(\frac{R_{sat}}{R_p}\right)^3 \left(\frac{T_p}{T_{sat}}\right)^2. \tag{11.13}$$

As an example of the use of (11.13), suppose the planet in question is Jupiter. Newton used Jupiter's fourth moon, Callisto, as the planet's satellite. The values of the parameters in (11.12) are:

$R_p = 5.2$ A.U.;
$R_{sat} = 0.0125$ A.U.;
$T_{sat} = 16.7$ days;
$T_p = 4{,}332.7$ days.

So,
$$\frac{M_{Jupiter}}{M_{Sun}} = \left(\frac{0.0125}{5.2}\right)^3 \left(\frac{4{,}332.7}{16.7}\right)^2 = 0.000935 = \frac{1}{1{,}070}.$$

In the various editions of the *Principia* Newton gave various values for this calculation, ranging from $1/1{,}033$ to $1/1{,}190$. This method for calculating $\frac{M_{planet}}{M_{Sun}}$ could be used, *in Newton's time*, only for the cases of Jupiter, Saturn, and the Earth, because they were the only planets then with visible satellites. If we repeat this calculation for Earth and its satellite we have

$R_p = 1$ A.U.;
$R_{sat} = 0.0026$ A.U.;
$T_{sat} = 27.3$ days;
$T_p = 365.25$ days,

and so
$$\frac{M_{Earth}}{M_{Sun}} = \left(\frac{0.0026}{1}\right)^3 \left(\frac{365.25}{27.3}\right)^2 = 3.15 \cdot 10^{-6} = \frac{1}{317{,}460}.$$

This is actually pretty close to the currently accepted value of
$$\frac{M_{Earth}}{M_{Sun}} = \frac{1}{329,000}.$$

Now, let's return, once again, to the problem of calculating the mass of the Moon. We still haven't solved *that* problem. With all of our earlier analyses, you can probably now see that we could answer this question if the Moon itself had a small satellite. Of course, it doesn't—or it didn't until the start of the space age. As part of the *Apollo 11* lunar expedition in July 1969, the spaceship *Columbia* went into lunar orbit while the lunar landing craft *Eagle* made moonfall. The orbital period of *Columbia* was widely reported on television as being a little less than two hours. Not so widely reported was the fact that this was all that was need to estimate the mass of the Moon.

Our earlier assumption of $M >> m$ is again good, with M as the mass of the Moon and m as the mass of the orbiting spacecraft. If we write R as the radius of the (assumed) circular orbit and T as the orbital period, then Kepler's third law says

$$T^2 = \frac{4\pi^2 R^3}{GM}. \tag{11.14}$$

To a good approximation, R is only a little bit larger than the radius of the Moon (the spacecraft is said to be in a low or *grazing* orbit), and so if we write ρ as the average density of the Moon, then

$$M = \frac{4}{3}\pi R^3 \rho. \tag{11.15}$$

Putting (11.15) into (11.14) we have the interesting result that R cancels and that

$$\rho = \frac{3\pi}{GT^2} = \frac{3\pi}{6.67 \cdot 10^{-11} T^2} \text{kg/m}^3 = \frac{3\pi}{6.67 \cdot 10^{-11} T^2} \cdot \frac{10^3 \text{ g}}{10^6 \text{ cm}^3}$$

$$= \frac{3\pi}{6.67 T^2} \cdot 10^8 \text{ g/cm}^3 = \frac{1.413 \cdot 10^8}{T^2} \text{ g/cm}^3. \tag{11.16}$$

The value of T, for a lunar grazing orbit, is $T = 108$ minutes $= 6{,}480$ seconds, and (11.15) gives $\rho = 3.37$ g/cm^3, significantly less than the average density of the Earth.

Well before the 1969 lunar orbiting event the Earth had already enjoyed a similar experience, with the October 1957 launch of *Sputnik 1*. The orbit of that Russian satellite was not really a grazing circular one, with *Sputnik*'s height above the Earth's surface varying from 142 miles to 591 miles. A truly low-orbit satellite would just skim over the surface of the planet, but of course that isn't possible with a planet that has an atmosphere. Still, we can use *Sputnik 1*'s orbital period of 96 minutes = 5,760 seconds in (11.16) to get a lower bound estimate on the average density of the Earth (lower because Kepler's third law tells us that the smaller the orbital radius the smaller the orbital period and so, from (11.16), the larger is ρ). Thus, from *Sputnik 1*,

$$\rho_{Earth} > \frac{1.413 \cdot 10^8}{(5,760)^2} \, g/cm^3 = 4.26 \, g/cm^3,$$

which is consistent with Cavendish's result that $\rho_{Earth} = 5.54 \, g/cm^3$. Notice, too, that even *Sputnik 1*'s crude density estimate still says $\rho_{Moon} < \rho_{Earth}$. Finally, we can turn (11.16) on its head and ask what it requires for the grazing orbit period to give Cavendish's density? That is, let's set $\rho = 5.54 \, g/cm^3$ and solve for T. The result is:

$$T = \sqrt{\frac{1.413 \cdot 10^8}{5.54}} \text{ seconds} = 5,050 \text{ seconds} = 84.2 \text{ minutes}.$$

If you recall (6.8), which told us how long it would take Hesiod's anvil to fall from the surface of the Earth to the center of the Earth (Hell), under the assumption of a constant-density Earth, then

$$\text{time to reach Hell} = \frac{1}{2}\sqrt{\frac{3\pi}{4G\rho}} = \frac{1}{4}\sqrt{\frac{3\pi}{G\rho}}.$$

The total travel time from Earth's surface to the center, then onward to the other side of the planet, and then all the way back to the starting point is therefore four times this ($\sqrt{3\pi/G\rho}$), which is of course just (11.16) solved for the orbital period of a surface-grazing satellite. That is, the so-called *tunneling time* through a constant-density Earth is precisely equal to the low-orbit-satellite period. Interesting! We'll do more inside the Earth in the next discussion.

We can apply these same ideas to regions of the universe that are far, far away from the Earth. Consider, for example, *pulsars* (a misnomer for *pulsating stars*), which are rapidly spinning objects created by the supernova explosions of certain dying stars. Via a number of different mechanisms, a pulsar emits a well-defined beam of electromagnetic radiation that rotates (as the pulsar spins) through space, much like a beam of light shines from the spinning lens of a lighthouse. The rotation rate of the pulsar beam is equal to the spin rate of the pulsar. Imagine such a beam that happens to have Earth along its direction of propagation. Each time the beam sweeps over Earth, an energy pulse can be detected, and since the beam sweeps over Earth at a regular rate, the detected energy is in the form of a periodic sequence of pulses. There is such a pulsar, for example, in the Crab Nebula (about 6,000 light-years from Earth, in the constellation Taurus) that has a period of 33 milliseconds, which means the Crab pulsar spins completely around its rotation axis thirty times a second. That is, in the twenty-four hours Earth takes to spin on its axis *once*, the Crab pulsar spins well over two-and-a-half *million* times!

Now, let's suppose "the radiation originates from some type of plasma layer rotating with the observed period, taken to be greater than or equal to the orbital period at the surface."[8] We can use (11.16) to again calculate a lower bound on the pulsar's average density. For the Crab pulsar we get

$$\rho_{Crab} > \frac{1.413 \cdot 10^8}{(0.033)^2} \text{ g/cm}^3 = 1.3 \cdot 10^{11} \text{ g/cm}^3.$$

This density is far greater than even the enormous density of a white dwarf star ($\sim 10^6$ g/cm^3), the ultimate fate of our own sun in a few billion years. Since ρ_{Crab} is actually on the order of nuclear matter ($\sim 10^{14}$ g/cm^3), pulsars are today believed to be spinning neutron stars.

CP. P11.1:

Imagine a satellite launched into orbit around the Earth such that the satellite is directly above the equator and orbiting in the same direction that the Earth below is rotating. The satellite is said to be *geostationary* if it "hovers" over the same spot. That is, to an observer on Earth's surface the satellite appears to be motionless in the sky. Estimate the height of such a satellite above the surface of the planet. (Such satellites are in orbit and are used as relay stations to send telephone, radio, and television signals around the Earth). *Hint*: Use Kepler's third law and the approximation that the Moon orbits the Earth at a distance of 238,850 miles from the Earth's center in 27.3 days.

Notes and References

1. One of Hooke's modern champions has written a vigorous analysis supporting Hooke's claim that he deserved better treatment from Newton; see Michael Nauenberg, "Hooke, Orbital Motion, and Newton's *Principia*" (*American Journal of Physics*, April 1994, pp. 331–350). Hooke was certainly not a quack. His greatest mistake was to be a contemporary of Newton, whose star was so bright that even an intellectually gifted person such as Hooke could be diminished by comparison. One of Hooke's personality quirks was a firm belief that if he even *thought* of something, then if anybody else had the same idea they must have somehow stolen it from him. Whatever Hooke may have thought about the inverse square law in the solitude of his own mind, the fact remains that he never published anything about those thoughts. Newton, on the other hand, published the *Principia*. Today we remember Hooke mostly for Hooke's law, an empirical statement about how "ideal" springs stretch and compress in response to an applied external force, a far cry from the glory of universal gravitation.

2. A nice tutorial discussion on the announcements and acceptance of the three laws up to Newton's time is in J. L. Russell, "Kepler's Laws of Planetary Motion: 1609–1666" (*The British Journal for the History of Science*, June 1964, pp. 1–24).

3. Newton's work in the *Principia* emphasizes starting with Kepler's laws and deriving from them the inverse square, central gravity force. In recent years a curious debate boiled up over whether or not Newton had solved the so-called inverse problem—that is, going in the other direction and starting with an inverse square, central force. Had Newton shown that Kepler's elliptical orbits are the only possible solutions? This debate was never taken very seriously by most physicists, engineers, or mathematicians, but it did spur a great deal of high-level historical scholarship. The debate started when the late professor of physics at Oberlin College Robert Weinstock (1919–2006) made claims that Newton has wrongly received credit for numerous discoveries, not just the inverse problem. Weinstock was not a nut case; he enjoyed a considerable, well-deserved reputation as a skillful mathematical physicist. His historical

arguments, however, were received with far less enthusiasm. I personally think Professor Weinstock was on pretty shaky ground, but you can decide for yourself; you can find a lengthy presentation of both his views and those of his critics in *The College Mathematics Journal* (May 1994, pp. 178–222).

4. The eccentricities of the planetary orbits vary from 0.0068 (Venus) to 0.206 (Mercury); Pluto's orbit has an even greater eccentricity of 0.25, but of course Pluto isn't a planet anymore! The eccentricity of Earth's orbit is 0.0167. From (11.2) we have the ratio of Earth's semiminor orbital axis to the semimajor orbital axis as $\frac{b}{a} = 0.99986$; to the naked eye, such an elliptical orbit would be indistinguishable from a circular orbit.

5. This part of the discussion was motivated by Franklin Miller, Jr., "Kepler's Third Law and the Mass of the Moon" (*American Journal of Physics*, January 1966, pp. 53–57).

6. This point is vigorously made in Robert Osserman, "Kepler's Laws, Newton's Laws, and the Search for New Planets" (*American Mathematical Monthly*, November 2001, pp. 813–820).

7. This part of the discussion was motivated by Robert Garisto, "An Error in Isaac Newton's Determination of Planetary Properties" (*American Journal of Physics*, January 1991, pp. 42–48). I have made some simplifications to Professor Garisto's impressively detailed analyses; I hope he will forgive me.

8. H. L. Poss, "Pulsar Periods and Kepler's Third Law" (*American Journal of Physics*, January 1970, pp. 109–110).

Gravity Inside the Earth

> The acceleration of gravity on the Earth's surface is greater at the poles than it is at the equator. Why?
> — Question on a freshman physics exam

12.1 Newton's Experiment

Most people, when faced with the above opening question, almost immediately think of the Earth's rotation, probably because of the common familiarity just about everybody has (by age five) with the tendency of a merry-go-round to toss a rider off when it spins. For the Earth, this "tendency" appears as an acceleration component opposite to that of Earth's gravity. My wife, who has never had a physics course in her life—she was an art history major in college—came up with this explanation in less than five seconds after I ran the opening test question by her. A nonrotating planet with a spherically symmetric density should have a surface gravity that is the same everywhere, but if it is rotating, the merry-go-round effect will be maximum at the equator, zero at the poles, and vary smoothly from maximum to minimum as we move up (or down) in latitude from the equator toward either pole. The magnitude of this effect is not hard to calculate using Figure 12.1, which shows a mass m on the surface of a rotating planet (with radius R) at latitude λ.

The mass is traveling along a circular path with radius $R\cos(\lambda)$, and so in one rotation of duration T the mass travels a distance of

GRAVITY INSIDE THE EARTH   187

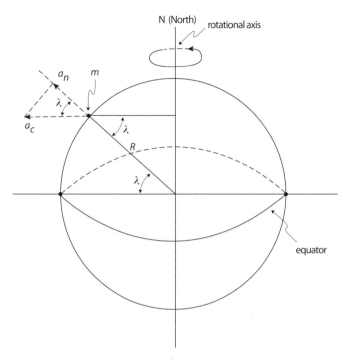

Figure 12.1. How the rotation of a planet reduces its surface gravity.

$2\pi R \cos(\lambda)$; its speed v is therefore

$$v = \frac{2\pi R \cos(\lambda)}{T}.$$

The centrifugal acceleration (directed away from the planet *in the plane* of m's circular path) is thus

$$a_c = \frac{v^2}{R \cos(\lambda)} = \frac{4\pi^2 R \cos(\lambda)}{T^2}.$$

The normal component of this acceleration, a_n, which is *opposite* to the inward centripetal acceleration of gravity (that is, the merry-go-round effect reduces the net *radial* acceleration of the planet's surface gravity), is given by

$$\frac{a_n}{a_c} = \cos(\lambda),$$

or
$$a_n = \frac{4\pi^2 R \cos^2(\lambda)}{T^2}.$$

Obviously, $a_n = 0$ at the poles ($\lambda = \pm 90°$) and $4\pi^2 R/T^2$ at the equator ($\lambda = 0°$). Inserting numbers, using $R = 6{,}371$ kilometers, the value of a_n at the equator is

$$\frac{4\pi^2 (6{,}371 \cdot 10^5 \text{ cm})}{(86{,}400 \text{ s})^2} = 3.4 \text{ cm/s}^2,$$

and this is the reduction in Earth's gravity at the equator from what it is at either pole.

This isn't a huge difference in the net polar and equatorial accelerations of gravity—the nominal textbook value is 981 cm/s^2, and so we're talking about an effect with a *maximum* of one-third of 1% of the nominal value—but it is not difficult to measure. In fact, our calculation is a correct one and goes a long way toward answering the exam question. But not *all* the way. The actual difference in the net polar and equatorial accelerations of gravity is found by experiment to be even greater than 3.4 cm/s^2; the measured difference is 5.2 cm/s^2. So, where does the extra 1.8 cm/s^2 come from? From our (incorrect) assumption of a uniform-density, spherical Earth. And that's the rest of this discussion.

As 1679 drew to a close, years before relations between the two irrevocably broke over the priority squabble concerning the inverse square law of gravitation, Hooke and Newton exchanged letters containing specific and implied speculations on the nature of gravity *inside* the Earth. The exchange started when Newton (in a letter dated November 28, 1679) described an interesting experiment that he thought would allow detection of the Earth's daily rotation on its axis. The fact that the Earth rotates is, of course, literally as obvious as are night and day, but the fascinating thing about Newton's proposal is that it would allow an observer in a sealed, windowless laboratory to also conclude that Earth rotates. (Today, this is most dramatically demonstrated not by Newton's experiment but by the Foucault pendulum, invented in 1851 by the French physicist Léon Foucault [1819–1868]. It's a popular science exhibit in numerous museums around the world.)

GRAVITY INSIDE THE EARTH _ _ _ _ **189**

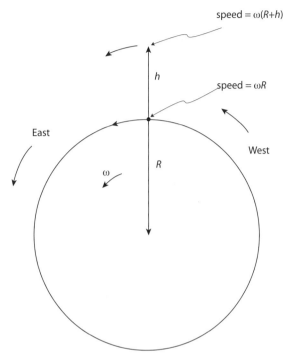

Figure 12.2. Newton's experiment.

Newton's experiment can be understood with the aid of Figure 12.2, which shows the view seen by an observer above the North Pole looking downward along the Earth's rotation axis as the planet spins below him. (To keep things as simple as possible, imagine that the experiment is being performed on the equator.) The Earth, with radius R, is rotating counterclockwise from west to east at angular speed ω. Newton proposed dropping an object (he suggested a bullet) from a tower of height h and simply observing where it landed. Most people, when asked about this, think of the object as falling straight down toward the center of the Earth while the Earth rotates beneath the falling object. Thus, according to this (erroneous) image, the object will land to the *west* of the foot of the tower.

Not so, said Newton. The rotation speed of the foot of the tower is ωR, while the rotation speed of the top of the tower is $\omega(R+h)$, which is also the rotation speed of the dropped object at the instant it is released—and that, argued Newton, will continue to be the object's rotation speed all through its fall. That is, the dropped object is, all through the fall, moving to the east faster than the foot of the tower by the speed difference ωh. Thus, the object will land to the east of the foot of the tower. As Newton wrote to Hooke, the falling object would "outrun" the foot of the tower and "shoot forward to the east." Newton did not actually calculate the distance the object would be to the east of the tower foot, but it isn't difficult to estimate it.

Assuming that over the entire fall that the acceleration of gravity is a constant (g), then starting from rest (in the downward direction) the duration of the fall (T) is

$$h = \frac{1}{2}gT^2,$$

or

$$T = \sqrt{\frac{2h}{g}}.$$

Since the falling object is moving east faster than the foot of the tower is moving east by the speed difference ωh, then the eastward displacement is given by

$$E_d = \omega h \sqrt{\frac{2h}{g}} = \omega \sqrt{\frac{2h^3}{g}}. \quad (12.1)$$

When Hooke read Newton's letter he got quite enthusiastic about this experiment and convinced the Royal Society to commission its Curator of Experiments to perform it (you'll recall from the previous discussion that the Curator was, conveniently, Hooke himself!). Hooke performed the experiment in London and not on the equator, and so a final correction factor of $\cos(\lambda)$ is required in (12.1), where λ is the latitude of the tower ($\lambda = 51.5$ degrees at London). (Take a look back at Figure 12.1 to see where the $\cos(\lambda)$ factor comes from.) You won't be surprised to learn that Hooke's results were less than definitive, once you put numbers in (12.1). The value of ω is found from the Earth's single rotation through 2π radians each day (once every 86,400 seconds):

$$\omega = \frac{2\pi}{86,400} \frac{\text{rad}}{\text{s}} = 7.3 \cdot 10^{-5} \frac{\text{rad}}{\text{s}},$$

and so, for the $h = 27$ foot drop Hooke used (a small ball was dropped into a box full of tobacco pipe clay, so the ball would stick), (12.1) gives

$$E_d = 7.3 \cdot 10^{-5} \sqrt{\frac{2(27)^3}{32.2}} \cos(51.5°) \text{ feet} = 0.00159 \text{ feet} = 0.019 \text{ inches},$$

that is, less than one-fiftieth of an inch! It is almost impossible that Hooke could possibly have detected Newton's predicted eastward displacement.

While Newton's experiment is an interesting concept in its own right, that initial letter to Hooke mostly fascinates historians of science for a different reason. After outlining the proposed experiment, almost as an afterthought Newton went on (as he wrote years later to his friend Edmond Halley in a letter dated May 27, 1686) to speculate about what would happen if the dropped object could continue its fall to *inside* the Earth: "I carelessly described the Descent of the falling body in a spirall to the center of the earth" (Newton was, of course, assuming that the matter of the Earth was not resisting the motion of the falling object). Hooke replied to Newton on December 9, 1679, saying that Newton's "spirall" was wrong and that the path of the falling body instead "would resemble An Elleipse." Hooke almost surely wrote that because he was thinking of the inverse square law of gravity operating inside the Earth as well as outside. He didn't come right out and specifically invoke the inverse square law, however, almost surely because (like Newton in 1679) he didn't have any idea of how to handle gravity for the case of distributed matter inside the Earth.

12.2 Gravity Inside the Earth

Newton replied to Hooke on December 13, 1679, in a letter in which he determined the path of the falling body under the assumption of a specific force law—that gravity is constant throughout the interior of the Earth. Newton's calculations were correct for this law, but the force law itself was as faulty as Hooke's (but take a look back at Challenge Problem 10.2); as Newton explained in his letter to Halley, he was just taking "the simplest case for computation, which was that of Gravity uniform in a medium not Resisting." My point here is that, in 1679, neither Hooke nor Newton, and for that matter nobody else in the world, had any idea at all of the physical nature of the regions deep beneath their feet. And yet, just seven years later, with Newton's publication of his two superb theorems in the *Principia*, he had all that is needed to calculate all sorts of wonderful things concerning the insides of Earth. Before we are done in this discussion, for example, I'll show you how to calculate the pressure at the center of the planet. I think we'd all agree it's got to be pretty big—but how big?

Recall that Newton's second superb theorem says that the gravitational force inside a uniform density, differentially thin spherical shell is zero. This tells us that, as we descend into the interior of a solid sphere of matter with a spherically symmetric density, the inverse square gravitational force we'll experience depends only on the matter still *beneath* us. The matter that is *above* us (the spherical shell we have passed through) contributes nothing to the gravitational force on us. For a sphere with radius R and a density ρ, then, the gravitational force F on a mass m varies with r (the distance from the center of the sphere), as follows:

$$F = \begin{cases} \frac{GMm}{r^2} = \frac{G\frac{4}{3}\pi R^3 \rho m}{r^2}, & r \geq R \\ \frac{G\frac{4}{3}\pi r^3 \rho m}{r^2} = G\frac{4}{3}\pi \rho m r, & r \leq R. \end{cases}$$

When $r = R$ (the surface of the sphere) we write $F = mg(R)$, where $g(R) = g(r = R)$ where $g(r)$ is the acceleration of gravity at distance r from the center of the sphere. So,

$$G\frac{4}{3}\pi R \rho = g(R)$$

is the acceleration of gravity at the surface, and therefore

$$g(r) = \begin{cases} g(R)\frac{R^2}{r^2}, & r \geq R \\ g(R)\frac{r}{R}, & r \leq R. \end{cases} \quad (12.2)$$

Figure 12.3 shows a plot of (12.2), and the portion for $0 \leq r \leq R$ is what many physicists and engineers think (because this is the typical freshman physics class result) is the actual behavior of Earth's interior gravity. It isn't (the acceleration of gravity, in the real Earth, is actually greater in most of the interior volume of the planet than it is at the surface). And it's a far cry from a constant interior gravity;

a constant-density Earth would have an interior gravity that linearly decreases with decreasing distance from the center, which certainly isn't Newton's original (1679) idea of constant gravity. So, what would the density have to be to give Newton's constant gravity of 1679? Newton himself never asked this question, but we can answer it as follows. Suppose we are distance r from the center of a planet with radius R, and that $\rho(r)$ is the (assumed) spherically symmetric density. The mass $M(r)$ beneath us is given by

$$M(r) = \int_0^r 4\pi x^2 \rho(x)\,dx, \; 0 \leq r \leq R, \tag{12.3}$$

where x is simply a dummy variable of integration. The gravitational force F on a mass m at distance r is then

$$F = \frac{GM(r)m}{r^2} = 4\pi Gm \frac{1}{r^2} \int_0^r x^2 \rho(x)\,dx = mg(r),$$

where $g(r)$ is the acceleration of gravity at distance r from the center of the planet. That is,

$$g(r) = 4\pi G \frac{1}{r^2} \int_0^r x^2 \rho(x)\,dx, \; 0 \leq r \leq R. \tag{12.4}$$

Now, suppose that, with k some constant, the density is given by

$$\rho(r) = \frac{k}{r}, \; 0 \leq r \leq R. \tag{12.5}$$

Then, (12.4) reduces to

$$g(r) = 4\pi G \frac{1}{r^2} \int_0^r x^2 \frac{k}{x}\,dx = 4\pi G \frac{1}{r^2} k \int_0^r x\,dx = 4\pi G k \frac{1}{r^2} \left(\frac{1}{2} x^2\right)\Big|_0^r = 2\pi G k,$$

which is a constant, that is, we have Newton's constant interior gravity. If we use $r = R$ (the surface of the planet) in (12.5), we have $k = R\rho(R)$, and so Newton's constant gravity is achieved with a density of

$$\rho(r) = \frac{R}{r}\rho(R), \; 0 \leq r \leq R, \tag{12.6}$$

where $\rho(R)$ is the density at the planet's surface.

This density function is not physically possible, even if we ignore the infinite density at the center of the planet. Just approaching the

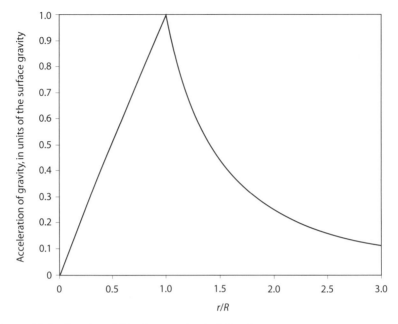

Figure 12.3. Gravity of "freshman physics" Earth.

center requires the density to become unreasonable. For example, when $r = R/1{,}000$, which for the Earth is still thousands of meters short of the center, (12.6) says the density would have to be a thousand times greater than the surface density. Interestingly enough, however, (12.6) doesn't require anything dramatic at all for the *average* density. That is, if ρ_a is the average density of Newton's constant interior gravity planet, then

$$\rho_a = \frac{\text{total mass of planet}}{\text{volume of planet}} = \frac{\int_0^R 4\pi r^2 \rho(r)\, dr}{\frac{4}{3}\pi R^3} = \frac{4\pi \int_0^R r^2 \frac{R}{r}\rho(R)\, dr}{\frac{4}{3}\pi R^3}$$

$$= \frac{3R\rho(R)}{R^3} \int_0^R r\, dr = \frac{3\rho(R)}{R^2} \left(\frac{1}{2}r^2\right)\Big|_0^R = \frac{3}{2}\rho(R).$$

That is, the average density of Newton's planet is only 50% greater than the (finite) surface density, even though the density becomes unbounded as we approach the core.

TABLE 12.1:
The Insides of the Earth

Region	Inner radius (miles)	Outer radius (miles)	ρ (g/cm^3)
crust/oceans	3,944	3,960	2.4
upper mantle	3,542	3,944	3.6
lower mantle	2,163	3,542	4.9
outer core	759	2,163	10.9
inner core	0	759	12.9

The actual variation of gravity inside the Earth is given by neither the linear (12.2) nor by Newton's constant gravity. Surprisingly, it wasn't until two centuries after the *Principia* that the modern quantitative model of the Earth's interior structure began to be developed. In 1896 the German physicist Emil Wiechert (1861–1928) proposed an Earth with a dense solid iron core surrounded by a less dense stony shell (the mantle). In 1899 the Irish geologist Richard Oldham (1858–1936), using the seismological records of an 1897 earthquake in India, determined that there was indeed a density discontinuity inside the Earth representing the change from a low-density mantle to a high-density core. It is now believed, too, that there is a distinct density transition in the mantle itself, that is, there is a lower (higher-density) mantle and an *upper* (lower-density) mantle. In 1910 the Croatian Andrija Mohorovičić (1859–1936) found seismic wave evidence for yet another density discontinuity above the mantle. This is now famously known as the Moho discontinuity between the outermost crust of the Earth (about sixteen miles thick on average, but varying from as thin as three miles more than twenty miles) and the upper mantle. It wasn't until as recently as 1926 that the English mathematical physicist Harold Jeffreys (1891–1989) determined that the Earth's core is actually fluid. Ten years later the Danish seismologist Inge Lehmann (1888–1993) proposed a solid inner region to the core, which is now the accepted geophysical doctrine. That is, there is a liquid outer core and a solid inner core. The interior structure of the Earth is pretty complicated![4] There are, in fact, five major divisions to the Earth's interior, listed in Table 12.1 with their inner and outer radii, along with their average densities.

The complex inner structure of the Earth that is evident in Table 12.1 is behind one of the more striking aspects of the real Earth's gravity. As one descends into the Earth the acceleration of gravity, at least for a while, actually increases rather than decreasing as the freshman physics graph of Figure 12.1 would lead you to (incorrectly) believe. Here's why. Let's make the following definitions:

$M(r)$ = mass of Earth inside a radius of r, $r \leq R$;
$\rho(r)$ = Earth's density at r;
$\rho_a(r)$ = Earth's average density inside a radius of r.

Then, as in (12.3),

$$M(r) = \int_0^r 4\pi x^2 \rho(x) dx$$

and so, using Leinbiz's rule to differentiate the integral (take a look again at the Challenge Problem in Discussion 5),

$$\frac{dM(r)}{dr} = 4\pi r^2 \rho(r). \tag{12.7}$$

Also,

$$M(r) = \frac{4}{3}\pi r^3 \rho_a(r). \tag{12.8}$$

Now, the gravitational force F on a mass m' at distance $r \leq R$ from the center of the Earth is

$$F = \frac{GM(r)m'}{r^2} = m'g(r),$$

or

$$g(r) = \frac{GM(r)}{r^2}.$$

So, differentiating this $g(r)$, we get

$$\frac{dg}{dr} = \frac{r^2 G \frac{dM(r)}{dr} - 2GrM(r)}{r^4},$$

or

$$\frac{dg}{dr} = \frac{G}{r^2}\frac{dM(r)}{dr} - \frac{2GM(r)}{r^3}. \tag{12.9}$$

Putting (12.7) and (12.8) in (12.9),

$$\frac{dg}{dr} = \frac{G}{r^2} 4\pi r^2 \rho(r) - \frac{2G}{r^3} \cdot \frac{4}{3}\pi r^3 \rho_a(r),$$

or

$$\frac{dg}{dr} = 4\pi G \left[\rho(r) - \frac{2}{3}\rho_a(r) \right]. \tag{12.10}$$

Look carefully at what (12.10) tells us. The acceleration of gravity will decrease as r decreases—that is, $\frac{dg}{dr} > 0$—only if $\rho(r) > \frac{2}{3}\rho_a(r)$. In particular, if we are on the surface (where $r = R$), then gravity will decrease as we descend into the Earth only if $\rho(R) > \frac{2}{3}\rho_a(R) = \frac{2}{3} \cdot 5.54 \text{ g/cm}^3 = 3.69 \text{ g/cm}^3$. But this is not the case! The continental crust (2.7 g/cm^3), the oceanic crust (3 g/cm^3), and the oceans (1 g/cm^3) average together in some way to give a $\rho(R)$ that cannot possibly exceed 3 g/cm^3. You simply can't have an *average* density greater than a *maximum* density. From Table 12.1 we see that the average near-surface density is even less: 2.4 g/cm^3. So, as we start our descent from the surface, (12.10) says we have $\frac{dg}{dr} < 0$, which means that as r decreases, we must have $g(r)$ increase, and that is precisely what is actually observed.

The most striking aspect of Table 12.1 is the very distinct difference in densities between the upper three regions and the lower two. If the first three are lumped together to form what we'll simply call the mantle, and if the inner two regions are lumped together into what we'll call the core, then we'll get a mathematically simpler model that nevertheless will still exhibit much of the actual behavior of gravity inside the Earth. This is called the two-density model of the Earth.[6] Specifically, we'll take the lumped core to be a sphere of radius $R_c = 2{,}163$ miles with an average density of $\rho_c = 11.0 \text{ g/cm}^3$, and the lumped mantle to be a shell with inner radius 2,163 miles and outer radius $R = 3{,}960$ miles with an average density of $\rho_m = 4.44 \text{ g/cm}^3$. These two densities are the volume-weight averages of the densities given in the regions listed in Table 12.1.[7]

For $r \leq R_c$ the gravitational force F on a mass m' distance r from the Earth's center is

$$F = m'g(r) = \frac{GM(r)m'}{r^2} = \frac{G\frac{4}{3}\pi r^3 \rho_c m'}{r^2}$$

where $M(r)$ is, as before, the mass of the Earth inside a radius of r. Thus,

$$g(r) = \frac{4}{3}\pi G \rho_c r = \left(\frac{4}{3}\pi\right) 6.67 \cdot 10^{-11} \frac{\text{m}^3}{\text{kg} \cdot \text{s}^2} \cdot 11 \frac{\text{g}}{\text{cm}^3} r$$

$$= \left(\frac{4}{3}\pi\right) 6.67 \cdot 10^{-11} \frac{(10^2 \text{ cm})^3}{10^3 \text{ g} \cdot \text{s}^2} \cdot 11 \frac{\text{g}}{\text{cm}^3} r.$$

At $r = R_c = 2{,}163$ miles $= 3.481 \cdot 10^8$ cm this gives $g(R_c) = 1{,}070$ cm/s^2. That is, if we start at the center of the Earth and move radially outward toward the surface, the acceleration of gravity will increase linearly with r until, at the outer radius of the lumped core, the acceleration is $1{,}070$ cm/s^2. If C is some constant, we can write this behavior in a particularly convenient form with r normalized to Earth's radius, R, as

$$g(r) = C \frac{r}{R}, \quad 0 \leq r \leq R_c.$$

Since

$$\frac{R_c}{R} = \frac{2{,}163}{3{,}960} = 0.546,$$

then

$$g(R_c) = C \frac{R_c}{R} = 0.546 C = 1{,}070,$$

which gives $C = 1{,}960$, and so, at last,

$$g(r) = 1{,}960 \frac{r}{R} \text{ cm/s}^2, \quad 0 \leq \frac{r}{R} \leq 0.546. \tag{12.11}$$

For $R_c \leq r \leq R$, we have

$$F = m' g(r) = \frac{G m' \left[\frac{4}{3}\pi R_c^3 \rho_c + \frac{4}{3}\pi (r^3 - R_c^3) \rho_m\right]}{r^2},$$

and so

$$g(r) = \frac{4}{3}\pi G \left[\rho_c \frac{R_c^3}{r^2} - \rho_m \frac{R_c^3}{r^2} + \rho_m r\right] = \frac{4}{3}\pi G R_c \left[(\rho_c - \rho_m)\left(\frac{R_c}{r}\right)^2 + \rho_m \frac{r}{R_c}\right].$$

This becomes, when numbers are inserted,

$$g(r) = 97.256 \left[(11 - 4.44) \left(\frac{R_c}{r} \right)^2 + 4.44 \frac{r}{R_c} \right]$$

$$= 638 \left(\frac{R_c}{r} \right)^2 + 432 \frac{r}{R_c} \, \text{cm/s}^2, \, R_c \leq r \leq R.$$

(As a check, notice that at $r = R_c$, we have $g(R_c) = 638 + 432 = 1,070 \, \text{cm/s}^2$, just as we should.) Now, since $R_c = 0.546 \, R$, we arrive at

$$g(r) = 638 \left(\frac{0.546R}{r} \right)^2 + 432 \left(\frac{r}{0.546R} \right),$$

or, finally,

$$g(r) = 190 \left(\frac{R}{r} \right)^2 + 791 \left(\frac{r}{R} \right) \text{cm/s}^2, \, 0.546 \leq \frac{r}{R} \leq 1. \quad (12.12)$$

(As a check, notice that at the Earth's surface, that is, when $r = R$, (12.12) says $g(R) = 190 + 791 = 981 \, \text{cm/s}^2$, the usual textbook value for Earth's surface gravity.)

It is interesting to compare (12.11) and (12.12) with the acceleration of gravity inside a uniform-density Earth. From (12.12) the interior gravity for such a Earth—what I'll call a "freshman physics" Earth —is

$$g(r) = 981 \frac{r}{R} \, \text{cm/s}^2, \, 0 \leq \frac{r}{R} \leq 1. \quad (12.13)$$

Figure 12.4 shows (12.11) and (12.12) plotted as a solid line, while (12.13) is the dashed line. There is a marked difference between the two gravity plots. You'll notice, however, that the simplification from five to two inner regions has lost some detail; in particular, the calculation we did earlier of an increase in gravity (near the Earth's surface) as one travels radially inward is not present in Figure 12.4. In the plot there is a slight decrease. Does that mean we really haven't gained much (if anything), despite all the calculating that went into (12.11) and (12.12)? The answer is most definitely *no*, and to illustrate why I say that, I'll finish this discussion with a calculation that if we use (12.13) will result in significant error, whereas if we use (12.11) and (12.12), the analysis will give a result in excellent agreement with modern geophysics.

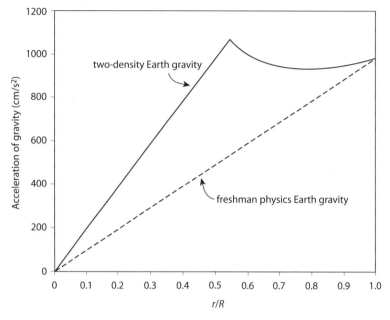

Figure 12.4. Gravity inside the Earth.

12.3 Pressure at the Center of the Earth

To calculate the pressure at the center of the Earth, we'll need what is called the *differential equation of static equilibrium*. Despite the impressive name, it is actually not at all difficult to derive. To simplify our thinking, let's assume that the Earth behaves like a fluid that has realized a steady-state condition. By that I mean let's suppose that all the matter in the Earth, even if once flowing and moving about, has settled down into a static equilibrium because all energy of motion has been dissipated away via friction. Imagine, then, a small cubic block of matter with its bottom face parallel with Earth's surface, at distance r from the Earth's center, and the top face at distance $r + \Delta r$. The top and bottom faces of the block each have surface area ΔA. The pressure on the top surface is $P(r + \Delta r)$, while the pressure on the bottom face is $P(r)$. The mass of the block is Δm, where, since $\Delta A \Delta r$ is the volume of the block,

$$\Delta m = \rho(r) \Delta A \Delta r.$$

The magnitude of the downward-directed force on the block is the force on the top face plus the gravitational force on the block:

$$\text{downward force} = P(r+\Delta r)\Delta A + g(r)\rho(r)\Delta r\,\Delta A.$$

The magnitude of the upward force on the block is simply the force on the bottom face:

$$\text{upward force} = P(r)\Delta A.$$

If we assume (as we have) that the block is in static equilibrium, then it isn't moving (and it certainly isn't accelerating!), and so the net force on Δm must be zero. (The lateral forces on all the vertical sides of the block sum to zero, as well; if they didn't, the block would move sideways, and then we would be faced with the awkward problem of explaining where that kinetic energy came from). Thus, the two force magnitudes, downward and upward, must be equal, and so

$$P(r+\Delta r)\Delta A + g(r)\rho(r)\Delta r\,\Delta A = P(r)\Delta A,$$

or

$$P(r+\Delta r) + g(r)\rho(r)\Delta r = P(r),$$

or, as $P(r+\Delta r) = P(r) + \Delta P$,

$$P(r) + \Delta P + g(r)\rho(r)\Delta r = P(r),$$

or

$$g(r)\rho(r)\Delta r = -\Delta P,$$

or

$$\frac{\Delta P}{\Delta r} = -g(r)\rho(r).$$

In the limit as $\Delta r \to 0$, we replace the deltas with differentials and so arrive at our fundamental preliminary result, the promised differential equation of static equilibrium:

$$\frac{dP}{dr} = -g(r)\rho(r). \tag{12.14}$$

The right-hand side of (12.14) is clearly always negative, because both $g(r)$ and $\rho(r)$ are *always* positive, and so $\frac{dP}{dr} < 0$, which makes physical sense. That is, just as you'd expect, because r decreases as one approaches the center of the planet, the pressure P continually increases.

We can integrate (12.14) easily for the case of the freshman physics Earth, where $\rho(r)$ is a constant ($\rho = 5.54$ g/cm^3) and $g(r)$ is given by (12.2). That is,
$$g(r) = g(R)\frac{r}{R},$$
where $g(R) = 981$ cm/s^2. Then
$$\int_0^{P_c} dP = P_c = -\frac{\rho g(R)}{R}\int_R^0 r\,dr,$$
where P_c is the pressure at the Earth's center ($r = 0$), and the pressure is zero at the Earth's surface ($r = R$). Continuing,
$$P_c = -\frac{\rho g(R)}{R}\left(\frac{1}{2}r^2\right)\Big|_R^0 = \frac{\rho g(R)R}{2}$$
$$= \frac{5.54\text{ g/cm}^3 \cdot 981\,\frac{\text{cm}}{\text{s}^2} \cdot 3{,}960 \cdot 1609 \cdot 10^2\text{ cm}}{2} = 1.73 \cdot 10^2\,\frac{\text{g}}{\text{cm}\cdot\text{s}^2}.$$

The unit of pressure in the metric system is the pascal (Pa), with the units of kg/m·s^2 (equal to one newton of force per square meter). That is,
$$1\text{ Pa} = \frac{10^3\text{ g}}{10^2\text{ cm}\cdot\text{s}^2} = 10\,\frac{\text{g}}{\text{cm}\cdot\text{s}^2},$$
and so
$$P_c = 1.73 \cdot 10^{11}\text{ Pa}. \tag{12.15}$$

To give you a more intuitive feel for the magnitude of P_c, we can convert (12.15) to pounds per square inch with the conversion factor
$$101{,}000\text{ Pa} = 1\text{ atmospheric pressure} = 14.7\text{ pounds/in}^2.$$

Thus, the pressure at the center of the freshman physics Earth is
$$P_c = \frac{1.73 \cdot 10^{11}}{1.01 \cdot 10^5} \cdot 14.7\text{ pounds/in}^2 = 25.2 \cdot 10^6\text{ pounds/in}^2.$$

More than twenty-five million pounds per square inch—now that's a big squeeze! But not big enough. Modern geophysical theory puts the actual value of P_c at more than twice this value.

We can eliminate most of our error by using the two-density model of the Earth in (12.11) and (12.12). Returning to (12.14), its integral now becomes

$$\int_0^{P_c} dP = -\int_R^0 g(r)\rho(r)dr = P_c,$$

or, if we change the variable in the second integral to $u = r/R$,

$$P_c = -R\int_1^0 g(u)\rho(u)\,du = R\int_0^1 g(u)\rho(u)du,$$

where

$$g(u) = \begin{cases} 1,960u, & 0 \le u < 0.546 \\ \frac{190}{u^2} + 791u, & 0.546 \le u \le 1 \end{cases}$$

and

$$\rho(u) = \begin{cases} 11, & 0 \le u \le 0.546 \\ 4.44, & 0.546 \le u \le 1. \end{cases}$$

So,

$$P_c = R\left[\int_0^{0.546} (1,960u)11\,du + \int_{0.546}^1 \left(\frac{190}{u^2} + 791u\right)4.44du\right],$$

which, after doing the easy integrals, reduces to

$$P_c = 5,147.9R = 5,147.9 \cdot 3,960 \cdot 1.609 \cdot 10^5 \text{ g/cm} \cdot \text{s}^2$$
$$= 3.28 \cdot 10^{12} \text{ g/cm} \cdot \text{s}^2 = 3.28 \cdot 10^{11} \text{ Pa}.$$

This is almost twice the value computed in (12.15) and is within 10% of the modern textbook value of $3.62 \cdot 10^{11}$ Pa for the pressure at the center of the real Earth.

12.4 Travel Inside the Earth

As mentioned earlier in this book (see note 5 of Discussion 10), the exotic idea of human travel and amazing discovery deep inside the Earth was a romantic one for writers of nineteenth-century (and even earlier) fiction. It was the Indiana Jones adventure of the day. Only with Jules Verne's 1864 *A Journey to the Center of the Earth* did even a bit of reality begin to appear. A sort of scientific transition on the topic of deep-Earth travel occurred in the very late nineteenth century with the publication of stories involving characters discussing gravity-powered travel times based on real physics. At the same time, however, these same writers brushed aside all other technical issues, such as heat and pressure, with little concern. For example, in the January 1898 issue of *St. Nicholas Magazine* there began a four-part serial called "Through the Earth," written by Clement Fezandié (1865–1959).[8] (Later that same year the story appeared as a hardcover book.) Set a century in the future, it was the tale of a rapid-transit gravity-powered vacuum tunnel, thirty feet in diameter, joining New York City and a point about two hundred miles out at sea off the southwestern coast of Australia. It was to cost only a hundred million dollars.[9] The travel time to the Earth's center is correctly given as twenty-one minutes, but the thermal nature of the Earth's core was dismissed as "no problem" via the proper use of tunnel wall refrigeration! The story is still interesting, though, because while the solutions may be silly, many of the actual engineering difficulties of building such a tunnel are explicitly stated.

There is no direct evidence, but I would be surprised if the pioneer American rocket scientist Robert Goddard (1882–1945) didn't read "Through the Earth" as a youngster; the young Goddard was absolutely fascinated by the concept of a vacuum tunnel rapid transit system, and reading "Through the Earth" might well have put him onto the concept. While still an undergraduate at Worcester Polytechnic Institute (1904–1908), for example, he wrote a fictional treatment of a vacuum tunnel transport system,[10] and a less romantic essay appeared as an unsigned editorial in the November 20, 1909, issue of *Scientific American*. Goddard's tunnel transport system was not, however, a gravity-powered one but rather a passenger car to be driven by "the magic power of magnetism." The idea was to accelerate at a constant rate for the first half of the journey and then to decelerate at the same rate over the second half. As that editorial went on,

> Taking an acceleration of 11 feet per second each second, which would bring to rest, in 4 seconds, a train moving at 30 miles per hour (44 feet per second) ... the time from New York to Philadelphia (85 miles) and

from New York to Boston (190 miles) would be 6 minutes 44 seconds and 10 minutes 4 seconds respectively.[11]

In 1929 a strange novel by an author using the pen-name "Gawain Edwards" appeared called *The Earth-Tube*. It is the tale of a vaguely specified, hostile Asian country that, not having the airplane technology to directly challenge America, decides to instead tunnel through the Earth to launch a surprise attack. (Sure, that *would* be lots easier than simply building a better airplane.) The novel reads like Flash Gordon on Mars, and envelops the reader in a strongly racist atmosphere: the bad guys are clearly from either Japan or China. The "science" is generally preposterous, and all of the novel's miracle weaponry is developed by just two men working alone in a home lab.

The tunnel in *The Earth-Tube* is much bigger than Fezandié's—more than 500 feet in diameter compared to 30 feet—and it was traversed by a monster gravity-powered car a half-mile long. The total travel time through the Earth was given as 38.47 minutes, which is about right. As with Fezandié, however, the heat and pressure of the Earth's interior were no problem, because the invaders had discovered a mysterious metal ("undulal"), impervious to just about anything, with which to line the tunnel. An immense amount of this metal would be required, of course, but that was no problem, either—it was made (in some unspecified way) from mud!

The most interesting aspect of *The Earth-Tube* today is the identity of the author. Gawain Edwards was actually George Pendray (1901–1987), whose career was mostly as a public relations consultant. As one of the founders of the American Rocket Society in 1931, which in 1963 became the American Institute of Aeronautics and Astronautics, he came to know Goddard; he was, in fact, co-editor of Goddard's papers cited in note 10. The idea for his novel almost surely was sparked by conversations he had with Goddard, who never lost his youthful fascination with vacuum tunnel travel through the Earth. Indeed, as outrageous as the concept of a gravity-powered tunnel transport system might seem, it nevertheless remains a fixture in modern physics textbooks. It is just too romantic and exciting an idea to ignore.

It is also, with the aid of Newton's two superb gravity theorems, not at all difficult to analyze the gravity-powered tunnel for a

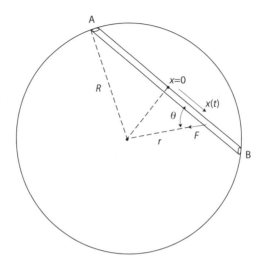

Figure 12.5. Geometry of a gravity tunnel.

constant-density planet. Consider Figure 12.5, which shows a tunnel connecting two arbitrary points A and B on the surface of a planet with radius R and constant density ρ. The axis of the tunnel defines the x-axis, with $x = 0$ locating the middle of the tunnel. The figure shows our traveler, with mass m, located at some point $x(t)$ at time t while on his way from A to B. The gravity force F acting on the traveler is always, of course, directed straight toward the center of the planet. Let's write the distance from the center of the planet to the traveler as r. Then we know the force F is due only to the planetary mass inside a sphere of radius r, that is, to the mass

$$M(r) = \frac{4}{3}\pi r^3 \rho.$$

The angle θ is, as shown, the angle the line joining the traveler's location to the planet's center makes with the tunnel axis, and it is geometrically clear that

$$\cos(\theta) = \frac{x}{r}.$$

Don't be misled by the way I've drawn Figure 12.5, with the trip from A to B as "downhill" all the way. It *is* gravitationally downhill on the first half of the trip, from A to the midpoint of the tunnel, in the sense that the traveler is continually getting closer to the center of the Earth. From the midpoint on, to B, however, it is gravitationally *uphill*, as the traveler is then continually moving away from the Earth's center. The return trip from B back to A is perfectly symmetrical in this sense, too, with B to $x = 0$ being "downhill" and $x = 0$ to A being "uphill."

It is the component of F *along* the tunnel axis (the x-axis) that powers our traveler on his journey from A to B. That force component is, from Figure 12.5,

$$F_x = F \cos(\theta) = F \frac{x}{r}.$$

Notice carefully that when $x < 0$, F_x is directed toward B, and that when $x > 0$, F_x is directed back toward A. During the first half of the journey ($x < 0$) F_x is in the direction of travel, while during the second half of the journey ($x > 0$) F_x is in the direction *opposite* to the direction of travel. That is, F_x is always directed toward the tunnel midpoint at $x = 0$. The traveler is accelerated by F_x during the first half of the trip and then decelerated by F_x during the second half, just as in Goddard's scheme. The force component of F that is normal to the tunnel axis, met by the reaction force of the suspension system of the vehicle in which sits our intrepid traveler, plays no role in our analysis.

So, putting all these expressions together, we have the gravitational force that powers the traveler through the tunnel as

$$F_x = G \frac{M(r)m}{r^2} \cdot \frac{x}{r} = G \frac{\frac{4}{3}\pi r^3 \rho m x}{r^3} = \frac{4}{3}\pi G \rho m x.$$

G is, of course, the universal gravitational constant. Thus, using Newton's second law of motion, we have

$$m \frac{d^2 x}{dt^2} = -F_x = -\frac{4}{3}\pi G \rho m x,$$

or, at last, we have the differential equation for travel through a gravity tunnel:

$$\frac{d^2 x}{dt^2} = -\frac{4}{3}\pi G \rho x. \tag{12.16}$$

If the minus sign bothers you, just remember that when $x < 0$ we must have the traveler's acceleration be positive (he is getting up to speed) and when $x > 0$ we must have the traveler's acceleration be *negative* (he is slowing down so as to arrive at B just as he reaches zero speed).

Now, we don't have to solve (12.16) from scratch because we already have—just take a look back at (6.4). Same equation! The general solution was given in (6.6) as

$$x(t) = A_1 e^{i\alpha t} + A_2 e^{-i\alpha t},$$

where $i = \sqrt{-1}$, $\alpha = \sqrt{\frac{4}{3}\pi G\rho}$, and A_1 and A_2 are constants that depend on the initial conditions of the problem. For a gravity tunnel, let's take the traveler's entire trip as requiring time T and the start of the trip, at A, as beginning at time $t = -\frac{T}{2}$. So, by symmetry the traveler is at the tunnel midpoint $x = 0$ at $t = 0$, and at the end of the trip the traveler is at B at time $t = \frac{T}{2}$. Using $x(0) = 0$ we have $0 = A_1 + A_2$ or $A_1 = -A_2$, and so

$$x(t) = A_1 e^{i\alpha t} - A_1 e^{-i\alpha t} = i2A_1 \sin(\alpha t).$$

Since the trip starts from rest, we have

$$\left.\frac{dx}{dt}\right|_{t=-\frac{T}{2}} = 0 = i\alpha A_1 2\cos(\alpha t)|_{t=-\frac{T}{2}}$$

$$= i\alpha A_1 2\cos\left(-\alpha\frac{T}{2}\right) = i\alpha A_1 2\cos\left(\alpha\frac{T}{2}\right).$$

Since whatever A_1 is (obviously it will be imaginary, since we know physically that dx/dt must be real), it must be that $\cos\left(\alpha\frac{T}{2}\right) = 0$, which means that $\alpha\frac{T}{2} = \frac{\pi}{2}$, and so

$$T = \frac{\pi}{\alpha} = \pi\sqrt{\frac{3}{4\pi G\rho}},$$

or, at last, the trip from A to B takes time

$$T = \frac{1}{2}\sqrt{\frac{3\pi}{G\rho}}. \tag{12.17}$$

Notice something wonderfully surprising about T in (12.17): there is no appearance in it of the distance between A and B. That means it doesn't matter what the distance is for any two points A and B connected by a straight gravity tunnel, the travel time is the same. Straight gravity tunnels in a constant density planet are said to be *isochronous*.[12]

The *speed* of the traveler, however, must be dependent on where A and B are if the time to travel tunnels of different lengths is a constant. Think about this, and then try your hand at CP. P12.3.

Plugging numbers into (12.17) to calculate the value of this constant travel time, for a constant-density Earth we have

$$T = \frac{1}{2}\sqrt{\frac{3\pi}{6.67 \cdot 10^{-11} \frac{m^3}{kg \cdot s^2} \cdot 5{,}540 \frac{kg}{m^3}}} = 2{,}525 \text{ seconds,}$$

or just over forty-two minutes. Since A and B can be any two points on Earth's surface, then in particular, they could be on a diameter of Earth (for example, pole to pole) and so the time to travel to the center of the Earth would be twenty-one minutes, the time given correctly in "Through the Earth." Compare these times with the value given in *The Earth-Tube* for a trip along an Earth diameter.

12.5 Epilogue

Finally, before ending, let me return to one issue that has been left unfinished: what about that last 1.8 cm/s^2 of gravity acceleration that we found, at the start of this discussion, that couldn't be accounted for by the Earth's rotation? That requires not only a nonuniform density for the Earth but also that a nonspherical shape for the Earth be allowed. The Earth is actually ellipsoidal, not spherical, with the poles about thirteen miles closer to the center than are points on the equator. If that is taken into account, then just about all of the difference between the polar and equatorial gravitational acceleration can be explained. Notice, of course, that the reason for the ellipsoidal shape is precisely because the Earth rotates, so rotation is the key idea once again. But I'll let you explore that connection on your own.[13]

CP. P12.1:

Suppose we have two planets, each with radius R and surface density ρ_0. The density of Planet 1 increases linearly to $2\rho_0$ as we descend toward its center, while the density of Planet 2 increases linearly to $3\rho_0$ as we descend toward its center. It is obvious that both the mass and the center pressure of Planet 2 will be greater than for Planet 1, but how much greater? Calculate the percentage increase in planetary mass and in center pressure of Planet 2 compared to Planet 1.

CP. P12.2:

If the radius of the Earth is 3,960 miles, what would be the maximum depth beneath the Earth's surface for a gravity tunnel linking New York City and Philadelphia? For a tunnel linking New York City and Boston? Use the surface distances between these cities cited in Goddard's 1909 *Scientific American* essay.

CP. P12.3:

What is the maximum speed of a traveler in a gravity-powered tunnel linking New York City and Philadelphia? New York City and Boston? Compare your second answer with the maximum speed achieved in Goddard's tunnel (see note 11). Also, for a pole-to-pole gravity tunnel (or for any tunnel that traverses a full diameter of the Earth), what is the maximum speed of the traveler? *Hint*: Use conservation of energy, with the Earth's center as the zero potential energy reference, and assume no energy loss mechanisms are present.

CP. P12.4:

Here's a different sort of transit time calculation, in the form of a pretty little problem first posed and solved by Galileo in a 1602 letter.[14] Figure 12.6 shows two masses, one at the very top (point A, directly above point C) of a circle with radius R in a vertical plane and the other at some arbitrary location (point B) on the circumference of that circle. The point B mass, when released, slides along a straight wire of length l to the bottom of the circle (point C).

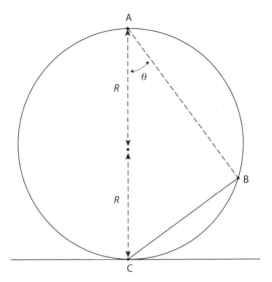

Figure 12.6. Galileo's problem.

If the two masses are released simultaneously, the mass at point A falling freely down to point C, then show that the two masses will reach point C simultaneously. Ignore air drag and friction, and take the acceleration of gravity as constant over the entire duration of the two transits. *Hint*: Remember from geometry that the triangle ABC is a right triangle for any B.

Notes and References

1. A very nice, detailed discussion of Newton's experiment (which to truly understand in depth requires a bit more analysis than I've given here) is in Archibald W. Handry, "Newton's Apple" (*The Physics Teacher*, May 2007, pp. 294–298).

2. Herman Erlichson, "Newton's 1679/80 Solution of the Constant Gravity Problem" (*American Journal of Physics*, August 1991, pp. 728–733).

3. Laurent Hodges, "Gravitational Field Strength Inside the Earth" (*American Journal of Physics*, October 1991, pp. 954–955).

4. Stephen G. Brush, "Discovery of the Earth's Core" (*American Journal of Physics*, September 1980, pp. 705–724).

5. Clyde S. Zaidins, "The Radial Variation of g in a Spherically Symmetric Mass with Nonuniform Density" (*American Journal of Physics*, January 1972, pp. 204–205).

6. Ralph Synder, "Two-density Model of the Earth" (*American Journal of Physics*, June 1986, pp. 511–513).

7. Here's where all these numbers come from. First, we calculate the fraction of Earth's total volume that each interior region represents, where if R_i and R_o are the inner and outer radii of the region, respectively, we have

$$\text{volume fraction} = \frac{\frac{4}{3}\pi R_o^3 - \frac{4}{3}\pi R_i^3}{\frac{4}{3}\pi R^3} = \left(\frac{R_o}{R}\right)^3 - \left(\frac{R_i}{R}\right)^3.$$

Then, using the values for R_i and R_o from Table 12.1, we get

Region	Volume fraction
crust/oceans	0.012
upper mantle	0.272
lower mantle	0.553
outer core	0.156
inner core	0.007

Notice that the average density of the entire Earth, from these volume fractions and the average density of each interior region, is

$$(2.4)(0.012) + (3.6)(0.272) + (4.9)(0.553) + (10.9)(0.156)$$
$$+ (12.9)(0.007) = 5.5 \text{ g/cm}^3,$$

which agrees well with Cavendish's value from 1798. Finally, we perform the region lumping calculations as follows. Looking at just the two core regions, the volume fraction of the core that is the outer core region is

$$\frac{0.156}{0.156 + 0.007} = \frac{0.156}{0.163} = 0.957,$$

and so, of course, the volume fraction of the core that is the inner core is 0.043. The average density of the *lumped core*, then, is

$$(10.9)(0.957) + (12.9)(0.043) = 11 \text{ g/cm}^3.$$

If you do the same thing for the upper three regions to form a *lumped mantle*, you can confirm that the volume fractions for the oceans/crust, upper mantle, and lower mantle are, respectively, 0.014, 0.325, and 0.66, and that these numbers give an average density for the lumped mantle of 4.44 g/cm^3.

8. Fezandié found later literary fame, at least for a while, in the new genre of "scientifiction," as the American publisher Hugo Gernsback called what would become today's science fiction. Fezandié's short stories of the Edison-like character he called "Dr. Hackensaw" appeared in Gernsback's magazines *Science and Invention* and *Amazing Stories*. Hackensaw was a kindly elderly scientist who could, on demand, cook up just about any gadget a story crisis required; among his many off-the-rack inventions were matter transmitters and invisibility. This sort of wonder character was a signature feature of the "superscience" stories popular in the 1920s and 1930s science fiction pulp magazines (see my book *Time Machines: Time Travel in Physics, Metaphysics, and Science Fiction*, New York, Springer-Verlag 1999); Hackensaw's spirit lived on in the television series *MacGyver* (1985–1992), whose hero, armed with only duct tape, a Swiss army knife, and his technological wits, overcame all obstacles.

9. How times have changed: To build the underground tunnels of Boston, in the monumental engineering fiasco called the Big Dig, required over $22 *billion*. Even though declared to be done at the end of 2007, the tunnels continue to leak enormous amounts of water onto their roadways, and multiple law suits over part of the ceiling falling down (killing a motorist) continue in the courts as I write. If the Big Dig contractors ever build a gravity tunnel, my advice would be either to find another way to make your trip or to stay home.

10. Goddard's fictional tunnel transport tale was called "The High-speed Bet" and is set in 1948. It is reprinted in volume 1 of *The Papers of Robert H. Goddard 1898–1924* (New York: McGraw-Hill, 1970, pp. 69–74). It contains a few hints on how Goddard proposed to evacuate the tunnel, and on the magnetic propulsion system (which would be something like that of modern magnetically levitated high-speed trains).

11. Goddard's travel times are easily confirmed by writing the distance traveled from rest in time t at constant acceleration a as $\frac{1}{2}at^2$. So, for two cities separated by distance d, the time T for the first half of the journey is $\frac{1}{2}aT^2 = \frac{d}{2}$ or, $T = \sqrt{\frac{d}{a}}$. The total journey time is $2T = 2\sqrt{\frac{d}{a}}$. For Goddard's New York City–Boston tunnel, for example, the travel time is $2\sqrt{\frac{190 \cdot 5,280}{11}}$ seconds = 604 seconds = 10 minutes 4 seconds, just as stated. This calculation does ignore the slight difference in the lengths of a direct tunnel and the surface separation. At midjourney the speed of Goddard's New York–Boston tunnel traveler would be an impressive $302 \text{ seconds} \cdot 11 \text{ ft/s}^2 = 3,322 \text{ ft/s} = 2,265$ mph. This speed was not calculated in the *Scientific American* editorial, perhaps because it would have raised obvious concerns and skepticism about the practicality of the whole idea.

12. There are other gravity tunnels that are *not* straight that are also isochronous in a uniform-density planet. See S. M. Lee, "The Isochronous Problem Inside the Spherically Uniform Earth" (*American Journal of Physics*, February 1972, pp. 315–318). Further, there are faster gravity tunnels that achieve *minimum* travel time, called *brachistochrone* tunnels. For some historical discussion of these minimum time solutions, whose proper study requires the calculus of variations, see my book, *When Least Is Best* (Princeton N.J.: Princeton University Press, 2007, pp. 229–230).

13. Marcelo Z. Maialle and Oscar Hipólito, "Acceleration of Gravity for the Earth Model as an Ellipsoidal Mass with Nonuniform Density" (*American Journal of Physics*, April 1996, pp. 434–436).

14. This important letter, written to Galileo's friend and mentor the Italian mathematician Guidobaldo dal Monte (1545–1607), was part of a larger exchange that has been lost. You can read about the historical context of the letter in the scholarly book by Domenico Bertolini Meli, *Thinking with Objects: the transformation of mechanics in the seventeenth century* (Baltimore: Johns Hopkins University Press, 2006, pp. 70–72). For more on why Galileo was interested in the physics of this Challenge Problem, see my book, *When Least Is Best* (pp. 200–210).

Quilts & Electricity

> The writer would prefer to find perfect squares by a theoretical method not demanding exhaustive computational searches, but must confess that so far he has seen no hint of such a process. Instead we have a pioneering example of the construction by computers of graphs satisfying specified nontrivial conditions.
> — Mathematician William Tutte, conceding victory to computers in "solving" a mathematical problem

13.1 Recreational Mathematics

Newtonian gravity is a fascinating subject, but I've been discussing it for a long while now, and I (and perhaps you, as well) would enjoy a break from it. So, let's do something completely different here for just a bit. Would you be surprised if I told you that there is an intimate connection between the artistic world of quilting and the mathematical world of electric circuit theory? *There is*. And I don't mean either the light bulb above or the electronic circuitry inside a modern quilt sewing machine.[1]

I'm talking about the fundamental conservation laws of energy and of electric charge.

The story I'm about to tell you begins with a question that appeared in a 1907 issue of *Our Puzzle Magazine*, published by the American puzzlemeister Sam Loyd (1841–1911). A square quilt of 169 identical square patches (that is, the quilt is thirteen patches by thirteen patches)

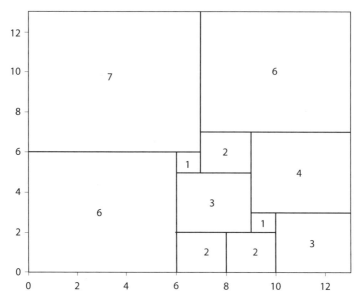

Figure 13.1. Mrs. Perkins's solution to Sam Loyd's puzzle.

is to be cut into a number of pieces which are themselves to be square by cutting along the stitch lines. In other words, you are not allowed to cut *through* a patch. How can this be done in the fewest number of subsquares? The answer is shown in Figure 13.1 (eleven subsquares), and it is unique. The numbers in the subsquares are the edge lengths (in units of the original patches), not areas.

Loyd's puzzle was repeated by the English puzzle genius Henry Dudeney (1857–1930) in his 1917 book, *Amusements in Mathematics*; Dudeney's version gave the problem its modern name, the one that is used to this day by mathematicians—dividing a square into subsquares, some of which may be equal in size, produces a *Mrs. Perkins's Quilt*.[2] (There was no real Mrs. Perkins. The name was simply an "it sounds funny" whimsy by Dudeney, who named a character in his puzzle statement Mrs. Potipher Perkins.)

The delicate nature of Figure 13.1 is made clear when one focuses on all of the constraints it satisfies. For example, if you pick any point at the top edge of the figure and work your way straight down, the sum of all the numbers you encounter as you go from one subsquare to the next must always total 13. That is, proceeding from left to right with a starting top edge point, $7+6=13$, $7+1+3+2=13$, $6+2+3+2=13$, $6+4+1+2=13$, and $6+4+3=13$. We'll call these relations the *vertical constraints*. Similarly, if you pick any point at the left edge of the figure and work your way straight across, the sum of all the numbers you encounter as you go from one subsquare to the next must always total 13. That is, proceeding from top to bottom with a starting left edge point, $7+6=13$, $7+2+4=13$, $6+1+2+4=13$, $6+3+4=13$, $6+3+1+3=13$, $6+2+2+3=13$. We'll call these relations the *horizontal constraints*.

In 1925, Zbigniew Moroń (1904–1971), a Polish secondary school teacher who had studied mathematics at the University of Lwoów, became interested in the so-called dissection problem, with the additional constraint that all the subsquares had to be of different sizes. Such a dissection (often called a *tiling*) is said to be *perfect*. The number of subsquares is called the *order* of the dissection; the order of Figure 13.1, for example, is 11. Moroń was unable to find a perfect decomposition of an initial *square* into subsquares, but he did discover an example of an "almost square" ($33 \cdot 32$) divided into squares with no two squares equal; this is shown in Figure 13.2. That is, Moroń had found a perfect rectangle of order 9.

Because of Moroń's failure (and that of all others exploring the problem) it was conjectured by the Russian mathematician Nikolai Luzin (1883–1950) that "squaring a square" (as a perfect square decompostion is called) was not possible. In 1939, however, the German mathematician Roland Sprague (1894–1967) showed that Luzin was wrong by succeeding at last in squaring a square, achieving that impressive feat with fifty-five subsquares (no two equal!) fitted together to form a huge square of size $4,205 \cdot 4,205$. Moroń later claimed that in the period 1925–1928 he had not only proved that a minimum of nine different squares is needed to dissect a "dissectable" rectangle (as in Figure 13.2), but also that *he* had been the first to find Sprague's perfect dissection solution to a square. He did not publish these results, however, and in mathematics the glory goes only to the first to publish.

In 1978 Adrianus Duijvestijn (1927–1998), an electrical engineer and mathematician at Twente University of Technology in the Netherlands, used a computer to discover a perfect squared square

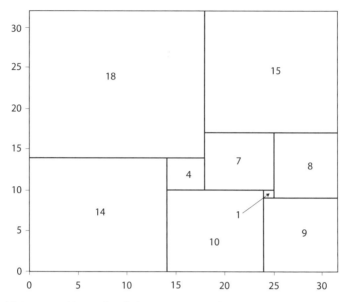

Figure 13.2. Moroń's perfect "almost square" decomposition.

of order 21 (with size 112 by 112), which is in fact the only order 21 perfect squared square—see Figure 13.3. It has been shown, too, that there is no perfect squared square of lower order. Notice that in Figure 13.3 there is no subset of the squares that, themselves, form a squared rectangle. If such a squared rectangle is present in a dissection, the squared square is called *compound* (like Sprague's); Duijvestijn's solution, however, is what is called *simple*, the most impressive status (to mathematicians) a solution can have.

As impressive as Sprague's 1939 result appeared at the time, even more important was work already completed (1936–1938) in England by four young undergraduates at Trinity College, Cambridge. Three of them—Rowland Brooks (1916–1993), Cedric Smith (1917–2002), and Arthur Stone (1916–2000)—were mathematics majors, while the fourth, William Tutte (1917–2002), from whom the opening quotation comes,[3] was a chemistry major (although he ended up as a distinguished Canadian professor of mathematics, first at the University of Toronto and then at the University of Waterloo). Here I'll simply refer to the four as BSST.

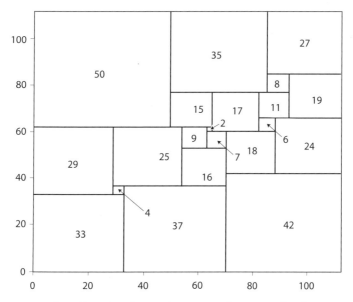

Figure 13.3. Duijvestijn's simple, perfect squared square of order 21 (112 · 112).

Aware of Moroń's work, BSST were also familiar with Dudeney's very first puzzle book, *The Canterbury Tales* (1917), which includes a puzzle ("Lady Isabel's Casket") that appeared to them to suggest the impossibility of a squared square (Luzin's conjecture). When BSST tried to prove the impossibility, however, they failed, and for a good reason: the "suggestion", as Sprague would later demonstrate in 1939, is false.[4] In the course of their efforts, though, they made some curious discoveries. For example, they found that if one simply scales Moroń's "almost square" solution of order 9 in Figure 13.2 up by a factor of 3 (to 99 · 96), then, in addition to the obvious scaled-up version of Moroń's original solution, there is now a second, different perfect solution (of order 12), shown in Figure 13.4. BSST found many other perfect rectangles and "almost squares," and also proved that there is no perfect rectangle of order less than 9, and that there are just two such rectangles of order 9 (Moroń's dissection in Figure 13.2 is one, and BSST published the other). But by far their greatest discovery is the totally unexpected, intimate connection of the dissection problem to electric circuit theory. Here's how that works.[5]

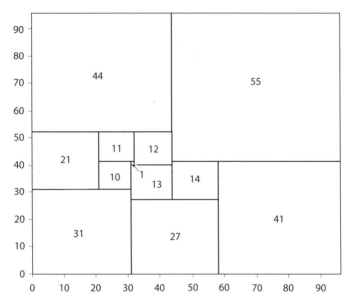

Figure 13.4. Solution to the scaled-up-by-3 33 · 32 rectangle of Figure 13.2.

13.2 Electric Quilts

Imagine that we have a vertical rectangular plate of electrically conductive material with uniform thickness—perhaps a slab of copper, for example—as shown in Figure 13.5. The slab has width W and height H. The top edge is connected to the positive terminal of a perfect (this means zero internal resistance) voltage source of value V volts, and the bottom edge is connected to the negative terminal. We further imagine both the top and bottom edges are coated with a perfect conducting film. We can define the bottom edge, of which every point is at the same potential (because of the perfect conductive film), to be our zero potential reference. Thus, the top edge, of which every point is at the same potential (because of the perfect conductive film), is at potential V. Now, imagine a horizontal line across the entire width of the rectangular plate. The potentials at all points along that line are equal, and the potential decreases from V to 0 as the horizontal line moves downward from the top edge to the bottom edge.

As shown in Figure 13.5 by the downward-directed arrows, there is a uniformly distributed current in the rectangular plate from the top

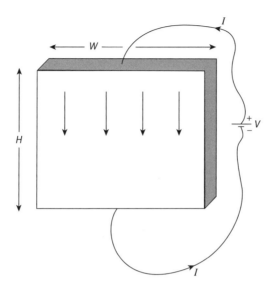

Figure 13.5. A rectangular copper plate and an ideal battery.

edge to the bottom edge; if the total current flowing from the voltage source's positive terminal is I, then the current that flows through any horizontal line segment in the plate of length l is $I\frac{l}{W}$. The total resistance R of the rectangular plate, as "seen" by the voltage source, is directly proportional to H and inversely proportional to W. This is so because the farther apart are the voltage source's terminals (H) the greater the resistance, and the more parallel paths available for the current to flow through (W) the less the resistance. If k is some proportionality constant that is a property of the specific nature of the material of the plate and of its thickness, then

$$R = k\frac{H}{W}.$$

Notice that if we have not just a rectangular plate but in fact we have a square plate ($H = W$), then this says

$$R = k,$$

a result independent of the actual physical size of the square! This conclusion is, I think, not at all obvious by inspection of the physical setup itself. Without the loss of any generality, we can always manufacture our plate in such a way that $R = 1$ ohm: simply pick some real conductive material (say, copper), and then adjust the plate thickness until $R = 1$ ohm. These two observations are absolutely crucial to what follows, so be sure you are quite clear on both points.

Now, imagine that we take the rectangular plate and cut it into subsquares according to some squared rectangle pattern. *For each pair of vertically adjacent squares (this means they share a common horizontal boundary), we connect the bottom edge of the upper subsquare to the top edge of the lower subsquare with a perfectly conducting wire.* Since the current flowing in the original plate is everywhere vertical, the dissection has no effect; that is, the same current continues to flow vertically through each subsquare, just as it did before the dissection. Because of our earlier "crucial" observations, we know that each and every one of the individual subsquares of the dissection has the same resistance ($R = 1$ ohm) from its top edge to its bottom edge, independent of the actual physical size of each subsquare. We can therefore replace our plate with a network of one-ohm resistors (each subsquare becomes a resistor) connected together in a way dictated by the common horizontal edges shared by the various subsquares.

For example, consider again Moroń's decomposition of Figure 13.2, where I've now labeled (in Figure 13.6) the various depths of the horizontal edges as a, b, c, d, e, and f as we travel from the top of the plate (a) to the bottom of the plate (f). The voltage at a given level is the same across the entire width of the (now cut) plate. Notice that I've labeled the current into the top of the plate as thirty-three amperes, and so, as argued before, there is a current of eighteen amperes in the top left subsquare and a current of fifteen amperes in the top right subsquare. And those currents divide up even further as they travel downward, e.g., the fifteen amperes divides into seven amperes and eight amperes in the two squares immediately beneath the top right subsquare. The corresponding resistor network is shown in Figure 13.7.

The number next to each of the one-ohm resistors in Figure 13.7 is the current in that resistor, with the arrowheads showing the direction of the currents. Notice carefully that by the very construction of Figure 13.7, Kirchhoff's current law is satisfied at every node (vertex) of the resistor network. That is, the sum of the currents into each node is zero

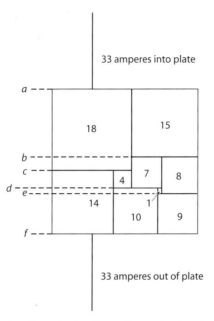

Figure 13.6. Voltage levels at the horizontal cuts in the plate.

(this is, of course, simply a statement of the conservation of electric charge). Since the current in the plate divides up according to the *horizontal* edge lengths of the subsquares, we see it is the horizontal constraints I mentioned earlier that are behind this satisfaction of the current law. Notice, too, that Kirchhoff's voltage loop law is also satisfied—the sum of the voltage drops around any closed loop is zero—because of the *vertical* constraints. That is, starting from any horizontal edge we can travel from that edge to another horizontal edge to another horizontal edge to and so on, and if we finally return to the starting edge, the net voltage change is zero. This is a statement of the conservation of energy. We can actually be specific about the voltage drops across each subsquare because we know each subsquare has a resistance of one ohm. The voltage drop from top edge to bottom edge is simply (by Ohm's law) numerically equal to the current in the subsquare (times one ohm), which says the voltage drop across each subsquare is simply its edge length. So, for example, the voltage drop from a to b in Figure 13.6 is eighteen volts if the input current is thirty-three amperes at the top of the copper plate. The entire voltage

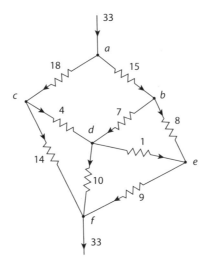

Figure 13.7. The resistor network corresponding to Figure 13.6.

drop across the plate from top to bottom is then equal to the value of H when the input current is $I = W$ amperes. So, for Figure 13.6, $V = 32$ volts.

Two final comments on the resistor network associated with a tiling. First, if we take the tiling of a rectangle and rotate it 90 degrees, we obviously change nothing electrically, but now the original horizontal and vertical edges have reversed roles (the voltage source is now connected to what used to be the vertical edges of the copper plate); this gives a new, different resistor network for the same tiling. And second, the equivalent resistance of any associated network of one-ohm resistors for a tiling must be H/W ($= V/I$) ohms (look back at the start of this discussion and you'll see I've already made this argument, but it's worth repeating). So, for example, the resistance from a to f in Figure 13.7 must be $32/33 = 0.9697$ ohms.[6] BSST continued on in their work to expand and greatly elaborate on the electrical analogy to the study of the squaring problem, but I'll stop here and simply refer you to their paper (see note 4). I should mention, however, that one of the modern technological applications of the squaring of rectangles (a pure mathematical problem) is the determination of the "best" way to layout electronic circuitry on microchips![7]

To finish my discussion of the squaring problem I'll show you three quite clever impossibility proofs, the second making clever direct use of the electrical connection.

13.3 Three Impossibility Proofs

For the first impossibility proof, I'll start with an easy, all-mathematical demonstration. Suppose the dimensions of our rectangle to be tiled are such that W/H is irrational. Then it is impossible to tile the rectangle with a finite number of equal subsquares (of any size). We can prove this by contradiction. That is, suppose we *can* tile the rectangle with m rows of n tiles each (m and n are integers), with each tile having edge length l. This says that $W = nl$ and $H = ml$, but *that* says

$$\frac{W}{H} = \frac{nl}{ml} = \frac{n}{m},$$

which is rational (by definition), in contradiction to the given irrationality of W/H. Thus, our assumption that the tiling exists (that there are integers m and n) must be false, and we are done.

That was pretty easy,[8] but we can actually prove a second, much stronger result if we bring some physics into the argument. The assertion is that if W/H is irrational, then it is impossible to tile our rectangle with a finite number of tiles even if we allow them to be of unequal sizes. Our first impossibility result is a special case of this more general statement. This stronger assertion might seem to be of much greater difficulty to show, but that isn't so. All we need to observe is that if our rectangle could be tiled in some way, there would exist an associated network of one-ohm resistors. The equivalent resistance of this network, as shown at the end of the last section, is H/W, which is given as irrational. So, our proof of impossibility is equivalent to showing that it is impossible to connect a finite number of one-ohm resistors together to form an equivalent resistance that is irrational. We can show that as follows.

Given any network, we can write Kirchhoff's node current and voltage loop equations, where the resistor currents and node voltages appear as the unknowns. All the resistors in the network are one ohm, and so all the coefficients in all the equations (which are linear equations) are rational. We can solve the equations for the currents and the voltages using just rational operations (addition, subtraction, multiplication, and division). That means the solutions will all be rational, too; that is, for an applied rational plate voltage $V(=H)$ the resulting plate current $I(=W)$ will also be rational, and so the plate resistance H/W will be rational and not irrational. Thus, our initial assumption that there *is* a network (that our rectangle with irrational H/W can be tiled) must be false, and we are done.

I'll finish up here by showing you a beautiful third impossibility result that BSST showed at the end of the 1940 paper. I think it is a perfect example of how much of mathematical reasoning is of a *non*-numerical nature. To quote BSST, "There is no perfect cube (or parallelepiped)." That is, it is impossible to cut a rectangular parallelepiped into a finite number of cubes such that no two cubes are equal. This is a remarkable statement—there is an infinity of such perfect dissections in two dimensions, but none in three dimensions. The proof of this assertion is equally remarkable in its simplicity.

BSST start off their proof with the throwaway comment: "It is easily seen that in any perfect rectangle, the smallest element [subsquare] is not on the boundary of the rectangle."' Their impossibility proof in three dimensions depends on this initial result in two dimensions, but it is so "easily seen" that they don't bother to prove it. *I* had to think about it for a while, however, so let me first establish this preliminary result. There are two distinct cases to be considered: (1) the smallest subsquare as a possible *corner* subsquare and (2) the smallest subsquare as a possible *side* subsquare. We can argue as follows.

(1) A corner subsquare will have two of its edges in the interior of the rectangle. Along each of those interior edges a larger subsquare must be positioned, and those two larger subsquares *would by necessity overlap*;

(2) A side subsquare will have three of its edges in the interior of the rectangle. Let's call them the *parallel* edge (parallel to the touched rectangle boundary) and the two *normal* edges (normal to the touched rectangle boundary). Along each of the two normal edges of the smallest subsquare a larger subsquare must be positioned. That means

there is not enough room for even a single (also larger) subsquare to be positioned along the interior parallel edge of the smallest subsquare without overlap occurring.

Since overlapping is not allowed, we've demonstrated that the smallest subsquare cannot be *anywhere* along the rectangle's boundary, and we are done. With this preliminary result established, we can now finish the proof of the claim that "there is no perfect cube (or parallelepiped)."

Assume we *have* successfully cubed a parallelepiped with a finite number of subcubes all of a different size. Then the bottom face will, by necessity, be a squared rectangle. From our preliminary result we know that the smallest subsquare in that squared rectangle cannot be on the boundary of the bottom face of the parallelepiped. That says that the smallest cube that occurs on the bottom face of the parallelepiped cannot touch any of the upright sides of the parallelepiped. But that says the smallest cube on the bottom must be *surrounded*, on all four of its upright sides, by larger subcubes. And that in turn says that the upper face of that smallest subcube is "walled in." To cover that upper face will require subcubes even smaller than the one we are presently focused on (certainly a bigger subcube wouldn't fit, and we can't use a subcube of the same size because we aren't allowed to repeat subcube sizes!).

And now we can repeat the above argument. Among all the subcubes that are positioned on that upper face, the smallest must be surrounded on all four of its upright sides by even bigger subcubes, which again creates a walled-in space that can be filled only by even smaller subcubes! This process has generated what mathematicians call an *infinite regress*, and we can cube the original parallelepiped only if we use an infinity of ever smaller subcubes. Our starting assumption of a finite decomposition with no two subcubes equal is *not possible*, and our proof is done.

CP. P13.1:

One way to discover a squared rectangle is to sketch an assumed tiling pattern of subsquares and, from that sketch, write out some equations that satisfy the horizontal and vertical constraints. Then,

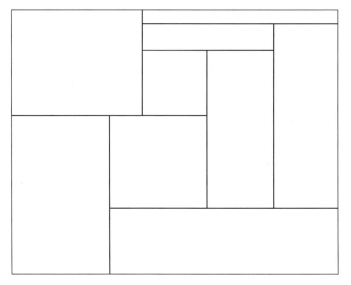

Figure 13.8. How big are the subsquares?

from those equations, you attempt to find positive integer solutions. See if you can do this for the tiling pattern of Figure 13.8, *with no two subsquares of the same size*. That is, you are to find a perfect squaring of order 9. Notice that I've purposely presented the tiling pattern so that none of the assumed subsquares actually *looks* square—I did that so you can't use a ruler. You will have to do some analysis![9]

CP. P13.2:

If you look carefully at Figures 13.1, 13.2, 13.3, and 13.4, you'll notice that the smallest subsquare in each case (which we showed earlier cannot be on the boundary of the original rectangle) is surrounded by exactly four other (all larger, of course) subsquares. This is not so, in general, for larger interior subsquares. Prove that this will always be true for any tiling, i.e., explain why the smallest subsquare of any tiling will always have exactly four adjacent neighbors (*adjacent* means sharing all or part of an edge). *Hint*: Prove by contradiction, that is, assume there are five (or more) adjacent neighbor subsquares, and show how that assumption quickly leads to an impossible situation.

Notes and References

1. When my wife, Pat, who is an accomplished quilter (see the end of the next note), sits down at her quilting workstation and flips on the power to her computer-controlled Bernia 440, she looks like a starship captain on her command bridge. Behind her is a bookcase stuffed with quilting technique and tech reference manuals. Quilting today isn't your great-grandmother's needle, thread, and rocking chair activity anymore.

2. The English-born mathematician John Conway (born 1937), now on the mathematics faculty at Princeton University, has shown that if N is the edge length of a square, then the smallest number of subsquares possible in a Mrs. Perkins's quilt, $f(N)$, satisfies the double inequality $\log_2(N) \leq f(N) \leq 6\sqrt[3]{N}+1$. For the Loyd/Dudeney problem $N=13$, and so this says $3.7 \leq f(13) \leq 15.1$; the solution in Figure 13.1, with $f(13) = 11$, is consistent with Conway's inequalities. See J. H. Conway, "Mrs. Perkins's Quilt" (*Proceedings of the Cambridge Philosophical Society*, July 1964, pp. 363–368). An actual Mrs. Perkins's quilt, based on Figure 13.1, is shown in the photograph of Figure 13.9 (both the quilt and the photograph courtesy of Pat Nahin).

3. W. T. Tutte, "The Quest of the Perfect Square" (*American Mathematical Monthly*, February 1965, pp. 29–35).

4. BSST found, *before* Sprague's publication, a perfect squared square of order 39 and size $4,639 \cdot 4,639$, but Sprague was the first to get into print. BSST didn't publish their work until 1940; see their paper, "The Dissection of Rectangles into Squares" (*Duke Mathematical Journal* 7, 1940, pp. 312–340). Although published seventy years ago, the BSST paper continues to inspire to this day; see, for example, the recent paper by Frederick V. Henle and James M. Henle, "Squaring the Plane" (*American Mathematical Monthly*, January 2008, pp. 3–12).

5. The rest of this discussion was inspired by an essay written by Professor Tutte: see Chapter 17 ("Squaring the Square") in Martin Gardner, *The 2nd Scientific American Book of Mathematical Puzzles and Diversions* (New York: Simon and Schuster, 1961, pp. 186–209, 250). See also Sherman K. Stein, *Mathematics: The Man-made Universe* (San Francisco: W. H. Freeman, 1963, chaps. 6 and 7).

6. This is actually quite a remarkable result. I doubt there is an electrical engineer or physicist on the planet who could just *look* at Figure 13.7 and tell you that the equivalent resistance R is $32/33 = 0.9697$ ohms. Rather, most engineers would simply sigh with resignation and write out Kirchhoff's equations and grind through the grubby math to find the input current I

230 _ _ _ _ DISCUSSION 13

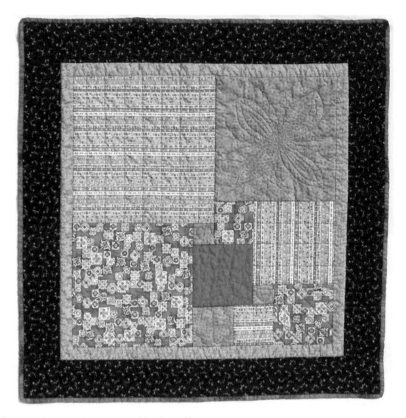

Figure 13.9. Pat's Mrs. Perkins's quilt.

as a function of the applied voltage V (and then calculate $R = V/I$). I don't want to do that (it's a lot of work), but just to convince you that all the discussion here is indeed okay, I used a commercially available electronic circuit simulator (ELECTRONICS WORKBENCH, or EWB) that implements what is called *schematic capture*. That is, it allows one to literally "draw" a circuit on the computer screen (you click and drag, with a mouse, various electrical components out of virtual parts bins of endless capacity and hook those components together as you please). When I did that for Figure 13.7, and then applied a thirty-three-ampere constant current source (the circle at the left with an upward arrow in it) to the network, the result was (are you surprised?) a thirty-two-volt drop as measured by an EWB voltmeter, as shown in Figure 13.10. (The default internal resistance of the voltmeter is one million ohms—EWB would let me set it even higher if I wanted—and so the voltmeter current can be ignored.) I keep EWB idling on my computer

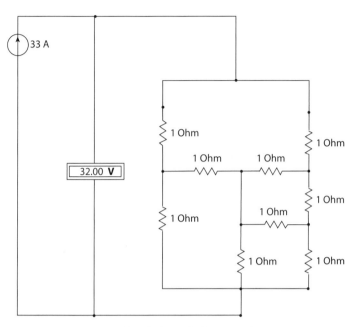

Figure 13.10. Computer simulation of Figure 13.7.

desktop, along with MATLAB, just for situations such as this (when I'm too lazy to grind out the equations). In Discussion 17 I'll show you yet another way, using probability theory, to "solve" resistor circuits.

7. Keishi Sakanushi, Yoji Kajitani, and Dinesh Mehta, "The Quarter-State-Sequence Floorplan Representation," (*IEEE Transactions on Circuits and Systems—I. Fundamental Theory and Applications* March 2003, pp. 376–386). This paper references the early 1930s work of the Japanese scientist (it isn't clear if he was a physicist or a mathematician) Michio Abe. After publishing two papers (one in English) on the squaring problem in journals of mathematical physics, Abe vanished from history, and I've had no luck in finding out anything about him. Perhaps a reader can help?

8. Proof by contradiction is a powerful technique, one that is ranked high by mathematicians in elegance and beauty. Here's another example of it in a nonmathematical setting. Suppose A and B play the game of tic-tac-toe, with A moving first. (I'll assume that this game is so well-known that I don't have to give the rules here.) Then we have the following assertion: *There exists a strategy by which A can always force either a win or a draw.* To prove this we'll assume the

opposite of what we wish to show, and then demonstrate that assumption leads to a contradiction. What is the opposite of "A can always force either a win or a draw"? The answer is, A can *never* force a win or a draw or, equivalently, B can always force a win. So, we start our proof by assuming B can always force a win. The game starts with A making some (any) move. After making that move, A then considers himself the second player and in possession of the winning strategy. He can do this because the move he just made can only help him as the game proceeds. And so he, A, the actual first player to move, can force a win. But this contradicts our initial assumption. So, that assumption must be false, and it is the alternative, that it is A who can force either a win or a draw, that is true. Notice that while this proof shows the existence of a strategy for A, we have not had to say anything about the details of that strategy. Mathematicians consider that absence of specificity about the details of what we have proven to be the most elegant part of the proof!

9. This method obviously becomes less useful as the order of the assumed tiling pattern increases, because then the number of resulting constraint equations also increases. In fact, just describing solutions of high order presents an obvious problem, and as the order increases it becomes more and more difficult even to sketch a readable tiling pattern. There is, fortunately, a clever notational convention that answers this difficulty and allows the unambiguous description of any squared rectangle, no matter how high the order. The idea is to identify, starting at the top left of the rectangle, adjacent subsquares with level (flush) top edges. The edge lengths of these adjacent subsquares are placed in brackets. One then proceeds all through the rectangle doing this, from left to right and top to bottom, and then simply writes out all the resulting brackets one after the other. The result is the so-called *Bouwkamp code* of the squared rectangle, named after the Netherlands mathematician Christofel Bouwkamp (1915–2003). For example, the Bouwkamp code of the original Mrs. Perkins's quilt of Figure 13.1 is

[7, 6], [2, 4], [6, 1], [3], [1, 3], [2, 2],

while the Bouwkamp code of Duijvestijn's simple perfect squared square of Figure 13.3 is

[50, 35, 27], [8, 19], [15, 17, 11], [6, 24], [29, 25, 9, 2],

[7, 18], [16], [42], [4, 37], [33].

Random Walks

> "Sal, we gotta go and never stop going till we get there."
> "Where we going, man?"
> "I don't know but we gotta go."
> — Exchange between two wandering bums in Jack Kerouac's 1950s beat generation novel, *On the Road*

14.1 Ronald Ross and the Flight of Mosquitoes

In September 1904, at the International Congress of Arts and Science in St. Louis, Sir Ronald Ross (1857–1932) gave an interesting talk (reprinted a year later in *Science*)[1] on how to mathematically model the spread of mosquitoes from a breeding pool. As the vectors of such devastating diseases as yellow fever and malaria, understanding how these insects migrate was an important practical problem of the day. Yellow fever in particular had some years before laid waste to the French effort to build a Panama Canal. Ross was not a mathematician by training—he was a British medical doctor who had received the 1902 Nobel Prize in Medicine for his work on malaria—but his simple mathematical analysis is still of interest.

Ross prepared his audience for the mathematics to come with these words:

> Suppose that a mosquito is born at a given point, and that during its life it wanders about, to and fro, to left or to right, where it wills, in search of food or of mating, over a country which is uniformly attractive to it. After a time it will die. What are the probabilities that its dead body

will be found at a given distance from its birthplace? That is really the problem which governs the whole of this great subject of the prophylaxis of malaria.

He had earlier hinted at the probabilistic mathematics he was about to introduce when he said,

> The answer depends upon the distance which a mosquito can traverse, not during a single flight, but during its whole life; and so upon certain laws of probability, which must govern its wanderings to and fro upon the face of the earth.

Ross began by imagining that the life of a mosquito can be divided into n stages; during each stage the insect can fly a total distance of l. He further imagined, for mathematical simplicity, that while individual insects each begins its life by flying away from its birthplace at a random angle (which is almost surely different for each mosquito), once picked that angle remains fixed for each mosquito. (Fixing our attention now on one mosquito in particular, we lose no generality in imagining its line of flight is along the x-axis. Multiple mosquitoes, taken together, will of course be distributed over a *two*-dimensional plane.) During each of its stages of life, imagine that our mosquito decides, twice, at random, whether to fly "forward" or "backward" by a distance of $l/2$. Ross took the forward and backward choices to be independent and equally likely (each with probability 1/2). So, for example, if $t=0$ is the time of birth of our mosquito, then at time $t=1$ the insect is either at $x=l/2$ with probability 1/2 or at $x=-l/2$ with probability 1/2. At $t=2$ the insect is at $x=l$ with probability 1/4, or at $x=-l$ with probability 1/4, or back at $x=0$ with probability 1/2. And so on. Since the mosquito makes $2n$ decisions during its lifetime, the unit of time is simply that lifetime divided by $2n$ (Ross argued that n could be taken as large as one pleased, and so of course l could be made as small as one pleased). Ross never gave a name to this sequential decision process, but today mathematicians call it a *one-dimensional random walk*.

Ross next invoked a very simple combinatorial argument to compute the probabilities the mosquito's "dead body" would be found at $x=0$, $x=\pm l, x=\pm 2l, \ldots, x=\pm nl$, where it should be clear that those are the *only* locations at which it *could* be found; that is, since the mosquito makes an even number of flight direction choices ($2n$), its final location must be an even multiple of $l/2$. For example, for the mosquito to be at either nl or at $-nl$, *all* of its $2n$ decisions must have been in the same direction. Thus,

$$Prob(x = \pm l) = 2 \binom{2n}{2n} \left(\frac{1}{2}\right)^{2n} = \frac{2}{2^{2n}}.$$

For the mosquito to be at either $(n-1)l$ or at $-(n-1)l$, *all but one* of its $2n$ decisions must have been in the same direction, with one additional decision in the opposite direction. Thus,

$$Prob(x = \pm(n-1)l) = 2 \binom{2n}{2n-1} \left(\frac{1}{2}\right)^{2n} = \frac{4n}{2^{2n}}.$$

For the mosquito to be at either $(n-2)l$ or at $-(n-2)l$, *all but two* of its $2n$ decisions must have been in the same direction, with two additional decisions in the opposite direction. Thus,

$$Prob(x = \pm(n-2)l) = 2 \binom{2n}{2n-2} \left(\frac{1}{2}\right)^{2n} = \frac{2n(2n-1)}{2^{2n}}.$$

And so on, until we get to the $x = 0$ case. Then precisely half of the mosquito's decisions are in one direction and the other half are in the reverse direction. Thus,

$$Prob(x = 0) = \binom{2n}{n} \left(\frac{1}{2}\right)^{2n} = \frac{\frac{(2n)!}{(n!)^2}}{2^{2n}}.$$

To give a specific numerical example of these expressions, Ross imagined that the initial total number of mosquitoes is 1,024 (he set $n = 5$ in 2^{2n}). Then the expected number of mosquitoes reaching the various possible distances from the starting point is just the numerators of the above probability expressions. That is,

Distance from the breeding pool	Number of dead mosquitoes
0	252
l	420
$2l$	240
$3l$	90
$4l$	20
$5l$	2

Ross wrote of these numbers,

> Thus only 2 out of the 1,024 mosquitoes are ever likely to reach the extreme limits; while, on the other hand, no less than 912, or 89 percent, are likely to die somewhere [within $2l$] ... around the [pool] ... the large majority [perish] close to the point of origin.

Ross thought so much of this observation that he elevated it to a general principle: "The law here enunciated may, perhaps, be called *the centripetal law of wandering*. It ordains that when living units wander from a given point *guided only by chance* they will always tend to revert to that point." As far as I know, mathematicians and physicists have not adopted Ross's somewhat grandiose "Newtonian" terminology.

14.2 Karl Pearson Formulates a Famous Problem

In the summer of 1905, almost a year after Ross's address, the eminent English statistician Karl Pearson (1857–1936) wrote a letter to the British science journal *Nature* in which he asked for help on a mathematical problem. The problem was a sophisticated version of Ross's mosquito migration problem and one Ross himself had earlier asked for help on from Pearson. The letter is short, and so, because of its historical importance, I will quote it here in full:[2]

> Can any of your readers refer me to a work wherein I should find a solution of the following problem, or failing the knowledge of any existing solution provide me with an original one? I should be extremely grateful for aid in the matter.
>
> A man starts from point O and walks l yards in a straight line; he then turns through any angle whatever and walks another l yards in a second straight line. He repeats this process n times. I require the probability that after these n stretches he is at a distance between r and $r + \delta r$ from his starting point O.
>
> The problem is one of considerable interest, but I have only succeeded in obtaining an integrated solution for *two* stretches. I think, however, that a solution ought to be found, if only in the form of a series in powers of $1/n$, when n is large.

This letter is historically important because it gave the name to one of the most beautiful subjects in probability theory (itself considered by many to be one of the most beautiful areas of mathematics[3]), the theory of the *random walk* (in this case, a true two-dimensional random walk, unlike Ross's mosquito walk, which, despite appearances, is really just one-dimensional).

A reply[4] was published in the very next issue of *Nature* as a letter signed only as *Rayleigh*. No more identification was needed, in fact, as the author was known to all as the brilliant mathematical and experimental physicist John William Strutt (1842–1919), who sat in the House of Lords as Lord Rayleigh and who just the previous year had received the Nobel Prize in Physics for his discovery of argon. (Rayleigh was one of those rare individuals who could have been legitimately awarded multiple Nobel Prizes many of his other discoveries, as well.) Rayleigh's letter was even shorter than Pearson's:

> This problem, proposed by Prof. Karl Pearson..., is the same as that of the composition of n iso-periodic vibrations of unit amplitude and of phases distributed at random [Rayleigh then gives citations to his earlier work on this problem in the theory of sound, dating back to 1880]. If n be very great, the probability sought is
>
> $$\frac{2}{n} e^{-r^2/n} r \, dr.$$

This probability[5] is of the form $f_{\mathbf{r}}(r) \, dr$, where $f_{\mathbf{r}}(r)$ is called a *probability density function* (pdf):

$$f_{\mathbf{r}}(r) = \frac{2}{n} e^{-r^2/n} r \tag{14.1}$$

is famous today as *Rayleigh's pdf* (see Figure 14.1 for the case of $n = 6$, for example—this is not "very great," but you'll soon see why I picked this value), and there isn't a modern textbook on introductory probability theory that doesn't devote a fair amount of space to deriving and discussing its many applications.[6]

As a testament to the very rapid mail delivery in those long ago days, Pearson's reply[7] to Rayleigh appeared in yet the very next issue of *Nature*. Pearson's letter is quite interesting, as comments in it will motivate much of what appears later in this discussion. So, quoting

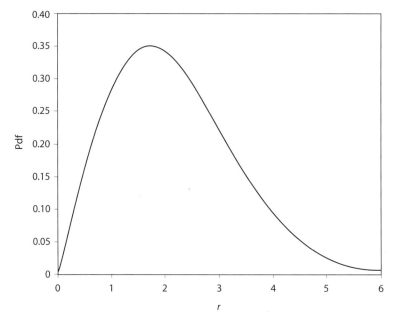

Figure 14.1. Rayleigh's pdf for $n = 6$.

Pearson again:

> I have to thank several correspondents for assistance in this matter. Mr. G. J. Bennett[8] finds that my case of $n = 3$ can really be solved by elliptic integrals, and, of course, Lord Rayleigh's solution for n very large is most valuable, and may very probably suffice for the purposes I have immediately in view [Pearson's interest was, of course, in the migration of very large numbers of mosquitoes, and also of birds]. I ought to have known it, but my reading of late years has drifted into other channels, and one does not expect to find the first stage in a biometric problem in a memoir on sound.

(And so, in that last sentence, we see yet another example of the "mutual embrace" of mathematics and physics that I mentioned in the Preface.) Pearson's letter continues:

> From the purely mathematical standpoint, it would still be very interesting to have a solution for n comparatively small. ... For $n = 2$, for example, [the pdf is] of the form of a double U, thus **UU**, the whole

being symmetrical about the centre vertical corresponding to $r = 0$, but each U itself being asymmetrical.... [I'll elaborate on just what Pearson meant by this in section 14.4].

The lesson of Lord Rayleigh's solution is that in open country the most probable place to find a drunken man who is at all capable of keeping on his feet is somewhere near his starting point!

Pearson's last sentence makes Ross's point that most of a large number of mosquitoes will die near their breeding pool. It also contributed the colorful alternative name of *drunkard's walk* to the random walk process, a name you can find in many modern probability books.

I'll return to the second letter from Pearson later in this discussion, but first let me remark that while Ross clearly described a one-dimensional random walk, and while Pearson gave the name *random walk* to a more sophisticated version of Ross's problem, they were definitely not the first to touch on such questions.[9] In 1865, for example, the Irish-born mathematician Morgan Crofton (1826–1915) proposed the following pretty little problem:

> A traveller starts from a point on a straight river, and travels [not crossing the river] in a random direction a certain distance in a straight line. Having quite lost his way, he starts again at random the next morning and travels the same distance; what is the chance of his reaching the river again in the second day's journey?

Crofton is remembered today as one of the early developers of geometric probability, and his solution shows how nicely the elementary concepts of that subject handle this question with ease. The answer is that the probability of returning to the river is 1/4—can you see how to calculate that? (I'll show you the solution at the end of this section, but see if you can figure out where the 1/4 comes from before reading the solution.)

A few years later, in 1873, the English mathematician J.W.L. Glaisher (1848–1928) posed a question very much more like what Pearson would ask thirty-two years later:

> Another question on Probability, the method of solution of which is not very easy to see, is this: **O** is a point in a plane, and any curve of given length l is drawn from **O** on that plane, required the chance that the

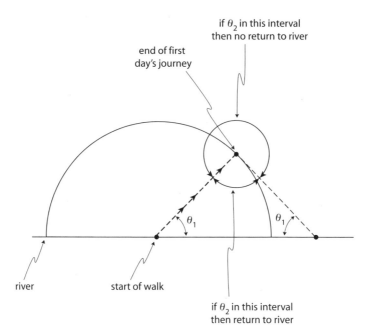

Figure 14.2. Crofton's problem.

other extremity will be at a distance $r < l$ from **O**; the curve is supposed to be capable of any form, crossing itself... any element ds, in fact, making any angle whatever with the next element ds. The question may be stated... thus: an insect is placed on a table, find the chance that it will be found at time t at a distance r from the point it started; if the insect be supposed to hop we have the simpler case of which that of the crawling insect is the limit.

Glaisher's problem, while much more mathematically sophisticated than Crofton's (and Pearson's, too), suffered the same fate of both in failing to stimulate additional study.

Okay, have you solved Crofton's problem? If not, here's how to do it. After the first day's journey the traveler could be anywhere on a semicircle centered on his starting point, as shown in Figure 14.2. Now, focus on one possible first day's journey, shown in the figure as the dashed line at angle θ_1 in the interval 0 to $\pi/2$ (θ_1 can actually be anywhere in the interval 0 to π, but for now let's work with 0 to $\pi/2$). For that angle, any second angle θ_2 in an interval of width $\pi - 2\theta_1$ will result in a return to the river (see the figure). Since $\boldsymbol{\theta}_1$ is uniformly distributed from 0 to π, $\boldsymbol{\theta}_1$ is a random variable with probability density $f_{\boldsymbol{\theta}_1}(\theta_1) = 1/\pi$. $\boldsymbol{\theta}_2$ is also a uniform random variable, but over the interval 0 to 2π, and so has probability density $f_{\boldsymbol{\theta}_2}(\theta_2) = 1/2\pi$. So, writing $P(return|\boldsymbol{\theta}_1 = \theta_1)$ as the so-called *conditional probability* of returning to the river *for the particular angle* θ_1, we have

$$P(return|\boldsymbol{\theta}_1 = \theta_1) = \frac{\pi - 2\theta_1}{2\pi},\ 0 \leq \theta_1 \leq \frac{\pi}{2}.$$

To calculate $P(return)$ we need to integrate out the dependency on θ_1, that is

$$P(return) = \int_0^{\pi/2} P(return|\boldsymbol{\theta}_1 = \theta_1)\, f_{\boldsymbol{\theta}_1}(\theta_1)\, d\theta_1$$

$$= \int_0^{\pi/2} \frac{\pi - 2\theta_1}{2\pi} \cdot \frac{1}{\pi}\, d\theta_1 = \frac{1}{2\pi}\int_0^{\pi/2} d\theta_1 - \frac{1}{\pi^2}\int_0^{\pi/2} \theta_1\, d\theta_1$$

$$= \left(\frac{1}{2\pi} \cdot \frac{\pi}{2}\right) - \left(\frac{\theta_1^2}{2\pi^2}\right)\bigg|_0^{\pi/2} = \frac{1}{4} - \frac{\pi^2}{8\pi^2} = \frac{1}{4} - \frac{1}{8}$$

$$= \frac{1}{8},\ 0 \leq \theta_1 \leq \frac{\pi}{2}.$$

Since the situation for an initial angle θ_1 in the second quadrant ($\pi/2 \leq \theta_1 \leq \pi$) is symmetrical with the first quadrant case we just analyzed, our total probability of return is double this, that is,

$$P(return) = \frac{1}{4}.$$

14.3 Gambler's Ruin

Drunkard's walk is but one colorful name given to the one-dimensional random walk. Another, a name that doesn't even sound like a random walk at all (but is), is *gambler's ruin*. This is a favorite of mathematicians as a classroom example, but for us the real reason for its interest is that it can be solved exactly, in analytical form. That will be useful for us when I show you yet another, approximate way to solve random walk problems; we can check how well the approximate solution for gambler's ruin agrees with the theoretically exact solution. There are numerous variations of gambler's ruin, but I'll treat one of the simplest, a version that can be traced back to a problem posed by the French mathematician Blaise Pascal (1623–1662) in 1654.

Suppose two people, A and B, agree to play the following game. Together, initially, they have five one-dollar bills, with three of them A's and the other two belonging to B. They also have a penny that, when flipped, shows heads with probability p and tails with probability q (with $p+q=1$, as we'll assume landing "on edge" never occurs). A and B start flipping the penny, where we'll also assume the individual flips are independent of one another. Each time the coin shows heads A gets one of B's dollar bills, and each time it shows tails A gives one of his dollar bills to B. The game ends when one of them first has all the money. What is the probability A is *ruined*, that is, loses all of his money to B? This is mathematically equivalent to saying A starts at $x=3$, moves right one step on heads and left one step on tails, and then asking what is the probability A's position becomes $x=0$ (A is ruined) before it becomes $x=5$ (A wins all of B's money, and so it is B who is ruined)?

We can solve for the required probability as follows, beginning with this definition:

$u(i)$ = probability A is ruined, given he starts with i dollars. (14.2)

Then the following equations can be immediately written:

(1) $u(0) = 1$,
(2) $u(1) = qu(0) + pu(2)$,
(3) $u(2) = qu(1) + pu(3)$,
(4) $u(3) = qu(2) + pu(4)$,
(5) $u(4) = qu(3) + pu(5)$,
(6) $u(5) = 0$.

The first and last equations are obvious: (1) says A is ruined with certainty if he starts broke, and (6) says A cannot be ruined if he in fact has all the money (five dollars) from the get-go. The other equations are only slightly more subtle. Equation (4), for example, says that the probability A is subsequently ruined (if he starts with three dollars) is the probability he loses on the next flip (probability q) times the probability he is then ruined with his now two dollars, plus the probability he wins on the next flip (probability p) times the probability he is then ruined with his now four dollars. The other equations follows from the same sort of argument.

We can now solve these coupled equations—in particular, for $u(3)$—using the following clever approach. First, let's rewrite equations (2) through (5) by multiplying their left-hand sides by $1 (= p + q)$. Then, rearranging terms, we have

$$q[u(1) - u(0)] = p[u(2) - u(1)],$$
$$q[u(2) - u(1)] = p[u(3) - u(2)],$$
$$q[u(3) - u(2)] = p[u(4) - u(3)],$$

and

$$q[u(4) - u(3)] = p[u(5) - u(4)].$$

If we define

$$c = u(1) - u(0),$$

then we have

$$u(2) - u(1) = \frac{q}{p}[u(1) - u(0)] = \frac{q}{p}c,$$

$$u(3) - u(2) = \frac{q}{p}[u(2) - u(1)] = \left(\frac{q}{p}\right)^2 c,$$

$$u(4) - u(3) = \frac{q}{p}[u(3) - u(2)] = \left(\frac{q}{p}\right)^3 c,$$

and

$$u(5) - u(4) = \frac{q}{p}[u(4) - u(3)] = \left(\frac{q}{p}\right)^4 c.$$

So, remembering that $u(0) = 1$, we have

(1') $u(1) = u(0) + c = 1 + c$,
(2') $u(2) = \frac{q}{p}c + u(1) = \frac{q}{p}c + 1 + c$,
(3') $u(3) = \left(\frac{q}{p}\right)^2 c + u(2) = \left(\frac{q}{p}\right)^2 c + \left(\frac{q}{p}\right)c + 1 + c$,
(4') $u(4) = \left(\frac{q}{p}\right)^3 c + u(3) = \left(\frac{q}{p}\right)^3 c + \left(\frac{q}{p}\right)^2 c + \left(\frac{q}{p}\right)c + 1 + c$,
and
(5') $u(5) = \left(\frac{q}{p}\right)^4 c + u(4) = \left(\frac{q}{p}\right)^4 c + \left(\frac{q}{p}\right)^3 c + \left(\frac{q}{p}\right)^2 c + \left(\frac{q}{p}\right)c + 1 + c$.

Let's now assume $p \neq q$, that is, that the penny is not a fair coin but rather is biased (the case of a fair coin—when $p = q = 1/2$, will be separately and easily handled in just a bit). Since $u(5) = 0$, (5′) says

$$\left[\left(\frac{q}{p}\right)^4 + \left(\frac{q}{p}\right)^3 + \left(\frac{q}{p}\right)^2 + \left(\frac{q}{p}\right) + 1\right]c + 1 = 0,$$

or

$$c = -\frac{1}{\left(\frac{q}{p}\right)^4 + \left(\frac{q}{p}\right)^3 + \left(\frac{q}{p}\right)^2 + \left(\frac{q}{p}\right) + 1}, \frac{q}{p} \neq 1. \quad (14.3)$$

The denominator of (14.3) is a geometric series, easily summed, to give

$$c = -\frac{1 - \left(\frac{q}{p}\right)}{1 - \left(\frac{q}{p}\right)^5}, \frac{q}{p} \neq 1. \quad (14.4)$$

Now you can see why we took $p \neq q$, for $p = q$ (14.4) gives the indeterminate result 0/0. We can handle the case of $p = q$ by returning to (5′) and setting $p = q$ there before summing a geometric series. Then, $5c + 1 = u(5) = 0$ or,

$$c = -\frac{1}{5}, \, p = q = \frac{1}{2}.$$

This special case, of a fair coin, results in the probability solutions

$u(1) = 1 + c = 4/5,$
$u(2) = c + 1 + c = 1 + 2c = 3/5,$
$u(3) = c + c + 1 + c = 1 + 3c = 2/5,$
$u(4) = c + c + c + 1 + c = 1 + 4c = 1/5.$

So, the answer to our original question is that, flipping a fair penny, A will be ruined with probability $u(3) = 0.4$ if he starts with three dollars and B starts with two dollars.

Looking back at (1′) through (4′), you can see, in general, that

$$u(k) = \left[\left(\frac{q}{p}\right)^{k-1} + \left(\frac{q}{p}\right)^{k-2} + \cdots + \left(\frac{q}{p}\right) + 1\right]c + 1.$$

Again, the expression in brackets on the right-hand side is a geometric series, easily summed to give (using (14.4) for the case of an unfair

($p \neq q$) coin)

$$u(k) = \frac{1-\left(\frac{q}{p}\right)^k}{1-\left(\frac{q}{p}\right)} \cdot \left[-\frac{1-\left(\frac{q}{p}\right)}{1-\left(\frac{q}{p}\right)^5}\right] + 1,$$

or, at last,

$$u(k) = 1 - \frac{1-\left(\frac{q}{p}\right)^k}{1-\left(\frac{q}{p}\right)^5}, \quad k = 1, 2, 3, 4, \ p \neq q. \tag{14.5}$$

This solution to the gambler's ruin problem is credited to the Swiss mathematician James Bernoulli (1654–1705); it appears in his *Ars Conjectandi* (1713). For $k = 3$,

$$u(k) = 1 - \frac{1-\left(\frac{q}{p}\right)^3}{1-\left(\frac{q}{p}\right)^5} = \frac{p^2 q^3 - q^5}{p^5 - q^5}, \quad p \neq q. \tag{14.6}$$

If $p = 0.55$, for example (and so $q = 0.45$), we have

$$u(3) = 0.2859$$

as the probability A is ruined if he starts with three dollars and B starts with two dollars, and the penny is not fair. Figure 14.3 shows what $u(3)$ looks like for $0 \leq p \leq 1$.

14.4 The Monte Carlo Method

The analytical solutions I showed you earlier to Crofton's problem and the gambler's ruin problem are elegant and beautiful. But they are also, to different degrees, *clever*. What if you are not clever? Maybe that's not a personal concern for these two particular problems, but be absolutely certain in your mind that, no matter how clever you are, there will always be some probabilistic problem that will be beyond your analytical reach. What do you do then? Well, I suppose the first thing to try is to find somebody to help you who is cleverer than you are. But that's such a depressing task! Here's another approach—if the problem that has you stumped is based on a well-defined physical process, then try to develop a probabilistic computer simulation of that process. Because of the randomness involved this is called the

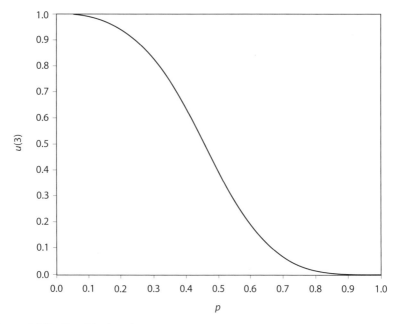

Figure 14.3. Gambler's ruin.

Monte Carlo approach, and it is now a recognized branch of mathematical physics.[10] Let me first show you how it works by applying it to the Crofton and gambler's ruin problems, and then we'll return to Pearson's problem and see what we can do there with some Monte Carlo coding.

First, Crofton's problem. The key observation is that, for a return to the river, the net total motion of the traveler *in the y-direction* (perpendicular to the river) must be nonpositive. The traveler's motion in the x-direction (parallel to the river) is irrelevant in determining if a return to the river has occurred. In the notation of Figure 14.2, then, the condition for a return is $l\sin(\theta_1) + l\sin(\theta_2) \leq 0$, where l is the distance the traveler can journey in a day. The l's, of course, cancel away, and so we are left with the simple condition of $\sin(\theta_1) + \sin(\theta_2) \leq 0$ for a return. A very short simulation code is the following (while it is technically MATLAB, I have limited myself to widely used control statements that are found in practically all scientific programming languages; even MATLAB's rand, which generates numbers uniformly distributed from 0 to 1, is a pretty standard command). This code, called crofton.m, is transparent in its operation; it simply generates ten million random journeys and keeps track of how many of them return to the river. It took me two minutes to write and type it but less than eight seconds for the code to execute on my little home computer. The simulation found that 2,500,964 (the final value of the variable river) returned to the river, giving an estimate of

$$P(return) = 0.2500964.$$

This is pretty close to the theoretical answer we previously calculated to be 0.25.

---

crofton.m
```
c=2*pi;
river=0;
for loop=1:10000000
    theta1=pi*rand;
    theta2=c*rand;
    if sin(theta1)+sin(theta2)<=0
        river=river+1;
    end
end
river
```

The gambler's ruin problem is just as easy to simulate, and the MATLAB code ruin.m does the job. It simulates 50,000 games between A and B for each of the uniformly spaced ninety-nine values of p from 0.01 to 0.99, and then creates a plot of $u(3)$ versus p (see Figure 14.4 and then compare it to the theoretical plot in Figure 14.3). In particular, the simulation estimates for $p = 0.5$ and 0.55, respectively, are $u(3) = 0.397$ and 0.2911. Both of these Monte Carlo estimates are fairly close to the theoretical values calculated in the previous section (0.4 and 0.2859, respectively). To write ruin.m took about five minutes, but it took only three seconds for the code to execute.

And now, to complete this discussion, let's go back to Pearson's plea for help. If personal computers (and a scientific programming

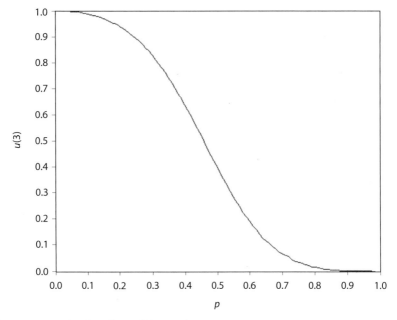

Figure 14.4. Simulated gambler's ruin.

language like MATLAB) had been available in 1905, he would have been able to answer his own question, at least numerically, with a Monte Carlo simulation. The surprisingly short code pearson.m is all that is required; after asking for the number of *unit*-length steps ("stretches" in Pearson's word) in an individual random walk, the code performs two million such walks (all starting

ruin.m
```
prob=linspace(0.01,0.99,99);pAruined=zeros(1,99);
for loop1=1:99
    p=prob(loop1);totalruined=0;
    for loop2=1:50000
        ruined=0;u3=3;
        while u3>0&&u3<5
            flip=rand;
```
(continued)

(continued)

```
                    if flip<p
                        u3=u3+1;
                    else
                        u3=u3-1;
                    end
                end
                if u3==0
                    ruined=1;
                end
                totalruined=totalruined+ruined;
            end
            pAruined(loop1)=totalruined/50000;
        end
        plot(prob,pAruined,'k')
    xlabel('p','FontSize',16)
    ylabel('u(3)','FontSize',16)
    title('Figure 14.4 - Simulated Gambler's Ruin','FontSize',16)
```

from the origin) and determines where the walker is at the end of the last step. The *squared* distance the walker is from the origin—the value of the variable distance2—is stored in the first row of the $2 \cdot 2{,}000{,}000$ matrix history[11], while the actual distance—the square root of distance2—is stored in the second row of history.

We can now use the values stored in history to plot approximations to the probability density functions for the random variables that are the squared distance and the distance of the walker from the origin. As I'll show you soon, these density functions are easy to calculate exactly for the case of two-step walks (the single case that Pearson said he could do), and we can use those calculations to check the performance of the computer code. Once we are convinced pearson.m is giving us the right results for the two-step walk, we can run it with some confidence for longer walks, walks for which Pearson could not calculate the probability density functions.

This code, unlike either crofton.m or ruin.m, takes advantage of a special MATLAB command, histc (for *histogram count*). The command

$N=histc(X,EDGES)$ creates a vector N that has the same number of elements as does the vector EDGES, such that if EDGES(k) is the value of the kth element of EDGES then $N(k)$ will be the number of elements of the vector X that have values in the interval ($EDGES(k)$, $EDGES(k+1)$). The elements of EDGES must monotonically increase in value with increasing k. For example, suppose X is the first row vector of history, that is, the values of X are the squared distances of the two million individual walks. Suppose also that EDGES is a row vector whose elements determine the start and stop points (the *edges*) of 500 subintervals over the interval which a squared distance could occur; for a two-step walk this interval would then of course be (0,4). MATLAB could create EDGES with the command

EDGES = linspace(0,4,500)

or, more generally, for a walk of length *steps*,

EDGES = linspace(0,steps^2,500).

Now, if the random variable **X** has the probability density function $pdf(x)$, and if Δx is "sufficiently small" that we can assume the pdf is "constant" over the interval Δx, then

$$\int_{x}^{x+\Delta x} pdf(u)\,du = pdf(x) \cdot \Delta x = prob(x \leq \mathbf{X} \leq x + \Delta x).$$

If $N(k)$ is the number of values (out of two million) of the elements of **X** that fall in the interval $(x, x + \Delta x)$, then

$$pdf(x) \cdot \Delta x = \frac{N(k)}{2 \cdot 10^6}, k = 1, 2, \ldots, 500$$

and so

$$pdf(x) = \frac{N(k)}{2 \cdot 10^6 \Delta x}, k = 1, 2, \cdots, 500.$$

But the length of the subinterval Δx is just the length L of the interval in which the values of **X** could fall (for the distance squared L = steps^2, and for the distance L = steps), divided by 500 (the number of subintervals). So,

$$pdf(x) = \frac{N(k)}{2 \cdot 10^6 \frac{L}{500}} = \frac{N(k)}{4{,}000L}, k = 1, 2, \cdots, 500.$$

With this explanation, the code pearson.m should now be transparent.

---

pearson.m
```
steps=input('How many steps?');
c=2*pi;history=zeros(2,2000000);
for loop1=1:2000000
    x=0;y=0;
    for loop2=1:steps
        angle=c*rand;
        x=x+cos(angle);y=y+sin(angle);
    end
    distance2=x*x+y*y;
    history(1,loop1)=distance2;
    history(2,loop1)=sqrt(distance2);
end
edges2=linspace(0,steps^2,500);
n2=histc(history(1,:),edges2);
pdf2=n2/(4000*steps^2);
subplot(2,1,1)
plot(edges2,pdf2,'k')
title('Figure 14.5 - Two Step Walk', 'FontSize',16)
xlabel('distance squared','FontSize',16)
ylabel('distance squared pdf','FontSize',16)
subplot(2,1,2)
edges=linspace(0,steps,500);
n=histc(history(2,:),edges);
pdf=n/(4000*steps);
plot(edges,pdf,'k')
xlabel('distance','FontSize',16)
ylabel('distance pdf','FontSize',16)
```

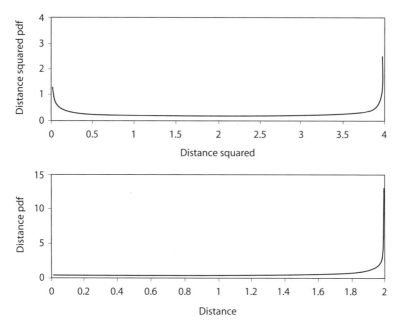

Figure 14.5. Two-step walk.

Figure 14.5 shows the estimated pdfs of the distance and the distance squared random variables for the two-step random walk. Notice that the pdf for the distance *squared* does indeed have the **U**-shaped form that Pearson mentions in his second *Nature* letter, but I have no idea what he meant by "a double U, thus **UU**, . . . symmetrical about the centre vertical corresponding to $r = 0$, but each U itself being asymmetrical." Further, the U-shaped pdf is for the distance *squared*, and not for the distance as he requested. The simulation code also gives what Pearson said he was actually after—the pdf for the distance of the walker from the origin—and it is definitely *not* U-shaped. Are these simulation pdfs in Figure 14.5 correct? let's calculate them and see.

When the first step (of unit length) is taken we can, with no loss of generality, take the direction as along the positive x-axis. Then, the second step of unit length leads to the geometry of Figure 14.6, with θ as a random variable uniformly distributed from 0 to 2π; **R** is a random variable representing the distance of the walker from the origin. From

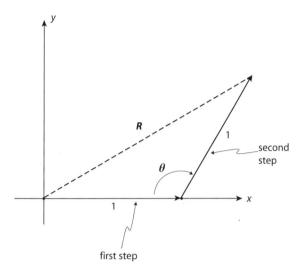

Figure 14.6. The geometry of the two-step walk.

the law of cosines we have the distance squared as

$$\mathbf{R}^2 = 2 - 2\cos(\theta) = \mathbf{Z} \qquad (14.7)$$

where I've written $\mathbf{Z} = \mathbf{R}^2$. To find the pdf of $\mathbf{Z}$, we'll first find the distribution of $\mathbf{Z}$ and then differentiate (see note 6 again). Thus, if $\mathbf{Z}$ is the distance squared random variable that takes on values z, we have the distribution

$$F_{\mathbf{Z}}(z) = P(\mathbf{Z} \leq z) = P(2 - 2\cos(\theta) \leq z) = \mathbf{P}\left(\cos(\theta) \geq \frac{2-z}{2}\right).$$

Looking at Figure 14.7 we see that, in the overall interval for θ of 0 to 2π, there are *two* subintervals (labeled a and b) in which $\cos(\theta) \geq \frac{2-z}{2}$. By symmetry the lengths of a and b are equal, and since θ is uniformly distributed these two subintervals have equal probability. So,

$$F_{\mathbf{Z}}(z) = 2\frac{\cos^{-1}\left(\frac{2-z}{2}\right)}{2\pi} = \frac{1}{\pi}\cos^{-1}\left(\frac{2-z}{2}\right).$$

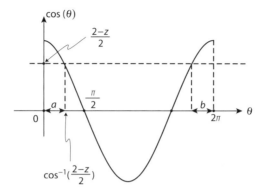

Figure 14.7. Calculating the distribution function for the distance squared of the two-step walk.

The pdf of **Z** (the distance-squared) is the derivative of the distribution and so

$$f_{\mathbf{Z}}(z) = \frac{d}{dz} F_{\mathbf{Z}}(z) = \frac{1}{\pi} \frac{d}{dz} \cos^{-1}\left(\frac{2-z}{2}\right),$$

or, using the well-known formula for the derivative of the inverse cosine (look in any good math handbook or calculus textbook, or just derive it for yourself—it's easy to do), and doing a line or two of easy algebra, we have the pdf for the distance squared as

$$f_{\mathbf{Z}}(z) = \begin{cases} \dfrac{1/\pi}{\sqrt{4z - z^2}}, & 0 \le z \le 4 \\ 0, & \text{otherwise}. \end{cases} \quad (14.8)$$

Notice that $\lim_{z \to 0} f_{\mathbf{Z}}(z) = \lim_{z \to 4} f_{\mathbf{Z}}(z) = \infty$, and also that the minimum value for $f_{\mathbf{Z}}(z)$ occurs when $z = 2$ and so $\min f_{\mathbf{Z}}(z) = 1/2\pi = 0.1592$. That is, the pdf of the distance squared *is* shaped like a U, just as Pearson said and as the simulation code **pearson.m** confirms. Further, the simulation gives a minimum value for the pdf as 0.1586 (the curve in Figure 14.5 is hard to read with accuracy, but a direct examination of the elements of the simulation's pdf2 vector is easy to do).

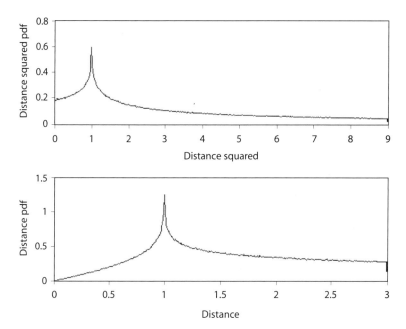

Figure 14.8. Three-step walk.

To calculate the pdf of the distance random variable **R**, which takes on values r, we have

$$F_{\mathbf{R}}(r) = P(\mathbf{R} \leq r) = P(\sqrt{2 - 2\cos(\theta)} \leq r) = P\left(\cos(\theta) \geq \frac{2 - r^2}{2}\right),$$

or, by the same symmetry argument used before,

$$F_{\mathbf{R}}(r) = 2 \frac{\cos^{-1}\left(\frac{2-r^2}{2}\right)}{2\pi} = \frac{1}{\pi} \cos^{-1}\left(\frac{2 - r^2}{2}\right).$$

And so the pdf of **R** is

$$f_{\mathbf{R}}(r) = \frac{1}{\pi} \frac{d}{dr} \cos^{-1}\left(\frac{2 - r^2}{2}\right),$$

or

$$f_{\mathbf{R}}(r) = \begin{cases} \frac{2/\pi}{\sqrt{4-r^2}}, & 0 \leq r \leq 2 \\ 0, & \text{otherwise.} \end{cases} \quad (14.9)$$

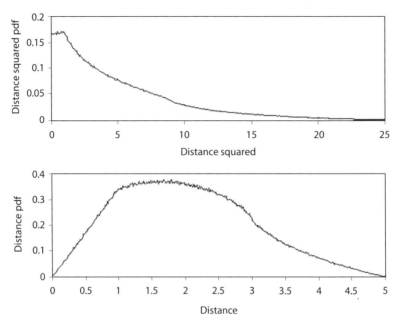

Figure 14.9. Five-step walk.

Notice that (14.9) says $\lim_{r \to 2} f_\mathbf{R}(r) = \infty$ and that as $r \to 0$ the pdf levels off and approaches the value of $1/\pi = 0.3183$ for $f_\mathbf{R}(0)$, just as the simulation curve in Figure 14.5 suggests (a direct examination of the elements of the *pdf* vector gives a value of 0.3174 for $f_\mathbf{R}(0)$). All of these calculations suggest the code pearson.m is working correctly.

So, what happens for the longer walks that Pearson could not work out analytically?[12] Figures 14.8, 14.9, and 14.10 show the distance squared and the distance pdfs for the three-, five-, and six-step random walks, respectively. You can see that while they are distinctly different, at the same time there is a definite pattern to how the pdfs change with an increasing number of steps. The distance pdfs are indeed looking more and more like Rayleigh's pdf (compare Figure 14.10 with Figure 14.1), while the distance squared pdfs "appear" to be approaching something like an exponential. And that "appearance" leads to the second challenge problem to end this discussion. But first, try your hand at this one.

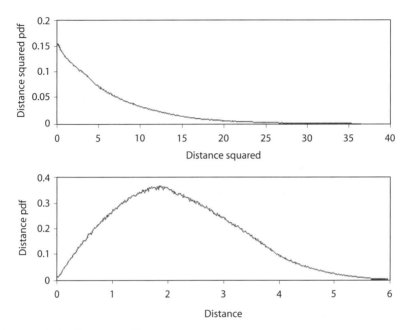

Figure 14.10. Six-step walk.

CP. P14.1:

A natural extension of Crofton's problem is to imagine the traveler's journey can be longer than two days. That is, on day 1 he walks distance l from the river at a random angle uniform in the interval 0 to $\pi/2$. Thereafter, on each of one *or more* additional days, he walks distance l (from the previous day's endpoint) at a random angle uniform from 0 to 2π. Write a Monte Carlo simulation that estimates the probability of returning at some time during the journey to the river. After partially validating your code by running it for the already solved case of two days, run it for the three-day case. Repeat for journeys of four, five, and six days.

CP. P14.2:

Show that if the random variable $\mathbf{X}$ (the distance) has the pdf given by (14.1), that is, is a Rayleigh random variable, then the random variable $\mathbf{Y} = \mathbf{X}^2$ (the distance squared) does in fact have an exponential pdf, yet another indication that pearson.m is working correctly.

Notes and References

1. "The Logical Basis of the Sanitary Policy of Mosquito Reduction" (*Science*, December 1, 1905, pp. 689–699).

2. "The Problem of the Random Walk" (*Nature*, July 27, 1905, p. 294).

3. Probability can be very practical, too. One of the first topics discussed in any introductory probability course, for example, is combinatorial theory, about which the famous Polish-born American mathematician S. M. Ulam (1909–1984) had this to say in his autobiography *Adventures of a Mathematician* (Scribner's 1976, p. 91): "When I became chairman of the mathematics department at the University of Colorado, I noticed that the difficulties of administering N people was not really proportional to N but to N^2. This became my first "administrative theorem." With sixty professors there are roughly eighteen hundred pairs of professors. Out of that many pairs it was not surprising that there were some whose members did not like one another." More precisely, Ulam's calculation is the binomial coefficient $\binom{N}{2} = \frac{N!}{(N-2)!2!} = \frac{N(N-1)}{2} = \frac{60 \cdot 59}{2} = 1,770$.

4. "The Problem of the Random Walk," p. 318.

5. If you're wondering why Rayleigh's probability expression in his letter to Pearson doesn't have any l-dependency, that's because Rayleigh has quietly assumed $l = 1$. Look back at his precise wording, where he speaks of "unit amplitude" oscillations.

6. Saying a random variable $\mathbf{X}$ has the pdf $f_{\mathbf{X}}(x)$ means that the integral of $f_{\mathbf{X}}(x)$ over an interval gives the probability $\mathbf{X}$ takes on a value in that interval. If the interval is $(-\infty, x)$ then that probability is called the *distribution function* of $\mathbf{X}$, written as $F_{\mathbf{X}}(x) = P(-\infty < \mathbf{X} \leq x) = P(\mathbf{X} \leq x)$: that is,

$$F_{\mathbf{X}}(x) = \int_{-\infty}^{x} f_{\mathbf{X}}(u)\, du.$$

In fact, the usual sequence of operations is just the reverse of this, that is, one usually first finds the distribution $F_{\mathbf{X}}(x)$ by other means and then calculates the density as

$$f_{\mathbf{X}}(x) = \frac{d}{dx} F_{\mathbf{X}}(x).$$

Rayleigh's pdf, in particular, appears in countless applications. Here's just one, of special interest to students of modern intercontinental warfare. If a large number of identical ICBM rockets are independently launched at a distant target, and if one measures the distance between the target and the actual impact point of each missile (the so-called *radial miss distance*), then the miss distances are the values of a random variable with a Rayleigh pdf. Each type of ICBM has a different Rayleigh pdf, parameterized on what is called the CEP (the initialism for the rather awkward phrase "circular error probability"). The CEP is the radius of the circle centered on the target within which half the missiles will impact (and outside of which the other half will land).

7. "The Problem of the Random Walk," p. 342.

8. Pearson's thanks for the $n = 3$ case were to Geoffrey Thomas Bennett (1868–1943)—the middle initial J. in Pearson's printed letter was a typo—who was a lecturer in mathematics at Emmanuel College.

9. Much interesting historical discussion on random walks is in the paper by Jacques Dutka, "On the Problem of Random Flights" (*Archive for the History of Exact Sciences* 32, 1985, pp. 351–375). See also S. Chandrasekhar, "Stochastic Problems in Physics and Astronomy" (*Reviews of Modern Physics*, January 1943, pp. 1–89).

10. You can find much more on the Monte Carlo approach in my two books, *Duelling Idiots* (2003) and *Digital Dice* (2008), both published by Princeton University Press. The two books together, in addition to their historical discussions, contain seventy or so problems, each with a complete, detailed solution in the form of a tested computer code written in MATLAB.

11. A personal note: between my master's degree and my doctoral degree, I worked for several years (1963–1972) as a digital system logic designer and programmer in the Southern California aerospace business. In those years, writing a computer program in which one casually dimensions a 2 by 2,000,000 matrix would have been considered a fantastic dream for all but those with access to one of the latest super-duper, top-secret computers operated by the National Security Agency. Today it is a routine task for anyone who owns a quite ordinary home computer running the commercial version of MATLAB. In the 1960s, computer memory was made from low-density integrated transistor circuitry, mechanically rotating magnetic drums, or expensive, bulky magnetic cores. To speak of 1,000-gigabyte memory storage units then would have been the signature of madness! Today, however, massive high-speed computer memory is literally cheaper than dirt. This revolution in data storage was recently recognized by the awarding of the 2007 Nobel Prize in Physics for the discoveries that made such massive computer memories commonplace.

12. E. Merzbacher, J. M. Feagin, and T-H. Wu, "Superposition of the Radiation from N Independent Sources and the Problem of Random Flights" (*American Journal of Physics*, October 1977, pp. 964–969). This paper presents computer simulation results that agree with the ones in this discussion (produced by simulating between 100,000 and 200,000 walks, as compared to the 2,000,000 walks used here), as well as the theoretical calculation of the pdf's of the distance squared for *all* finite N (the length, in steps, of the walk), following the method developed by the Dutch mathematician J. C. Kluyver (1860–1932) who solved Pearson's original problem only months after Pearson's first letter appeared in *Nature*, not just Rayleigh's limiting case of $N \to \infty$. In particular, the authors write, "In comparing our results with Pearson's tables and graphs [in a 1906 paper in which Pearson presented some numerical evaluations of Kluyver's formal integrals] we generally found satisfactory agreement—remarkable in view of the numerical technology available in 1906—but in a few instances we detected significant differences in detail. For example, while Pearson shows a constant value [in the distance-squared pdf for the five step walk of] 0.169 [over the interval 0 to 1], we noticed a slight rise." You can see that Figure 14.9 also suggests such a rise over that interval.

Two More Random Walks

> The Cat only grinned when it saw Alice....
> "Cheshire Puss," she began, rather timidly, as she did not at all know whether it would like the name: however, it only grinned a little wider. "Come, it's pleased so far," thought Alice, and she went on. "Would you tell me, please, which way I ought to go from here?"
> "That depends a good deal on where you want to go," said the Cat.
> "I don't much care where—" said Alice.
> "Then it doesn't matter which way you go," said the Cat.
> "—so long as I get somewhere," Alice added as an explanation.
> "Oh, you're sure to do that," said the Cat, "if you only walk long enough."
> — Lewis Carroll, *Alice in Wonderland*, chapter 6, "*Pig and Pepper*"

15.1 Brownian Motion

Perhaps the best known drunkard's walk is called *Brownian motion*, named after the Scottish botanist Robert Brown (1773–1858) who, in 1827, observed (under a microscope) the chaotic motion of tiny grains of pollen suspended in water drops. (This motion had been previously

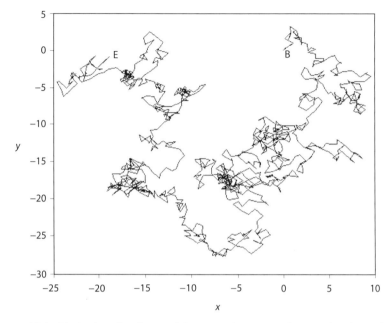

Figure 15.1. Typical path of a particle executing Brownian motion in a plane.

noted by others, but Brown took the next step of publishing what he saw in an 1828 paper in the *Philosophical Magazine*.) This motion exhibits a number of interesting physical characteristics: the higher the temperature and the smaller the suspended particle the more rapid the motion, while the more viscous the medium the slower the motion; the motions of even very close neighboring particles are independent of each other; the past behavior of a suspended particle has no bearing on its future behavior; and the likelihood of motion in a particular direction, at any instant of time, is the same for all directions. Figure 15.1 shows a typical (very kinky) 1,000-step two-dimensional path of a particle executing Brownian motion in a plane.[1] The beginning and the end of the path are marked with a B and an E, respectively.

Today, because of the work by the German-born physicist Albert Einstein (1879–1955) in a series of papers[2] published between 1905 and 1908 applying statistical mechanics to the question of Brownian motion, we understand that it is due to the random molecular bombardment experienced by the suspended particles. Brownian motion is, in fact, strong macroscopic experimental evidence for the reality of molecules, and therefore of atoms. Experiments done shortly after Einstein's theoretical studies that confirmed the theory would earn the French physicist Jean Perrin (1870–1942) the 1926 Nobel Prize in Physics. Some years before that, in the early 1920s, the American mathematician Norbert Wiener (1894–1964) made a deep analysis of the mathematics of Brownian motion, studying in particular what is going on in the limit of a particle moving in steps separated by vanishingly small time intervals. In that limit, Brownian motion becomes what mathematicians call a *Wiener stochastic process*, a concept of great continuing interest to modern physicists, mathematicians, and electrical engineers.

In particular, a "Wiener walk" is a continuous curve in space that, at almost every point, has no direction. Such "infinitely kinky" curves were known to mathematicians before Wiener's work,[3] but they were what mathematicians like to call *pathological*, and what other people would say are simply made up, extremely weird curves that are purposely weird just to show the extremes of what mathematics will allow. Wiener's work showed how such a strange thing could occur, in a natural way, in a *physical* context. In this discussion I'll not go anywhere nearly as deep as Wiener did, but there is a very pretty, elementary way of showing how time can be directly introduced into the probabilistic mathematical description of one-dimensional Brownian motion: it is the incorporation of time that makes a random process a stochastic process.

Our random walk will be much like Ross's mosquito walk in the previous discussion, that is, equally likely, independent steps of length Δs, back and forth along the x-axis ($\Delta s = l/2$ is Ross's walk). After N such steps, the particle, which *has to be somewhere*, could be at any one of the x-axis points $-N\Delta s$, $(-N+1)\Delta s$, ..., $-\Delta s$, 0, Δs, ..., $(N-1)\Delta s$, $N\Delta s$. To be precise, to be at the point with x-axis value $m\Delta s$, it is necessary that $(N+m)/2$ of the steps—it doesn't matter which ones— were in the positive direction and the other $(N-m)/2$ steps were in the negative direction. (Notice that m can be even only if N is even, and that m can be odd only if N is odd.) Since the total number of all possible, equally likely left/right step sequences is 2^N, and since the number of sequences for which $(N-m)/2$ steps are in the negative direction is $\binom{N}{\frac{N-m}{2}}$, then the probability the particle is at $x = m\Delta s$ after

N steps is

$$P(m\,\Delta s,\ N) = \begin{cases} \dfrac{\binom{N}{\frac{N-m}{2}}}{2^N} & N \geq 0,\quad m = -N, -N+1, \ldots, N-1, N \\ 0, \text{ otherwise.} \end{cases} \quad (15.1)$$

We can express (15.1) in a much different, and actually far more informative way, as follows. Expanding the binomial coefficient in the numerator, we have

$$\binom{N}{\frac{N-m}{2}} = \frac{N!}{\left[\frac{1}{2}(N+m)\right]!\left[\frac{1}{2}(N-m)\right]!}.$$

From Stirling's asymptotic formula for factorials,[4] that is from

$$n! \sim \sqrt{2\pi n}\, e^{-n} n^n,$$

we have

$$\ln(n!) \sim \left(n + \frac{1}{2}\right)\ln(n) - n + \frac{1}{2}\ln(2\pi).$$

So, taking the natural logarithm of (15.1) and using Stirling's approximation, we have

$$\ln P(m\,\Delta s,\ N) = \ln\binom{N}{\frac{N-m}{2}} - N\ln 2$$

$$= \ln N! - \ln\left[\frac{1}{2}(N+m)\right]! - \ln\left[\frac{1}{2}(N-m)\right]! - N\ln 2$$

$$\approx \left(N + \frac{1}{2}\right)\ln N - N + \frac{1}{2}\ln 2\pi - \left[\frac{1}{2}(N+m) + \frac{1}{2}\right]$$

$$\times \ln\left[\frac{1}{2}(N+m)\right] + \frac{1}{2}(N+m) - \frac{1}{2}\ln 2\pi - \left[\frac{1}{2}(N-m) + \frac{1}{2}\right]$$

$$\times \ln\left[\frac{1}{2}(N-m)\right] + \frac{1}{2}(N-m) - \frac{1}{2}\ln 2\pi - N\ln 2$$

$$= \left(N + \frac{1}{2}\right)\ln N - \frac{1}{2}(N+m+1)\ln\left[\frac{1}{2}N\left(1 + \frac{m}{N}\right)\right]$$

$$- \frac{1}{2}(N-m+1)\ln\left[\frac{1}{2}N\left(1 - \frac{m}{N}\right)\right] - \frac{1}{2}\ln 2\pi - N\ln 2.$$

Suppose now that we limit ourselves to the cases where $m \ll N$, that is, we will concentrate our attention on the probabilities that the particle has not wandered out nearly as far as it *might* possibly wander from its starting point at $x = 0$. Even though Stirling's approximation is just asymptotic, the fact that the numerator of (15.1) contains the *ratio* of factorials means our final approximation to $P(m \Delta s, N)$ will be pretty accurate even in the absolute sense *if* $m \ll N$. I'll show you how that works numerically in just a bit. Now, using the power series expansion approximation

$$\ln(1 \pm z) \approx \pm z - \frac{1}{2}z^2, |z| \ll 1,$$

we have

$$\ln\left[\frac{1}{2}N\left(1+\frac{m}{N}\right)\right] = \ln\left(\frac{1}{2}N\right) + \ln\left(1+\frac{m}{N}\right) \approx \ln(N) - \ln 2 + \frac{m}{N} - \frac{m^2}{2N^2}.$$

Similarly,

$$\ln\left[\frac{1}{2}N\left(1-\frac{m}{N}\right)\right] \approx \ln(N) - \ln 2 - \frac{m}{N} - \frac{m^2}{2N^2}.$$

So, for $m \ll N$,

$$\ln P(m \Delta s, N) = \left(N+\frac{1}{2}\right)\ln N - \frac{1}{2}(N+m+1)$$

$$\times \left[\ln(N) - \ln 2 + \frac{m}{N} - \frac{m^2}{2N^2}\right] - \frac{1}{2}(N-m+1)$$

$$\times \left[\ln(N) - \ln 2 - \frac{m}{N} - \frac{m^2}{2N^2}\right] - \frac{1}{2}\ln 2\pi - N\ln 2.$$

If you multiply this all out you'll find that a lot of terms cancel; in addition, making the final approximation that $\frac{m^2}{2N^2} \ll \frac{m^2}{2N}$ (since $N \gg m$) results in

$$\ln P(m \Delta s, N) = -\frac{1}{2}\ln N + \ln 2 - \frac{1}{2}\ln 2\pi - \frac{m^2}{2N} + \frac{m^2}{2N^2}$$

$$\approx \ln \frac{1}{\sqrt{N}} + \ln 2 + \ln \frac{1}{\sqrt{2\pi}} + \ln e^{-(m^2/2N)}$$

$$= \ln\left\{\frac{2}{\sqrt{2\pi N}}e^{-(m^2/2N)}\right\},$$

or, at last,

$$P(m\,\Delta s,\ N) \approx \sqrt{\frac{2}{\pi N}} e^{-(m^2/2N)},\ m \ll N. \tag{15.2}$$

We can numerically compare, for any given N and all possible m, the approximate (15.2) to the exact (15.1) to see how much damage all of our approximations have caused. Suppose, for example, that $N = 12$ (and so $m = 0, \pm 2, \pm 4, \ldots, \pm 12$ are the particle location possibilities). Then,

m	(15.1)—exact	(15.2)—approximation
0	0.2256	0.2303
2	0.1934	0.1950
4	0.1208	0.1183
6	0.0537	0.0514
8	0.0161	0.0160
10	0.0029	0.0036
12	0.00024	0.00057

As you can see, (15.2) is actually a pretty good approximation, even for $m \approx N$.

To finish our analysis, I'll next express the x-axis coordinate of our particle as $x = m\,\Delta s$. Thus, $m = x/\Delta s$, and so (15.2) becomes

$$P(x,\ N) \approx \sqrt{\frac{2}{\pi N}} e^{-x^2/2N(\Delta s)^2}. \tag{15.3}$$

In addition, suppose we now consider intervals along the x-axis of length Δx such that $\Delta x \gg \Delta s$; this means that there are *numerous* points within a Δx interval at which the Brownian particle *might* be after N steps. Note carefully—both Δx and Δs are small, but Δs is *very* small! Then the probability the particle is somewhere in the interval $(x, x + \Delta x)$ is

$$W(x, N) = \sum P(x,\ N) = \sqrt{\frac{2}{\pi N}} \sum e^{-x^2/2N(\Delta s)^2}, \tag{15.4}$$

where the sum is over all possible points on the x-axis where the particle *could be* located. To evaluate the sum, we need only make the following two observations:

(1) since Δx is small (but remember, Δs is even smaller), then the value of x in the exponential's exponent is nearly constant over the entire interval of width Δx;
(2) there are a total of $\Delta x/\Delta s$ points on the x-axis in an interval of width Δx, but only half of them are available as possible location points (which half depends, as mentioned earlier, on whether N is even or odd).

With these two observations, we can write (15.4) as

$$W(x, N) = \sqrt{\frac{2}{\pi N}} e^{-x^2/2N(\Delta s)^2} \cdot \frac{1}{2} \cdot \frac{\Delta x}{\Delta s},$$

or

$$W(x, N) = \frac{\Delta x}{\sqrt{2\pi N(\Delta s)^2}} e^{-x^2/2N(\Delta s)^2}. \tag{15.5}$$

For our last step in this analysis, which in fact is the crucial step that introduces time, let's suppose that the particle makes n steps per unit time. Then, starting from $t = 0$, the particle has made $N = nt$ steps at time t. This lets us write (15.5) as

$$W(x, t) = \frac{\Delta x}{\sqrt{2\pi nt(\Delta s)^2}} e^{-x^2/2nt(\Delta s)^2}$$

or, if we define the so-called *diffusion coefficient*[5] $D = \frac{1}{2}n(\Delta s)^2$, then we can write

$$W(x, t) = \frac{\Delta x}{2\sqrt{\pi Dt}} e^{-x^2/4Dt} \tag{15.6}$$

as the probability the particle is in the interval $(x, x + \Delta x)$ at time t. Since a probability is the product of a probability *density* and an interval (Δx), that is $W(x,t) = f_\mathbf{X}(x,t)\Delta x$ where $f_\mathbf{X}(x,t)$ is a pdf, we see from (15.6) that

$$f_\mathbf{X}(x,t) = \frac{1}{2\sqrt{\pi Dt}} e^{-x^2/4Dt}, \ t \geq 0 \tag{15.7}$$

is the pdf[6] that describes the location on the x-axis of a particle executing one-dimensional Brownian motion.[7] The pdf of (15.7) contains time and, obviously, it changes with time. Random processes

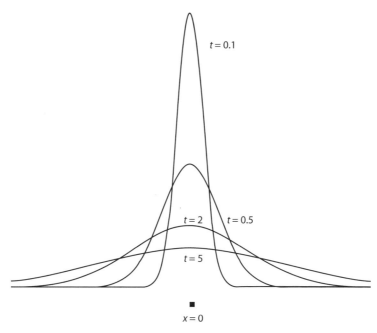

Figure 15.2. Gaussian pdfs for one-dimensional Brownian motion.

that have pdf's that do not change with time are said to be *stationary* (in time). So, the one-dimensional Brownian motion of a particle is a nonstationary random process, and it is a favorite of probability textbook authors looking for an example to illustrate that concept.

For a given value of time, the pdf of (15.7) is the well-known bell-shaped *Gaussian curve,* named for the great German mathematician Carl Friedrich Gauss (1777–1855), who studied it long before anyone had any idea it would occur in the theory of Brownian motion. For all non-negative values of t the pdf is symmetrical about $x = 0$. For values of t only slightly positive the bell-shaped pdf is tall and narrow, as the particle has not yet had enough time to wander very far from its starting point, but as time increases the bell-shaped pdf broadens and becomes less spiked. This reflects the fact that the total area under the pdf curve, which is the probability the particle is *somewhere*, is always unity. Figure 15.2 shows the relative shapes and sizes of (15.7) for four values of time. Notice that these curves clearly illustrate Karl Pearson's comment, in his second *Nature* letter (see the previous discussion), that at all times "in open country the most probable place to find a drunken man who is at all capable of keeping on his feet is somewhere near his starting point!" That is, for intervals of fixed width Δx, the one that has the largest probability of containing the particle is always the one centered on $x = 0$.

15.2 Shrinking Walks

In the random walks so far considered, the walker could, in the limit as the number of steps increases without bound, be found arbitrarily far from the starting point—the origin—of the walk. This is because the length of each step has been, so far, a constant. There is another interesting class of random walks, however, that is forever confined to a *finite* region of space even as the number of steps increases without limit. These are walks in which every step is shorter than the previous one. In particular, I'll discuss here what is called a one-dimensional *geometric walk*, in which the nth step has length s^n, where $n = 0$ is the first step and s is a positive number less than one. So, the lengths of the first four steps, for example, are $1, s, s^2$, and s^3. I'll take the steps as independent and equally likely to be in either direction along the x-axis. That is, if $\varepsilon_n = \pm 1$ with equal probability, then the location of the walker at the end of N steps is

$$x_N = \sum_{n=0}^{N-1} \varepsilon_n s^n. \tag{15.8}$$

The maximum distance of the walker from the origin occurs when $\varepsilon_n = +1$ (or -1) for all n:

$$|\max x_N| = \sum_{n=0}^{N-1} s^n = 1 + s + s^2 + s^3 + \cdots + s^{N-1},$$

which is a geometric series (hence the name of the walk), easily summed, to give

$$-\frac{1-s^N}{1-s} \leq \max x_N \leq \frac{1-s^N}{1-s}, \quad 0 < s < 1$$

and so, even as $N \to \infty$, the location of the walker is restricted to an interval of finite width $2/(1-s)$ centered on the origin. For example, if $s = 0.6$, then the resulting geometric walk will *always* be somewhere in an interval of length 5 centered on the origin, no matter how many steps long the walk may be.

As a variation on this sort of random walk, we could reinterpret the ε_n as equally likely to be 0 or $+1$, which physically represents a walk that, on the nth decision, either stays at the current location ($\varepsilon_n = 0$) or moves one step to the right ($\varepsilon_n = 1$). There is never a step to the left. Such a walk drifts to the right, but it too, of course, will always be confined to a finite interval, in this case to $0 < x < 1/(1-s)$, $0 < s < 1$. Making $\varepsilon_n = 0$ or -1 would give a walk drifting to the left.

Walks with nonconstant step size have been known to mathematicians since the 1930s, but in recent years they have also attracted the attention of physicists. For example, the authors of a paper[8] published in 2000 encountered such a walk during a "study of the state reduction (collapse) in multiple position or momentum observations of a quantum-mechanical particle." Well, that sounds pretty deep to me— I have no intention of jumping into *that* pool!—and my only point here is that walks with a *changing* (in this case, decreasing) step size can have physical significance. In fact, the other extreme, of an *increasing* step size, can occur in physics as well. For example, in a 1992 paper[9] it is shown that a particle executing Brownian motion in a fluid with a linear shear flow (imagine a wide river in which water flows ever faster downstream the farther out from one bank you move) performs a random walk with a step size that increases linearly with each new step. The authors call this walk a "stretched walk."

In the rest of this section I'll follow the lead of an even more recent (2004) paper that presents a fascinating mathematical discussion of geometric walks.[10] In particular, the authors, whom I'll refer to as K&R, were interested in the probability density functions of the distance x of the walker from the origin after an infinity of steps. That is, suppose we define the pdf for that distance as a function of the number of steps, and the step size, by $f_{\mathbf{X},s}(x, N)$. Then the question asked by K&R is, what happens as $N \to \infty$, that is, what is

$$\lim_{N \to \infty} f_{\mathbf{X},s}(x, N) = f_{\mathbf{X},s}(x) = ?$$

In general, this is not an easy question. But K&R do offer some quite interesting partial answers. For example,

(1) if $s < \frac{1}{2}$ the subset of the x-axis where $f_{\mathbf{X},s}(x)$ is non-zero is a Cantor set, that is, is fractal (which means, in practical terms, that we can *not* plot $f_{\mathbf{X},s}(x)$ for $s < \frac{1}{2}$);
(2) if $s = \frac{1}{2}$, then $f_{\mathbf{X},s}(x)$ is uniform; and
(3) if $\frac{1}{2} \leq s < 1$, then $f_{\mathbf{X},s}(x)$ is continuous.

These results are formally derived by K&R, but we can get an intuitive idea of what is going on with some Monte Carlo simulation, an approach K&R themselves use in the beginning of their paper. For example, suppose we take a look at the case of a right-drifting shrinking random walk with $s = 1/2$. We can't, of course, simulate such walks with an infinite number of steps, but just to get some idea of what happens, let's pick $N = 29$ steps (as do K&R), simulate 2,000,000 such random walks (K&R looked at 100,000,000 random walks), and record where the walk is at the end of the last step. Using the same sort of coding approach I used in the previous discussion, such a simulation produced Figure 15.3, and it does indeed suggest that in the limit $N \to \infty$ $f_{\mathbf{X},s}(x)$ is uniform over $0 < x < 2$ when $s = 1/2$ (remember, the simulation gives an *estimate* of $f_{\mathbf{X},s}(x, 29)$). The nature of $f_{\mathbf{X},s}(x)$ changes dramatically, however, for $s \neq 1/2$; for example, Figure 15.4 shows the result of simulating a twenty-nine-step random walk for $s = (\sqrt{5}-1)/2 \approx 0.618034$ (and as you'll soon see, that particular, perhaps curious-looking value for s was not chosen at random).

Before saying more about $s = (\sqrt{5}-1)/2$, let me make one final comment about the $s = 1/2$ case. K&R show, using some high-powered math (Fourier transforms), how to formally derive $f_{\mathbf{X},s}(x)$ for the values of $s = 2^{-\frac{1}{m}}$, where m is a positive integer. The very first value of m, $m = 1$, gives $s = 1/2$, and the formal result is a uniform pdf. If you are willing to settle for a less inclusive proof, however, one that works just for the single case of $s = 1/2$, there is a surprisingly simple proof of uniformity. If we put $s = 1/2$ in (15.8), then

$$x_N = \varepsilon_0 + \epsilon_1 \frac{1}{2} + \epsilon_2 \frac{1}{4} + \epsilon_3 \frac{1}{8} + \cdots + \epsilon_N \frac{1}{2^{N-1}}.$$

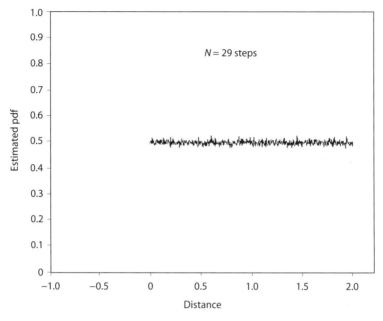

Figure 15.3. Shrinking walk for $s = 0.5$.

Interpreting all the ϵ_i's as either 0 or 1 (a drifting walk), this expression for the location of the walk at the end of the Nth step is simply an N-bit binary number in the interval 0 (all $\epsilon_i = 0$) to $2 - \frac{1}{2^{N-1}}$ (all $\epsilon_i = 1$). All possible combinations of values for all the ϵ_i give all possible N-step walks, with the numbers and the walks in a one-to-one correspondence. Since the binary numbers are uniformly distributed over the interval $(0, 2 - \frac{1}{2^{N-1}})$—I hope the uniformity is obvious!—then the walk lengths are uniform, too. In Figure 15.3 I used $N = 29$, and so there are a total of $2^{29} = 536{,}870{,}912$ such walks, while the simulation code that produced Figure 15.3 randomly examined just 2,000,000 walks. The wiggles you see in that figure are statistical sampling effects caused by looking at only a very small fraction of all possible twenty-nine-step walks.

Sampling error is just the first of our concerns with studying geometric walks by Monte Carlo simulation. Equally troublesome is that, as N increases, the step size becomes so small it eventually gets lost in roundoff error. There is a beautiful way, for one particular value of s, at least, to avoid this new problem. K&R give a broad overview of the approach, as follows, for walks that have both left and right moving steps:

> Unfortunately, a straightforward simulation of the geometric random walk is not a practical way to visualize fine-scale details of $[f_{\mathbf{X},s}(x,N)]$ because the resolution is necessarily limited by the width of the bin [what I called subintervals in Section 14.3 when discussing the MATLAB code **pearson.m**] used to store the [probability density]. We now describe an enumeration approach that is exact up to the number of steps in the walk. It is simple to specify all walks of a given number of steps. Each walk can be represented as a string of binary digits with 0 representing a step to the left and 1 representing a step to the right. Thus we merely need to list all possible binary numbers with a given number of digits N to enumerate all N-step walks.

This is true for *any* value of s. There still remains the roundoff problem, however, which K&R go on to explain:

> However, we need a method that provides the end point location without any loss of accuracy to resolve the fine details of $[f_{\mathbf{X},s}(x,N)]$. ... For large N, the accuracy in the position of a walk is necessarily lost by roundoff errors if we attempt to evaluate the sum for the end point location $x_N = \sum_{n=0}^{N-1} \varepsilon_n s^n$, directly.

To solve this problem for the particular value of $s = g = (\sqrt{5} - 1)/2$, K&R conclude this part of their paper with the following somewhat enigmatic (in my opinion) words:

> We may take advantage of the algebra of the golden ratio [that is, $1/g$] to reduce the N-*th* order polynomial in x_N to a first-order polynomial. To this end, we successively use the defining equation $g^2 = 1 - g$ to reduce all powers of g to the first order. When we apply this reduction to g^n, we obtain the remarkably simple formula $g^n = (-1)^n (F_{n-1} - g F_n)$, where F_n is the n-*th* Fibonacci number (defined by $F_n = F_{n-1} + F_{n-1}$ for $n > 2$, with $F_1 = F_2 = 1$). ... We now use this construction to reduce the location of each end point, which is of the form $\sum_{n=0}^{N-1} \varepsilon_n g^n$, to the much simpler

form $A + Bg$, where A and B are integers whose values depend on the walk. By this approach, each end point location is obtained with perfect accuracy.

Pretty intriguing stuff, alright, but what does it mean? K&R don't provide any of the details of their "reduction", and absent is any discussion of A and B, that is, they don't provide us with formulas with which to *calculate* A and B. What follows is my reconstruction of what they did, and (in my opinion) what K&R did is pretty darn clever!

To start, consider the second order difference equation

$$u(n+2) = u(n+1) + u(n), \ u(0) = u(1) = 1, \ n = 0, 1, 2, 3, \ldots, \quad (15.9)$$

which is of course the famous *Fibonacci equation*, named after the Italian mathematician Leonardo Pisano (circa 1170–1250), also known by the sobriquet *Fibonacci*.[11] Just as we can often solve differential equations by assuming an exponential solution, we can often solve difference equations by assuming a power solution. That is, let's assume

$$u(n) = ck^n$$

where c and k are some yet to be determined non-zero constants. Then, substituting this into (15.9), we have

$$ck^{n+2} = ck^{n+1} + ck^n,$$

or

$$k^2 - k - 1 = 0. \quad (15.10)$$

This quadratic is easily solved to give

$$k_1 = \frac{1+\sqrt{5}}{2} > 0 \text{ and } k_2 = \frac{1-\sqrt{5}}{2} < 0. \quad (15.11)$$

So, in general,

$$u(n) = c_1 \left[\frac{1+\sqrt{5}}{2}\right]^n + c_2 \left[\frac{1-\sqrt{5}}{2}\right]^n,$$

where, for complete generality, I've used a different c for each of the two k's. Using the given conditions $u(0) = u(1) = 1$, it is not difficult to show that the constants c_1 and c_2 are

$$c_1 = \frac{1+\sqrt{5}}{2\sqrt{5}}, \quad c_2 = \frac{\sqrt{5}-1}{2\sqrt{5}} = -\frac{1-\sqrt{5}}{2\sqrt{5}}.$$

Thus,

$$u(n) = \frac{1+\sqrt{5}}{2\sqrt{5}}\left[\frac{1+\sqrt{5}}{2}\right]^n - \frac{1-\sqrt{5}}{2\sqrt{5}}\left[\frac{1-\sqrt{5}}{2}\right]^n$$

or, at last,

$$u(n) = \frac{1}{\sqrt{5}}\left\{\left[\frac{1+\sqrt{5}}{2}\right]^{n+1} - \left[\frac{1-\sqrt{5}}{2}\right]^{n+1}\right\}, \quad n = 0, 1, 2, 3, \ldots.$$

(15.12)

As K&R indicate, the Fibonacci number $F_n = F(n)$ satisies (15.9), except that we normally start the indexing at $n = 1$ rather than at $n = 0$. All we need do, then, is replace $n+1$ in the exponents of (15.12) with $n = 1, 2, 3, \ldots$, and so

$$F(n) = \frac{1}{\sqrt{5}}\left\{\left[\frac{1+\sqrt{5}}{2}\right]^n - \left[\frac{1-\sqrt{5}}{2}\right]^n\right\}, \quad n = 1, 2, 3, \ldots. \quad (15.13)$$

The Fibonacci numbers are, of course, all integers $(1, 1, 2, 3, 5, 8, 13, \ldots)$, and I have always found it amazing that for every single one of the infinity of positive integer values of n, an expression like (15.13), with all of its irrational $\sqrt{5}$'s, produces an *integer*.[12]

Next, let's set $g = 1/k_1$, where we pick the value of $k > 0$ so that $g > 0$. Then, from (15.10), we have

$$\frac{1}{g^2} - \frac{1}{g} - 1 = 0, \quad (15.14)$$

or

$$g^2 = 1 - g, \quad (15.15)$$

which, you'll recall, K&R call "the defining equation." The physical interpretation of (15.15), which can be written as $1 - g - g^2 = 0$, is that g is that value of step size such that an initial step to the right (the "1"), followed by a second step to the left (the "$-g$"), followed by a third step to the left (the "$-g^2$"), returns the walk *exactly* to the origin. Notice that returning *exactly* to the origin in three steps is impossible with a constant step size.

It is easy to verify that

$$g = \frac{1}{\frac{1+\sqrt{5}}{2}} = \frac{\sqrt{5}-1}{2}$$

and that

$$g + 1 = \frac{1+\sqrt{5}}{2}.$$

Inserting these two results into (15.13), we have

$$F(n) = \frac{1}{\sqrt{5}}[(g+1)^n - (-g)^n],$$

or

$$F(n) = \frac{1}{\sqrt{5}}[(g+1)^n - (-1)^n g^n]. \tag{15.16}$$

Since we know from (15.14) that

$$\frac{1}{g^2} = \frac{1}{g} + 1,$$

then multiplying through by g gives

$$\frac{1}{g} = 1 + g. \tag{15.17}$$

Using (15.17) in (15.16) gives

$$F(n) = \frac{1}{\sqrt{5}}\left[\frac{1}{g^n} - (-1)^n g^n\right]. \tag{15.18}$$

And from (15.18) we have

$$F(n-1) = \frac{1}{\sqrt{5}}\left[\frac{1}{g^{n-1}} + (-1)^n g^{n-1}\right],$$

or

$$F(n-1) = \frac{1}{\sqrt{5}}\left[\frac{g}{g^n} + (-1)^n \frac{g^n}{g}\right]. \tag{15.19}$$

Consider now $F(n-1) - gF(n)$. That is, from (15.19) and (15.18) we can write

$$F(n-1) - gF(n) = \frac{1}{\sqrt{5}}\left[\frac{g}{g^n} + (-1)^n \frac{g^n}{g}\right] - \frac{g}{\sqrt{5}}\left[\frac{1}{g^n} - (-1)^n g^n\right]$$

$$= \frac{(-1)^n}{\sqrt{5}}\left[\frac{g^n}{g} + g^{n+1}\right] = \frac{(-1)^n}{\sqrt{5}} g^n \left[\frac{1}{g} + g\right].$$

Using (15.17), we know that

$$\frac{1}{g} + g = 1 + 2g,$$

and so

$$F(n-1) - gF(n) = \frac{(-1)^n}{\sqrt{5}} g^n (1 + 2g). \tag{15.20}$$

As we showed earlier,

$$\frac{1 + \sqrt{5}}{2} = g + 1,$$

and so

$$1 + \sqrt{5} = 2g + 2,$$

and so

$$\sqrt{5} = 1 + 2g.$$

Inserting this into (15.20) we have

$$F(n-1) - gF(n) = (-1)^n g^n,$$

or

$$g^n = (-1)^n [F(n-1) - gF(n)], \tag{15.21}$$

which is K&R's "remarkably simple formula."

Now we can calculate A and B. We have, once again, from (15.8), that the location of the end of our walk is at (with $s = g$)

$$x_N = \sum_{n=0}^{N-1} \varepsilon_n g^n.$$

That is, just to be quite clear, x_N is the endpoint of a geometric walk with N steps with a step size factor of g (with a first step of length $\varepsilon_0 = \pm 1$ with equal probability). So, writing x_N in detail, and using (15.21),

$$x_N = \varepsilon_0 + \epsilon_1 g + \epsilon_2 g^2 + \epsilon_3 g^3 + \epsilon_4 g^4 + \cdots + \epsilon_{N-1} g^{N-1}$$

$$= \varepsilon_0 + \epsilon_1 g + \epsilon_2 [F(1) - gF(2)] + \epsilon_3(-1)[F(2) - gF(3)]$$

$$+ \epsilon_4 [F(3) - gF(4)] + \epsilon_5(-1)[F(4) - gF(5)] + \epsilon_6 [F(5) - gF(6)]$$

$$+ \epsilon_7(-1)[F(6) - gF(7)] + \cdots + \epsilon_{N-1}(-1)^{N-1}$$

$$\times [F(N-2) - gF(N-1)]$$

$$= \{\varepsilon_0 + \epsilon_2 F(1) - \epsilon_3 F(2) + \epsilon_4 F(3) - \epsilon_5 F(4) + \epsilon_6 F(5) - \epsilon_7 F(6)$$

$$+ \cdots + \epsilon_{N-1}(-1)^{N-1} F(N-2)\} + g\{\epsilon_1 - \epsilon_2 F(2)$$

$$+ \epsilon_3 F(3) - \epsilon_4 F(4) + \epsilon_5 F(5) - \epsilon_6 F(6) + \epsilon_7 F(7)$$

$$+ \cdots - \epsilon_{N-1}(-1)^{N-1} F(N-1)\}$$

$$= \varepsilon_0 + \sum_{k=2}^{N-1} (-1)^k \varepsilon_k F(k-1) + g\left[\varepsilon_1 - \sum_{k=2}^{N-1} (-1)^k \varepsilon_k F(k)\right].$$

And this is, indeed, as K&R say, "the much simpler form $A + Bg$." That is, in the particular random geometric walk with $s = g$ (called by K&R the "golden walk," for the obvious reason), with both left- and right-moving steps, the endpoint of the walk after N steps is

$$x_N = A + Bg, \tag{15.22}$$

where

$$A = \varepsilon_0 + \sum_{k=2}^{N-1} (-1)^k \varepsilon_k F(k-1) \tag{15.23}$$

and

$$B = \varepsilon_1 - \sum_{k=2}^{N-1} (-1)^k \varepsilon_k F(k) \tag{15.24}$$

where all the ε's $= \pm 1$ with equal probability and the F's are the Fibonacci numbers of (15.13).

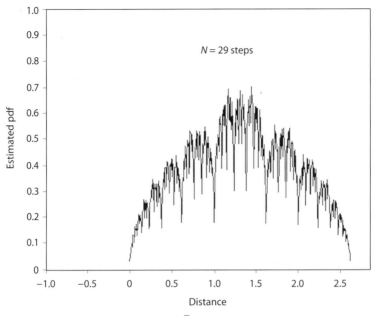

Figure 15.4. Shrinking walk for $s = \frac{(\sqrt{5}-1)}{2}$.

To see how (15.23) and (15.24) work, let's take a look, in two different ways, at golden walks with twenty-one steps. There are $2^{21} = 2{,}097{,}152$ such walks. First, as with Figure 15.4 (golden walks with twenty-nine steps), we can randomly simulate a lot of twenty-one-step walks and from them generate an estimate of the pdf for the location of the endpoints, that is, an *estimate* for $f_{\mathbf{X},g}(x, 21)$. And second, using (15.23) and (15.24), we can *compute* the endpoints of each and every one of the 2^{21} walks and thus arrive at the *exact* $f_{\mathbf{X},g}(x, 21)$. This second approach does not use a random number generator. The upper plot of Figure 15.5 shows the result of 2,097,152 simulated walks (this allows for the possibility of each and every one of the possible twenty-one-step walks to be generated). The lower plot in Figure 15.5 shows the result of directly computing the endpoint of each and every possible walk. The spatial resolution in both plots is 10^{-2}, that is, there are 524 subintervals with width 0.01 in the interval $\left(-\frac{1}{1-g}, \frac{1}{1-g}\right)$. The two plots are, to the eye at least, identical.

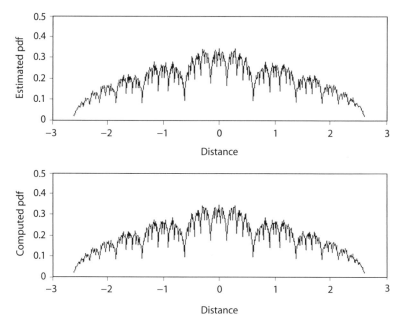

Figure 15.5. Simulated and computed golden walks.

CP. P15.1:

In the discussion on geometric walks, we've looked only at *shrinking* walks. What does the pdf of a *stretched* walk look like? As a start to answering this, you are to write a Monte Carlo simulation of a stretched random walk with $s = 1.01$. That is, the step sizes are $1, 1.01, 1.01^2, 1.01^3, \ldots$. Write the code for a simulation that estimates the pdf of the endpoint location of a twenty-one-step walk, with left- and right-moving steps equally likely. Plot the pdf with a spatial resolution of 10^{-2}. That is, since the maximum distance from the origin (the walk's starting point) is

$$1 + s + s^2 + s^3 + \cdots + s^{20} = \frac{s^{21} - 1}{s - 1} = \frac{(1.01)^{21} - 1}{1.01 - 1} = 23.24,$$

then the required number of subintervals (into which the individual endpoints of the simulated walks are placed) is $2\frac{23.24}{0.01} = 4{,}648$. It's always a good idea to have some idea of the nature of a solution before arriving at that solution, just as a partial check. Can you see, right now, that the value of the pdf at $x = 0$ must be zero? That is, it is impossible for this stretched walk to ever return to its starting point. Make sure your simulation agrees with this. If it does, that does not mean the simulation is correct, but if it doesn't agree, that does mean the simulation is incorrect!

Notes and References

1. Figure 15.1 was created by a MATLAB program I wrote for my book, *Duelling Idiots and Other Probability Puzzlers* (Princeton, N.J.: Princeton University Press, 2000 [corrected ed. 2002]). It's called **brownian.m**, and you can find the code on p. 210 of that book or on the press's math Web site (the code is available for download, for free) at http://press.princeton.edu/titles/6914.html. Superficially similar random behavior had been observed long before Brown's time. A half-century before Jesus, the Roman philosopher Lucretius wrote the following in Book 2 of his poem, *De rerum natura* (On the Nature of Things):

> when the sun's light and his rays penetrate and spread through a dark room: you will see many minute specks mingling in many ways throughout the void in the light of the rays, and as it were in everlasting conflict troops struggling, fighting, battling without any pause, driven about with frequent meetings and partings....

Some writers have claimed this is an early description of Brownian motion, but I suspect it has more to do with sunlight unequally heating the air and the resulting *macroscopic* thermal air currents rather than with the *microscopic*, molecular origin of true Brownian motion.

2. You can read, in English, these beautiful papers by Albert Einstein in *Investigations on the Theory of the Brownian Movement* (New York: Dover, 1956 [originally published in 1926]).

3. See, for example, my book, *When Least Is Best* (Princeton, N.J.: Princeton University Press, 2004 [corrected ed. 2007], pp. 40–45).

4. An asymptotic expansion has the property that while the expansion has an unbounded *absolute error, its relative error* approaches zero. That is, if $E(n)$ is an asymptotic expansion for some function $f(n)$, then

$$\lim_{n \to \infty} |E(n) - f(n)| = \infty$$

but
$$\lim_{n \to \infty} \frac{|E(n) - f(n)|}{f(n)} = 0.$$

For $f(n) = n!$, the decrease in the relative error of Stirling's approximation, as n increases, is remarkably fast. The approximations for $1! = 1$, $2! = 2$, and $5! = 120$ are 0.9221, 1.919, and 118.019, respectively (the percentage errors are 8%, 4%, and 2%, respectively), while for $10! = 3{,}628{,}800$ the approximation is 3,598,600 (a relative error of just 0.8%). For more on this, see Ian Tweddle, "Approximating n! Historical Origins and Error Analysis" (*American Journal of Physics*, June 1984, pp. 487–488), as well as the essay "Stirling's formula!" in the book by N. David Mermin, *Boojums All the Way Through: Commuicating Science in a Prosaic Age* (Cambridge: Cambridge University Press, 1990).

5. This is a mathematical definition. The diffusion coefficient is physically a simple function of two universal constants, the *gas constant* and *Avogadro's number* (both well-known to students of high school chemistry), the absolute temperature, and the viscosity of the suspension medium. While known today as a theoretician, Einstein was not just a symbol pusher; the physics of D appears in much detail in his papers on Brownian motion (see note 2).

6. If $f_{\mathbf{X}}(x, t)$ is a pdf, then it must be true, for any given $t > 0$, that $\int_{-\infty}^{\infty} f_{\mathbf{X}}(x, t)\, dx = 1$, since this is the *certain* probability that the particle is *somewhere* on the x-axis. And it certainly must be somewhere! That is, it must be true that $\int_{-\infty}^{\infty} e^{-x^2/4Dt}\, dx = 2\sqrt{\pi Dt}$. This is of course a purely mathematical assertion (with a clear physical interpretation), and so the question arises: can we mathematically prove that this is so? I still have a strong memory of discussing this very integral as a first-year graduate student at Caltech (1962–1963) with a fellow student in a course in which it came up in lecture one day. The professor just assumed everybody knew the value of the integral, and when I expressed surprise after class at that assumption, my fellow student (who, unlike me, had done his undergraduate work at Caltech), expressed surprise at my surprise. I was told every Techie alive knew that particular integral by the end of their freshman year. While I had no reason to doubt that claim, I soon realized that the main goal of my classmate was to wage a bit of psychological warfare on somebody who had come to Caltech from someplace else. Still, I also quickly realized that in this matter he was right—this integral *is* something every undergraduate in physics, engineering, and mathematics should indeed know how to do. Here's how. Write the integral

$$I = \int_{-\infty}^{\infty} e^{-x^2/4Dt}\, dx.$$

Since x is just a dummy integration variable, it is also true that

$$I = \int_{-\infty}^{\infty} e^{-y^2/4Dt}\, dy.$$

Thus, multiplying these two expression together, we get

$$I^2 = \int_{-\infty}^{\infty} e^{-x^2/4Dt}\, dx \int_{-\infty}^{\infty} e^{-y^2/4Dt}\, dy = \int_{-\infty}^{\infty}\int_{-\infty}^{\infty} e^{-(x^2+y^2)/4Dt}\, dx\, dy.$$

That is, I^2 is the integral of the function $e^{-(x^2+y^2)/4Dt}$ over the entire infinite xy plane, where $dxdy$ is the differential area in rectangular coordinates. Now, change variables to polar coordinates r and θ, where of course $x^2 + y^2 = r^2$ and the differential area is $r\,dr\,d\theta$. To cover the infinite plane we have $0 \leq r < \infty$ and $0 \leq \theta < 2\pi$. (For how to make coordinate transformations like this in general, see any good advanced calculus book and look in its index for a discussion of *Jacobian transformations*.) So,

$$I^2 = \int_0^{2\pi}\int_0^{\infty} e^{-r^2/4Dt}\, r\, dr\, d\theta = 2\pi \int_0^{\infty} re^{-r^2/4Dt}\, dr = 2\pi\left(-2Dt e^{-r^2/4Dt}\right)\big|_0^{\infty} = 4\pi Dt.$$

Thus, $I = 2\sqrt{\pi Dt}$, and the assertion is indeed correct. Is that slick, or what?

7. The density of (15.7) is perfectly well defined for all $t > 0$, but *at $t = 0$* it does seem to behave badly. Since the particle *is* at $x = 0$ *at $t = 0$* (by definition, with absolute certainty), then we must have

$$f_{\mathbf{X}}(x, 0) = \delta(x)$$

where $\delta(x)$ is Dirac's famous *impulse* or *delta* function, often written by engineers and physicists as

$$\delta(x) = \begin{cases} \infty, x = 0 \\ 0, otherwise \end{cases}$$

such that (see note 6 again)

$$\int_{-\infty}^{\infty} \delta(x)\, dx = 1.$$

Mathematicians are generally appalled at this sort of "engineer's math," but I believe it is okay to write

$$\lim_{t \to 0} \frac{1}{2\sqrt{\pi Dt}} e^{-x^2/4Dt} = \delta(x)$$

and then to talk of how $\delta(x)$ behaves *inside* an integral. You can find much more on Dirac's impulse in my book, *Dr. Euler's Fabulous Formula* (Princeton, N.J.: Princeton University Press, 2006, pp. 188–206).

8. A. C. de la Torre et al., "Random Walk With An Exponentially Varying Step" (*Physical Review E*, December 2000, pp. 7748–7754).

9. E. Ben-Naim, S. Redner, and D. ben-Avraham, "Bimodal Diffusion in Power-law Shear Flows" (*Physical Review A*, May 15, 1992, pp. 7207–7213).

10. P. L. Krapivsky and S. Redner, "Random Walks With Shrinking Steps" (*American Journal of Physics*, May 2004, pp. 591–598). This paper (which repeats much of the mathematics in the earlier paper I cited in note 8), while totally mathematical in nature, does mention another *physics* appearance of a variable-step random walk (in spectroscopy), in which the size of the nth step is $1/n^3$. The maximum distance such a walk can be found from the origin is then $\sum_{n=1}^{\infty} \frac{1}{n^3} = \zeta(3)$, which gives us yet another appearance of the zeta function in physics (see Discussion 7).

11. See my book, *An Imaginary Tale: The Story of $\sqrt{-1}$* (Princeton, N.J.: Princeton University Press, 1998, [corrected ed. 2007], p. 249).

12. Perhaps almost as amazing is that it wasn't until four centuries after Fibonacci that the French mathematician Albert Girard (1595–1632) first wrote the definition of the Fibonacci sequence in the form given by (15.9). And it took two more centuries for the French mathematician Jacques Binet (1786–1856) to discover (in 1834) the solution to (15.9) given in (15.13). There is evidence, however, that (15.13) was known a century earlier to the French-born mathematician Abraham de Moivre (1667–1754), who did all his great mathematical work after emigrating to England in 1688.

Nearest Neighbors

> Two cannibals are eating a badly cooked clown when one of them turns to the other, a frown on his face, and asks, "Does this taste just a little bit funny to you?"
> — Tasteless joke told by a mathematician to a physicist friend, after being driven slightly goofy by spending fifty fruitless years trying to calculate $\zeta(3)$

16.1 Cannibals Can Be Fun!

Well, with an opening like the one above you may be wondering if your author has himself been driven around the bend with all the random walks and MATLAB coding we've been doing. Actually, there is one more random walk I want to show you, with a totally unexpected application to physics, but I'll defer that until the next discussion. Here we'll take a break from random walks but still keep our feet in probability (and just a bit of MATLAB, too). Discussing the nearest neighbor problem in terms of cannibals is simply for fun, of course, but similar mathematical problems regularly occur in physics, for example, in astrophysics when discussing the distribution of stars in space, and I'll say just a bit more on that in the brief final section of this discussion. For now, though, back to cannibals.

Imagine $N+1$ cannibals who are randomly and independently scattered about a circular island with radius R. (What I mean by *random* is just this: when each cannibal is positioned on the island, he or she is as likely to be in the middle of one little patch of area as in any other little patch of area *of the same size*.) One of the cannibals happens to be at the exact center of the island, and I'll name him H (for *hungry*). H is a trained mathematician (culinary preferences having, of course, nothing to do with intellectual capability) who is interested in computing the average distance separating himself from the nearest of his N neighbors, whom of course H would like to have "join him for lunch." To calculate that average distance, H decides to first find the probability density function (pdf) of the random variable Z, which represents the distance from H to his nearest neighbor (Z's values are in the interval 0 to R). If we call this pdf $f_Z(z)$, then the average value of Z (also called the *expected value* of Z and written as E(Z)) is given by

$$E(Z) = \int_0^R z f_Z(z)\, dz, \qquad (16.1)$$

a formula derived in any elementary textbook on probability theory. Remember what $f_Z(z)$ means:

$$\int_a^b f_Z(z)\, dz = \text{probability} \quad a \leq Z \leq b.$$

Here's how to easily derive $f_Z(z)$. Imagine a circle of radius z centered on H. The probability that any individual one of H's neighbors is inside that circle (that is, the probability that that particular neighbor is no more than distance z from H)—because of the above area interpretation of what *random* means—is given by

$$\frac{\pi z^2}{\pi R^2} = \frac{z^2}{R^2}, \quad 0 \leq z \leq R.$$

It follows that the probability that that neighbor is *more* than distance z from H is

$$1 - \frac{z^2}{R^2}.$$

The probability, then, that *none* of H's N neighbors is within distance z of H (that is, *all* of H's neighbors are *more* than distance z away) is given (because of the given independence of the cannibals' locations) by

$$\left(1 - \frac{z^2}{R^2}\right)^N. \qquad (16.2)$$

This is, in other words, the probability that H's nearest neighbor is more than distance z from H. Since

$$\int_0^z f_{\mathsf{Z}}(u)\,du = \text{probability H's nearest neighbor is within distance } z \text{ of H,}$$

then

$$1 - \int_0^z f_{\mathsf{Z}}(u)\,du = \text{probability H's nearest neighbor is}$$

not within distance z of H,

or, from (16.2),

$$1 - \int_0^z f_{\mathsf{Z}}(u)\,du = \left(1 - \frac{z^2}{R^2}\right)^N.$$

Thus,

$$F_{\mathsf{Z}}(z) = \int_0^z f_{\mathsf{Z}}(u)\,du = 1 - \left(1 - \frac{z^2}{R^2}\right)^N,$$

where $F_{\mathsf{Z}}(z) = P(\mathsf{z} \le z)$ is the distribution function of z. To solve for the density $f_{\mathsf{Z}}(z)$ we differentiate the distribution[1] with respect to z to get

$$f_{\mathsf{Z}}(z) = \frac{d}{dz} F_{\mathsf{Z}}(z) = -\frac{d}{dz}\left(1 - \frac{z^2}{R^2}\right)^N = -N\left(1 - \frac{z^2}{R^2}\right)^{N-1}\left(\frac{-2z}{R^2}\right),$$

or

$$f_{\mathsf{Z}}(z) = \frac{2zN}{R^2}\left(1 - \frac{z^2}{R^2}\right)^{N-1}, \quad 0 \le z \le R. \qquad (16.3)$$

Figure 16.1 shows what the pdfs for some selected values of N look like for the case of $R = 1$.

Putting (16.3) into (16.1),

$$E(\mathsf{Z}) = \frac{2N}{R^2} \int_0^R z^2 \left(1 - \frac{z^2}{R^2}\right)^{N-1} dz. \qquad (16.4)$$

We can perform the integration as follows, with the aid of the binomial theorem. From high school algebra we have the wonderful identity

$$(x+y)^n = \sum_{k=0}^n \binom{n}{k} x^{n-k} y^k,$$

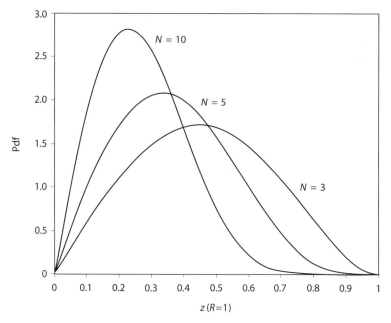

Figure 16.1. Some nearest neighbor pdfs.

or, if we set $x = 1$, $y = -z^2/R^2$, and $n = N - 1$, we can write

$$\left(1 - \frac{z^2}{R^2}\right)^{N-1} = \sum_{k=0}^{N-1} \binom{N-1}{k} \left(-\frac{z^2}{R^2}\right)^k.$$

Thus,

$$z^2 \left(1 - \frac{z^2}{R^2}\right)^{N-1} = R^2 \frac{z^2}{R^2} \sum_{k=0}^{N-1} \binom{N-1}{k} (-1)^k \left(\frac{z^2}{R^2}\right)^k,$$

or

$$z^2 \left(1 - \frac{z^2}{R^2}\right)^{N-1} = R^2 \sum_{k=0}^{N-1} \binom{N-1}{k} (-1)^k \left(\frac{z^2}{R^2}\right)^{k+1}. \quad (16.5)$$

Inserting (16.5) into (16.4),

$$E(\mathsf{Z}) = \frac{2N}{R^2} R^2 \sum_{k=0}^{N-1} \binom{N-1}{k} (-1)^k \int_0^R \left(\frac{z^2}{R^2}\right)^{k+1} dz,$$

or, as

$$\int_0^R \left(\frac{z^2}{R^2}\right)^{k+1} dz = \frac{1}{R^{2k+2}} \int_0^R z^{2k+2} dz = \frac{1}{R^{2k+2}} \left(\frac{z^{2k+3}}{2k+3}\right)\Big|_0^R = \frac{R}{2k+3},$$

then

$$E(Z) = 2NR \sum_{k=0}^{N-1} \binom{N-1}{k} \frac{(-1)^k}{2k+3}. \quad (16.6)$$

We can easily evaluate (16.6), either by hand or by writing a simple computer routine. For example, for $N = 1$ we have

$$E(Z) = 2R \sum_{k=0}^{0} \binom{0}{k} \frac{(-1)^k}{2k+3} = 2R \binom{0}{0} \frac{1}{3} = \frac{2}{3} R,$$

a result that sometimes surprises people who expect it to be $\frac{1}{2} R$. And for $N = 2$ we have

$$E(Z) = 4R \sum_{k=0}^{1} \binom{1}{k} \frac{(-1)^k}{2k+3} = 4R \left[\binom{1}{0} \frac{1}{3} - \binom{1}{1} \frac{1}{5} \right]$$

$$= 4R \left[\frac{5}{15} - \frac{3}{15} \right] = \frac{8}{15} R.$$

The following table shows the behavior of $E(Z)$ as a function of N; $E(Z)$ monotonically decreases with increasing N, as you might expect, but perhaps not as fast as you might think before doing this analysis.

We can "experimentally" check these calculations by performing a Monte Carlo computer simulation. That is, let's perform, say 100,000 times, the random placement of N cannibals inside a circle followed by the determination of the distance from H—at the circle's center—of that cannibal closest to H. An average of those 100,000 minimum values ought to give us a pretty good approximation to $E(Z)$, for a given N. To write the code that does this, however, requires a procedure that randomly picks N points uniformly distributed over the circle. One easy way to do this is to assume (with no loss of generality) that $R = 1$, and to then generate N random numbers uniformly distributed from 0 to 1, *and then to take their square roots*.[2] The following very simple MATLAB code near.m does the job, where all the commands should

N	E(Z)
1	$0.6667R$
2	$0.5333R$
3	$0.4571R$
4	$0.4063R$
5	$0.3694R$
6	$0.3410R$
7	$0.3183R$
8	$0.2995R$
9	$0.2838R$
10	$0.2703R$
20	$0.1945R$

be transparent with the following explanations: line 02 creates a row vector d with 100,000 elements; line 05 creates a row vector r of N random numbers uniformly distributed from 0 to 1 (and then line 06 replaces that vector with a vector whose elements are the square roots of the original vector's elements; line 07 stores the *smallest* element in the square root vector r in the current element (index j) of the d vector; line 10 sums all the elements of the d vector and divides by 100,000, that is, computes the average of the values stored in the elements of the d vector.

```
                near.m
        01  N=input('How many cannibals?');
        02  d=zeros(1,100000);
        03  for j=1:100000
        04      for k=1:N
        05          r=rand(1,N);
        06          r=sqrt(r);
        07          d(j)=min(r);
        08      end
        09  end
        10  sum(d)/100000
```

When run, near.m produced the following table of estimates for $E(Z)$ versus N.

N	$E(Z)$
1	$0.6676R$
2	$0.5332R$
3	$0.4565R$
4	$0.4059R$
5	$0.3698R$
6	$0.3419R$
7	$0.3175R$
8	$0.2991R$
9	$0.2836R$
10	$0.2697R$
20	$0.195R$

Comparing this table with the previous one based on (16.6), you can see that the agreement between theory and "experiment" is pretty good.

16.2 Neighbors Beyond the Nearest

What of H's second nearest neighbor, or his third one, or, indeed, his ith nearest neighbor? If, for example, the second nearest neighbor is a tastier possibility than the nearest neighbor, and if that second nearest neighbor isn't too much more distant, maybe it would be worth the extra travel distance! To work out the probability density functions for all these cases turns out, remarkably, to not be difficult at all (although perhaps just a bit tricky). This more general question should, of course, contain our earlier results as the special case for $i = 1$.

Let $Q(z)$ be the probability an individual cannibal is within distance z of H, which we know from the previous discussion is given by

$$Q(z) = \frac{z^2}{R^2}, \quad 0 \le z \le R.$$

Then, the probability that *exactly* k of his N neighbor cannibals are within distance z of H is given by the binomial probability law as

$$P_k(z) = \binom{N}{k} Q^k(z) [1 - Q(z)]^{N-k}.$$

Now, if the ith nearest neighbor is within distance z of H, then i *or more* neighbors are within distance z of H (if the ith nearest neighbor is within distance z, then $i - 1$ *even nearer* neighbors certainly are too, as well as perhaps one or more additional neighbors, all the way up to possibly *all* the neighbors). Thus, the total probability that the ith nearest neighbor is within distance z of H is given by

$$NN_i(z) = \sum_{k=i}^{N} P_k(z) = \sum_{k=i}^{N} \binom{N}{k} Q^k(z) [1 - Q(z)]^{N-k}.$$

Clearly, $NN_i(z)$ is a distribution function, that is, $NN_i(z) =$ Prob(distance from H of his ith nearest neighbor $\leq z$), and so the pdf of the distance of H's ith nearest neighbor is

$$nn_i(z) = \frac{d}{dz} NN_i(z).$$

Thus,

$$nn_i(z) = \sum_{k=i}^{N} \binom{N}{k} \left\{ k Q^{k-1} \frac{dQ}{dz} (1-Q)^{N-k} - Q^k (N-k)(1-Q)^{N-k-1} \frac{dQ}{dz} \right\}$$

$$= \frac{dQ}{dz} \left[\sum_{k=i}^{N} k \binom{N}{k} Q^{k-1} (1-Q)^{N-k} \right.$$

$$\left. - \sum_{k=i}^{N} (N-k) \binom{N}{k} Q^k (1-Q)^{N-k-1} \right]$$

or, explicitly writing out the first term of the first summation,

$$nn_i(z) = \frac{dQ}{dz} \left[i \binom{N}{i} Q^{i-1} (1-Q)^{N-i} + \sum_{k=i+1}^{N} k \binom{N}{k} Q^{k-1} (1-Q)^{N-k} - \sum_{k=i}^{N} (N-k) \binom{N}{k} Q^k (1-Q)^{N-k-1} \right].$$

(16.7)

In the second summation of (16.7), when $k = N$ the final term is zero, and so

$$nn_i(z) = \frac{dQ}{dz}\left[\begin{array}{l}i\binom{N}{i}Q^{i-1}(1-Q)^{N-i} + \sum_{k=i+1}^{N} k\binom{N}{k}Q^{k-1}(1-Q)^{N-k} \\ - \sum_{k=i}^{N-1}(N-k)\binom{N}{k}Q^k(1-Q)^{N-k-1}\end{array}\right].$$
(16.8)

In the first summation of (16.8) change index to $j = k-1$, that is, $k = j+1$. Then,

$$\sum_{k=i+1}^{N} k\binom{N}{k}Q^{k-1}(1-Q)^{N-k} = \sum_{j=i}^{N-1}(j+1)\binom{N}{j+1}Q^j(1-Q)^{N-j-1}.$$

Now, notice that

$$(j+1)\binom{N}{j+1} = (j+1)\frac{N!}{(j+1)!(N-j-1)!} = \frac{N!}{j!(N-j-1)!}$$

$$= (N-j)\frac{N!}{j!(N-j)!} = (N-j)\binom{N}{j}.$$

So,

$$\sum_{k=i+1}^{N} k\binom{N}{k}Q^{k-1}(1-Q)^{N-k} = \sum_{j=i}^{N-1}(N-j)\binom{N}{j}Q^j(1-Q)^{N-j-1},$$

but this is precisely the second summation in (16.8)! That is, the two sums in (16.8) cancel—this is, I think, a fantastic gift from the number gods—and we are left with the astonishingly simple result that the pdf of the random variable for the distance to H's ith nearest neighbor is

$$nn_i(z) = i\binom{N}{i}Q^{i-1}(1-Q)^{N-i}\frac{dQ}{dz},$$

or, since $Q(z) = z^2/R^2$,

$$nn_i(z) = i\binom{N}{i}\left(\frac{z^2}{R^2}\right)^{i-1}\left(1-\frac{z^2}{R^2}\right)^{N-i}\left(\frac{2z}{R^2}\right), \quad 0 \leq z \leq R. \quad (16.9)$$

The pdfs in (16.9) are *near* neighbor pdfs. If $i = 1$, then we are back to our original *nearest* neighbor pdf, and (16.9) reduces to

$$nn_i(z) = \frac{2zN}{R^2}\left(1-\frac{z^2}{R^2}\right)^{N-1},$$

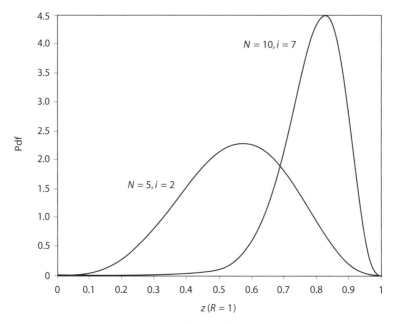

Figure 16.2. Two more nearest neighbor pdfs.

which is (as it had *better* be!) identical to (16.3), that is, $nn_1(z) = f_Z(z)$. Figure 16.2 shows the pdf curves for $i = 2$ and $N = 5$, and $i = 7$ and $N = 10$.

16.3 What Happens When We Have Lots of Cannibals

Imagine now that N is "large," but that we've allowed the island to also grow in size so that the *area density* of cannibals (number of cannibals per unit area) remains constant. That is, while N and R are increasing, the value of

$$\rho = \frac{N}{\pi R^2} \quad (16.10)$$

remains fixed. (I'll elaborate soon on just what "N is large" means.) The nearest neighbor pdf of (16.3) then becomes, with $R^2 = N/\rho\pi$,

$$f_Z(z) = 2\pi \rho z \left(1 - \frac{z^2 \rho \pi}{N}\right)^{N-1},$$

or, as $N \to \infty$ (recall Professor Grunderfunk's challenge problem from the Preface),

$$f_Z(z) = 2\pi\rho z e^{-\rho\pi z^2}. \tag{16.11}$$

Putting (16.11) into (16.1), we have for "large" N (and so of course $R \to \infty$) that

$$E(Z) = \int_0^\infty z f_Z(z)\,dz = 2\pi\rho \int_0^\infty z^2 e^{-\rho\pi z^2}\,dz.$$

From integral tables we have

$$\int_0^\infty x^m e^{-ax^2}\,dx = \frac{\Gamma\left(\frac{m+1}{2}\right)}{2a^{\left(\frac{m+1}{2}\right)}},$$

where Γ is the gamma function. For our problem $m = 2$ and $a = \rho\pi$, and so for "large" N with a given density of cannibals (ρ), we get the amazingly simple result

$$E(Z) = 2\pi\rho \frac{\Gamma\left(\frac{3}{2}\right)}{2(\rho\pi)^{3/2}} = \frac{\frac{1}{2}\Gamma\left(\frac{1}{2}\right)}{\sqrt{\rho\pi}} = \frac{\frac{1}{2}\sqrt{\pi}}{\sqrt{\rho\pi}} = \frac{1}{2\sqrt{\rho}}, \tag{16.12}$$

where I've used both the gamma function's recursive property of $\Gamma(x+1) = x\Gamma(x)$ and the particular value[3] $\Gamma\left(\frac{1}{2}\right) = \sqrt{\pi}$.

To see what N "large" means, let's just substitute (16.10) into (16.12) without worrying about the size of N. That is, let's write

$$E(Z) = \frac{1}{2\sqrt{\frac{N}{\pi R^2}}} = \frac{1}{2}\sqrt{\frac{\pi}{N}} R. \tag{16.13}$$

If we now compute $E(Z)$ from (16.13) then we can see how large N has to be to give values close to the correct values given in our first table calculated from (16.6).

By the time we get to $N = 10$, $E(Z)$ from (16.13) is less than 4% different from the exact value of $E(Z)$ from (16.6); that is, N doesn't really have to be very big at all to be "large."

N	$E(Z)$
1	$0.8862R$
2	$0.6267R$
3	$0.5117R$
4	$0.4431R$
5	$0.3963R$
6	$0.3618R$
7	$0.3350R$
8	$0.3133R$
9	$0.2954R$
10	$0.2802R$
20	$0.1982R$

16.4 Serious Physics

We can extend our cannibal problem to one of far more physical interest[4] if we move off our two-dimensional island surface and into three-dimensional space. Suppose now that we have N stars (considered as point objects) scattered uniformly throughout a spherical region of space with radius R. At the center of this region sits the $N+1$th star. Our question now is, what is the expected distance from the center star to its nearest neighbor star? The analysis goes through pretty much as it did with the cannibal problem—and that, of course, is why I did the cannibal analysis in the first place!

Let $f_Z(z)$ be the pdf of the random variable Z, which represents the distance between the center star and its nearest neighbor. We next imagine a sphere of radius z centered on the center star. The probability that any *individual one* of the N neighbor stars surrounding the center star is inside that sphere, that is, is within distance z of the center star, is

$$\frac{\frac{4}{3}\pi z^3}{\frac{4}{3}\pi R^3} = \frac{z^3}{R^3}, \quad 0 \leq z \leq R.$$

The distribution function of Z is $F_Z(z) = P(Z \leq z) = 1 - P(Z > z)$, where $P(Z > z)$ is the probability *all* N neighbor stars are more than distance

z from the center star. Since the probability any *individual one* of the N neighbor stars is more than distance z from the center star is $1 - \frac{z^3}{R^3}$, then $\left(1 - \frac{z^3}{R^3}\right)^N$ is the probability all N neighbor stars are more than distance z from the center star, and so $P(\mathsf{Z} > z) = \left(1 - \frac{z^3}{R^3}\right)^N$, and therefore

$$F_{\mathsf{Z}}(z) = \int_0^z f_{\mathsf{Z}}(u)\,du = 1 - \left(1 - \frac{z^3}{R^3}\right)^N.$$

At this point you should be able to complete the analysis and thus answer the following challenge problem question:

CP. P16.1:

Find an expression for $E(\mathsf{Z})$ for our star problem, and numerically evaluate $E(\mathsf{Z})$ for $N = 1$ to 10. Also, try your hand at:

CP. P16.2:

What is the probability that the *most distant* star from the center star is no closer than $0.9R$, as a function of N? Note that this probability should increase and approach one as N increases; does your expression have that behavior? How big does N have to be for this probability to be greater than 0.99?

Notes and References

1. See note 6 in Discussion 14.
2. For a complete discussion of this procedure for generating the radial distances of points uniformly distributed over a circle, see my book, *Digital Dice* (Princeton, N.J.: Princeton University Press, 2008, pp. 16–18). This procedure gives us only the radial distance of the points from the origin, of course, and says nothing about the *angular* dispersion of the points. But that's okay, because we don't need the angular information for H's problem. H doesn't care about what *direction* it is to his nearest neighbor, only how *far* he has to travel.

3. See note 4 in Discussion 7 for the definition of the gamma function, and my book, *An Imaginary Tale: The Story of $\sqrt{-1}$* (Princeton, N.J.: Princeton University Press, 2007, pp. 175–176, for both the recursive property of $\Gamma(x)$ and the calculation of the function's value at $x = 1/2$).

4. This entire mathematical discussion (except for the cannibals) was motivated by reading the physics papers by A. M. Stoneham, "Distributions of Random Fields in Solids: Contribution of the Nearest Defect" (*Journal of Physics C*, January 20, 1983, pp. 285–293), and M. Berberan Santos, "On the Distribution of the Nearest Neighbor" (*American Journal of Physics*, December 1986, pp. 1139–1141, and October 1987, p. 952).

One Last Random Walk

> Two roads diverged in a yellow wood,
> And sorry I could not travel both . . . .
> Two roads diverged in a wood, and I—
> I took the one less traveled by,
> And that has made all the difference.
> — Robert Frost, "The Road Not Taken" (1915)

17.1 Resistor Mathematics

In this discussion I'll show you a remarkable connection between random walks and electrical resistor networks. It is, at first glance, all too easy to dismiss the mathematics of resistors as trivial, but that would be a very big mistake. To set the stage for the rest of this discussion, then, let me immediately give you an example of the not so obvious ability of resistors to help us understand some very nontrivial mathematics. Consider the inequality

$$\frac{(a+b)(c+d)}{a+b+c+d} \geq \frac{ac}{a+c} + \frac{bd}{b+d}, a,b,c,d > 0.$$

Can you prove that this is so? Also, assuming the inequality is valid, under what conditions does equality hold? Here's how to answer these questions with resistors.

Consider the two resistor networks in Figure 17.1, where R and R' are the equivalent resistances between terminals 1 and 2 and 1' and 2', respectively. Clearly, the two networks are the same except for the

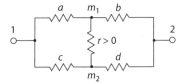

R = resistance between terminals 1 and 2

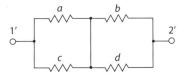

R' = resistance between terminals 1' and 2'

Figure 17.1. Resistors prove an inequality

value of the bridge resistor, but the unprimed network "becomes" the primed one if $r = 0$. Now, let me ask you if you find it "obvious" that $R \geq R'$, that is, if one reduces the value of *any* resistance in a network of resistors (in particular, $r > 0$ changes to $r = 0$), then the equivalent resistance must also decrease? The great Scottish mathematical physicist James Clerk Maxwell (1831–1879) thought so, writing in his legendary 1873 *A Treatise on Electricity and Magnetism* (a work often compared to Newton's *Principia*):

> This principle may be regarded as self-evident, but it may be easily shewn that the value of the expression for the resistance of a system of conductors between two points selected as electrodes, increases as the resistance of each member [individually] of the system increases.[1]

Today this is called Rayleigh's monotonicity law, and, as Maxwell said, it isn't hard to prove. But perhaps it is even more self-evident if we draw an analogy between water flowing through a network of pipes and electrons flowing through a network of resistors. As one famous modern book has expressed this,[2] "We just can't believe that if a water main gets clogged [that] the total rate of flow out of the local reservoir is going to increase." That is, the water flow (analogous to electrical current) will *decrease* because the resistance to the flow of water in the now partially clogged network of pipes has *increased*.

Using the well-known mathematics of resistors in series and in parallel, we can immediately write, for $r = \infty$,

$$R = (a+b) \parallel (c+d) = \frac{(a+b)(c+d)}{a+b+c+d},$$

while for $r = 0$ we have

$$R' = (a \parallel c) + (b \parallel d) = \frac{ac}{a+c} + \frac{bd}{b+d}.$$

By the monotonicity law we must have $R \geq R'$, because when writing our two resistance expressions we went from $r = \infty$ to $r = 0$. So, we instantly have our inequality. That's it!

As for when the inequality becomes equality, suppose that

$$\frac{b}{a+b} = \frac{d}{c+d}.$$

Then the voltages at m_1 and m_2 in Figure 17.1 must be the same (the resistor pairs a and b and c and d are what electrical engineers call a *voltage divider*). Thus, connecting m_1 and m_2 with *any* bridging resistance r will have no effect, since the current in r will be zero, because the voltage drop across r is zero. In particular, shorting m_1 and m_2 together, thereby creating the primed network, will have no effect.

Can you show, *analytically*, that equality results for the given condition, without making any reference to resistors and voltage dividers? (Fair warning: this can be a bit tricky, which is why you'll see this question again, at the end of this discussion, as a challenge problem.) Of course, without the resistor circuits to guide us, I don't think it would be at all evident what supposed condition for equality we should even try.

17.2 Electric Walks

Okay, down to business. Consider an arbitrary network of resistors, as shown in Figure 17.2, with any pair of nodes either not connected or, if connected, connected by a single resistor. (If two connected nodes are joined by multiple resistors in parallel, just replace them with their

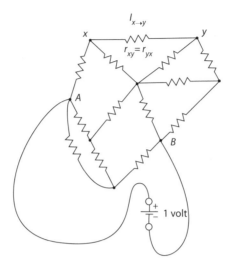

Figure 17.2. An arbitrary resistor network.

parallel equivalent. Resistors in series means, of course, that one or more additional nodes must be included in the set of all nodes.) Two particular nodes, labeled A and B, are connected to the terminals of an ideal (that is, zero internal resistance) one-volt source. If V_x is the voltage at node x, then in particular we write $V_A = 1$ and $V_B = 0$.

Now, consider two nodes, x and y, connected by the resistor r_{xy} (which of course equals r_{yx}). There are, in fact, possibly multiple nodes connected to node x, and so node y is just one of perhaps many (in Figure 17.2 there are three nodes connected to node x). Let's write $N(x)$ as the set of all nodes connected to node x, and since y is a member of that set, I'll write $y \in N(x)$.[3] We can write Kirchhoff's current law for the total current *out* of node x and *into* all the nodes that x is connected to as

$$\sum_{y \in N(x)} I_{x \to y} = 0 = \sum_{y \in N(x)} \frac{V_x - V_y}{r_{xy}} = V_x \sum_{y \in N(x)} \frac{1}{r_{xy}} - \sum_{y \in N(x)} \frac{V_y}{r_{xy}}. \quad (17.1)$$

Next, let's make the definition

$$c_x = \sum_{y \in N(x)} \frac{1}{r_{xy}}, \quad (17.2)$$

where we know, *physically*, that $c_x \neq 0$.[4] This allows us to write

$$0 = V_x c_x - c_x \frac{1}{c_x} \sum_{y \in N(x)} \frac{V_y}{r_{xy}},$$

or

$$0 = c_x \left(V_x - \sum_{y \in N(x)} \frac{V_y}{c_x r_{xy}} \right),$$

and so, finally, since $c_x \neq 0$,

$$V_x = \sum_{y \in N(x)} \frac{V_y}{c_x r_{xy}}, \qquad (17.3)$$

which is the result of a purely *electrical physics* analysis. Let's put (17.3) to one side for the present and start all over again with a second, purely *mathematical* analysis.

I'll start by first asking you to forget, just for now, all about electrical current being electrons moving through conductors because of forces acting on them from the electric fields produced by a voltage source. Now our typical conduction electron is to simply be imagined as a random walker moving node to node to ..., with each node transition to be made according to the following rule:

if an electron is at node x, then on its next move (to any one of the nodes that are connected to x) it will move to node $y \in N(x)$ with probability

$$p_{x \to y} \triangleq \frac{1}{c_x r_{xy}}. \qquad (17.4)$$

The choice of which of the nodes $y \in N(x)$ to move to is purely mathematical in nature, to be made randomly and independently of all the earlier node transitions the electron may have made; notice that now there is no electrical engineering tech talk of fields or forces. All we have are electrons wandering randomly, just like the hiker in Frost's poem, through a wood of nodes. Notice, too, that if we sum the probabilities of (17.4) over all possible y, we get

$$\sum_{y \in N(x)} p_{x \to y} = \sum_{y \in N(x)} \frac{1}{c_x r_{xy}} = \frac{1}{c_x} \sum_{y \in N(x)} \frac{1}{r_{xy}},$$

which becomes, from (17.2),

$$\sum_{y \in N(x)} p_{x \to y} = \frac{1}{c_x} c_x = 1.$$

This is the mathematics saying our wandering electron goes from node x to some other node y with certainty. The fact that all the transition probabilities sum to unity means that the electron always does *move*, and doesn't have a non-zero probability of remaining at node x.

Now, here's where all this is going. Let's define N_x as the probability that our electron, if at node x, will reach node A *before* reaching node B. Then,

N_x is the probability of reaching A before B (given it is at node x), which equals the probability of moving from node x to node y *times* the probability of reaching A before B (starting at y) summed over all possible y.

We can multiply probabilities this way because of the assumed independence of the node-to-node transitions. That is,

$$N_x = \sum_{y \in N(x)} p_{x \to y} N_y,$$

or, using (17.4),

$$N_x = \sum_{y \in N(x)} \frac{N_y}{c_x r_{xy}}. \tag{17.5}$$

Have you noticed that (17.3) and (17.5) are the same if we take $N_x = V_x$? That is, the *probabilities* N_x are numerically equal to the *node voltages* V_x! This probabilistic interpretation of the node voltages makes obvious sense, too, in the two special cases of node x as either A or B. That is,

$$N_A (= V_A) = 1,$$

which says starting at A, our electron will reach A before B with probability one (undeniably true), and

$$N_B (= V_B) = 0,$$

which says starting at B, our electron will reach A before B with probability zero (who would deny that?).

The two equations (17.2) and (17.4), as simple as they appear, are the basis for an incredible new way to "solve" dc resistor networks. To find the voltage at any node of the network, all we have to do is start a random walk at the node of interest, and then "decide" to which connected node to move. The "decision" is made by what amounts to the old children's party game of spin the bottle (which girl to kiss becomes which node to move to), with the node transition probabilities being given by (17.2) and (17.4). The walk continues until it reaches either A or B, with either condition terminating the walk. This is certain to happen eventually, although because of the randomness in the node-to-node transitions, it isn't a priori clear just how long it may take. It is perfectly possible, after all, for our random-walking electron to spend time simply bouncing back and forth between a pair of connected nodes. In practice, however, the entire process runs to completion on a computer *very* quickly. If we do this many times, all the while keeping track of how many walks terminate because they reached A, then we can estimate the probability of reaching A before B if the walk starts from the node of interest. But that probability *is* the voltage of the node of interest.

If we repeat the above for each of the nodes in the network (other than A or B, of course, the voltages of which we know from the start), then we'll have a complete solution to the circuit. If the actual input voltage for *your* circuit is different from one volt, then use the linearity of resistors to simply scale the one-volt solution up (or down) to get the node voltages for your circuit. And remember, at no time have we used the concepts of electric fields, forces, or moving electrons as electric current.

17.3 Monte Carlo Circuit Simulation

It is one thing to say (17.2) and (17.4) allow the Monte Carlo simulation of resistor circuits and quite another to actually produce a computer code that does the job. It's not that it is necessarily difficult to write such a code, just that it's hard to find a *published* code. More common is to find references to such codes, with only their results for particular circuits published. For example, in one paper[5] we are told of a code written in BASIC, specifically for the Wheatstone bridge circuit,[6] that "requires less than a page of code." Alas, the code itself was not presented. The author did say he would send a copy of his code to anyone who asked for it, but that was twenty years ago, and anyway, BASIC isn't the scientific programming language of choice any longer (if it ever was). We are told, however, that using "100 walks gives the exact answer ... to about 10% accuracy and 10,000 walks are needed for 1% precision. The 1% result requires only about a minute on a personal computer." A million walks should give about 0.1% accuracy.

So, here I'll show you how to write a MATLAB code that will solve *any* resistor circuit, not just a specific one. Further, it will use millions of random walks, and yet, even for circuits much more complicated than a Wheatstone bridge, it will require only a few seconds to run. (Of course, my 2007—when I'm writing this—personal computer is almost surely vastly more powerful than any 1990 machine was.) And finally, my MATLAB code will "require [almost] less than a page." Here it is, the code mccube.m. What follows after the code is how the code works.

---

mccube.m

```
01   numberofnodes=8;
02   data(3,:)=[3 1 1 4 1 8 1];
03   data(4,:)=[3 2 1 3 1 5 1];
04   data(5,:)=[3 4 1 1 1 6 1];
05   data(6,:)=[3 2 1 7 1 5 1];
06   data(7,:)=[3 1 1 8 1 6 1];
07   data(8,:)=[3 3 1 7 1 2 1];
08   for currentnode=3:numberofnodes
09       nynodes=data(currentnode,1);
10       cx=0;
11       for loop=3:2:2*nynodes+1
12           cx=cx+1/(data(currentnode,loop));
13       end
14       for loop=3:2:2*nynodes+1
```

(continued)

(continued)

```
15              pxy=1/(cx*data(currentnode,loop));
16              data(currentnode,loop)=pxy;
17          end
18          for loop=5:2:2*nynode+1
19              data(currentnode,loop)=data(currentnode,loop)
                    +data(currentnode,loop-2);
20          end
21          data(currentnode,loop)=1;
22      end
23      V=zeros(1,numberofnodes);
24      V(1)=1000000;V(2)=0;
25      for startnode=3:numberofnodes
26          for walk=1:1000000
27              currentnode=startnode;
28              while currentnode>2
29                  keeplooking=1;
30                  decision=rand;
31                  look=3;
32                  while keeplooking==1
33                      if decision<data(currentnode,look)
34                          keeplooking=0;
35                          currentnode=data(currentnode,look-1);
36                      else
37                          look=look+2;
38                      end
39                  end
40              end
41              if currentnode==1
42                  V(startnode)=V(startnode)+1;
43              end
44          end
45      end
46      for j=1:numberofnodes
47          V(j)/1000000
48      end
```

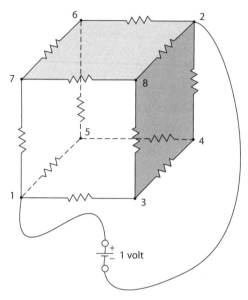

Figure 17.3. A 3-D resistor cube.

The hardest part of writing a circuit code is, I think, the development of a means by which to "tell" the code the *topology* of the circuit. That is, how do we tell a computer which nodes are connected to which nodes, and the values of the connecting resistors? There may be as many different ways to do that as there are analysts, but the method I used is quite simple and direct. I will be imposing the constraint that the nodes of the circuit be numbered sequentially from 1 (because MATLAB doesn't allow zero indexing of vectors and arrays), where node 1 is always the positive terminal of the voltage source (one volt) and node 2 is always the negative terminal of the voltage source (zero volts). Then all the other nodes are numbered, in any arbitrary order you wish, from 3. To be very specific, let's suppose our circuit is a three-dimensional cube of one-ohm resistors (hence the name mccube.m, for *Monte Carlo cube*, although the code will run any other resistor circuit just as well), as shown in Figure 17.3, where I've labeled each node with an integer running from 1 to 8. Our goal is to find the voltages at nodes 3 through 8.

This problem is a classic, a favorite on master's and even PhD oral exams, because electrical engineering and physics professors know

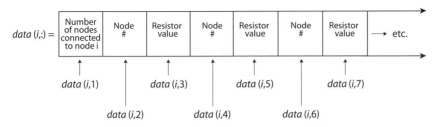

Figure 17.4. A row vector in the topology matrix **data**.

the easy way to analyze it, while most examinees—who are already in a state of nervous shock—immediately attempt to write Kirchhoff's equations for the cube and make a botch of it. It is something (unkind?) professors do because their professors did it to them!

The heart of mccube.m is the matrix **data**. Every node in whatever circuit we are going to study has a row in **data**. The general format of a row is shown in Figure 17.4, where data(i,:) is MATLAB's way of specifying the ith row (node i). The colon as the second index simply means we are talking of the *entire row*, not of an individual element in the row. The very first thing the code needs is the value of the variable numberofnodes (see line 01), which, as you have no doubt guessed, is the number of nodes in the circuit, including the voltage source nodes 1 and 2. In mccube.m this variable receives the value of 8 for the 3-D resistor cube. Then in lines 02 through 07, the rows of **data** are specified in the format of Figure 17.4. For example, line 02 says node 3 connects to three other nodes: to node 1 through a one-ohm resistor, to node 4 through a one-ohm resistor, and to node 8 through a one-ohm resistor. The rows for nodes 1 and 2 are not needed, since we are not going to launch any random walks from those nodes (we already know the voltages at those two nodes).[7]

Lines 08 through 22 then transform the topology matrix **data** into a probability transition matrix, using equations (17.2) and (17.4). What that means is the ith row of **data** will, when those lines of code have done their job, no longer specify the resistances through which node i is connected to other nodes; rather, the elements in **data** where the resistance values were will now contain numbers that determine the probabilities that random walker at node i will move to each of those other nodes. Because of the way I wrote the random walk section of the code (to be explained momentarily), these numbers are not the actual node transition probabilities but rather are *cumulative* probabilities. For example, the new row in data for node 3 will have become

data(3,:)=[3 1 0.3333 4 0.6666 8 1].

In the random walk portion of the code (lines 25 through 45) the variable decision is given a random value from 0 to 1 (line 30) and then the code sequentially uses decision to decide to which node to go to next. If, for example, the walk is currently at node 3 (currentnode=3) then data(3,:) says to go to node 1 if decision<0.3333, but if not then go to node 4 if decision<0.6666, but if not then go to node 8 (decision<1 is *certain* to occur at this point and that will be the case with a priori probability 0.3333 (I'm ignoring any roundoff error here, of course, a concern addressed in particular by line 21).

Line 23 defines the row vector V of length numberofnodes. $V(i)$ will, when the code finishes, be an estimate of the voltage at node i. The random walk algorithm, you'll recall, keeps track of how many times (out of one million walks) a walk starting from node i reaches node 1 before it reaches node 2. These numbers are stored in the vector V. To convert these numbers to probabilities (and, hence, to voltages) we simple divide by one million. That's why $V(1)$ and $V(2)$ are initially set (in line 24) to 1,000,000 and 0, respectively, since we know the voltages at node 1 and at node 2 are one volt and zero volts, respectively.

The random walk portion of the code is line 25 through line 45 and is, I think, pretty transparent; just keep the format of the data probability transition matrix in mind. An individual random walk, beginning at startnode (which itself will run from 3 to numberofnodes— see line 25), continues until line 28 determines a walk has reached either node 1 or node 2 (all other node numbers are of course greater than 2). Lines 41 through 45 determine which it is and, if it is node 1 then V(startnode) is incremented by one. A million such walks are launched from every node from 3 to numberofnodes. Finally, lines 46, 47, and 48 print the node voltage estimates.

So, at last, the big question: how well does mccube.m perform? To answer that, we need the actual node voltage values of the resistor cube to compare with the code's estimates. To get those exact values, let me now show you the easy way I mentioned earlier to analyze the 3-D resistor cube. To start, imagine that rather than a one-volt source connected to nodes 1 and 2, we have adjusted the voltage to be whatever it has to be so that the source *current* is one ampere. If we write $I_{x \to y}$ as the current flowing from node x to node y, as we did in (17.1), then by symmetry the source current splits equally at node 1, and so

$$I_{1 \to 3} = I_{1 \to 7} = I_{1 \to 5} = \frac{1}{3} A.$$

Further, and again by symmetry,

$$I_{3 \to 8} = I_{3 \to 4} = \frac{1}{2} I_{1 \to 3} = \frac{1}{6} A.$$

One can make the same argument for $I_{7 \to 6}$ and $I_{7 \to 8}$, too, that is

$$I_{7 \to 6} = I_{7 \to 8} = \frac{1}{2} I_{1 \to 7} = \frac{1}{6} A,$$

and so, too,

$$I_{5 \to 4} = I_{5 \to 6} = \frac{1}{2} I_{1 \to 5} = \frac{1}{6} A.$$

Now, in particular

$$I_{6 \to 2} = I_{7 \to 6} + I_{5 \to 6} = \frac{1}{6} + \frac{1}{6} = \frac{2}{6} = \frac{1}{3} A.$$

Thus, the voltage drop from node 1 to node 2 is

(drop from 1 to 7) + (drop from 7 to 6) + (drop from 6 to 2)

$$= \frac{1}{3} + \frac{1}{6} + \frac{1}{3} = \frac{5}{6} \text{ volts}$$

because all of the currents are in one ohm resistors.

To summarize: a voltage source of 5/6 volts causes a one ampere source current. So, linearly scaling (resistors are linear) we see that a one-volt source would result in a source current of 6/5 A. Again using the current splitting symmetry argument, we see that there is a current from node 1 to node 7 of

$$\left(\frac{6}{5}\right) \frac{1}{3} = \frac{2}{5} A,$$

which means there is a voltage drop of $2/5 = 0.4$ volts from node 1 to node 7 (and from node 1 to nodes 3 and 5, as well). Thus,

$$V(7) = V(3) = V(5) = 1 - 0.4 = 0.6 \text{ volts}.$$

Further, that $2/5\,A$ current flowing into node 7 splits (by symmetry) evenly, and so there is a current of $1/5\,A$ flowing from node 7 to node 8. This gives an additional voltage drop of $1/5 = 0.2$ volts from node 7 to node 8, and therefore

$$V(8) = 0.6 - 0.2 = 0.4 \text{ volts} = V(6) = V(4).$$

The following table sums all this up, as well as giving mccube.m's results, to four decimal places. The agreement is really pretty good; the execution time was less than six seconds.

Exact node voltage values	mccube.m
$V(3) = 0.6$	0.6004
$V(4) = 0.4$	0.4010
$V(5) = 0.6$	0.5997
$V(6) = 0.4$	0.4012
$V(7) = 0.6$	0.6001
$V(8) = 0.4$	0.4000

As a quick second example, consider the "pinwheel" circuit of Figure 17.5. It is not difficult to show[8] that the node voltage values are

$v(3) = v(4) = \frac{7}{18} = 0.3889$ volts;
$v(5) = v(6) = \frac{1}{6} = 0.1667$ volts;
$v(7) = \frac{1}{9} = 0.1111$ volts.

To input this circuit to the computer simulation code, all that is required is the modification of lines 01 through 06 to read

```
01    numberofnodes=7;
02    data(3,:)=[3 1 1 5 1 2 1];
03    data(4,:)=[3 1 1 6 1 2 1];
```

(continued)

(continued)

```
04    data(5,:)=[3 3 1 7 1 2 1];
05    data(6,:)=[3 4 1 7 1 2 1];
06    data(7,:)=[3 6 1 5 1 2 1];
```

as well as, of course, deleting line 07. In a bit less than three seconds the code produced the node voltage estimates

$v(3) = 0.3889$ volts;
$v(4) = 0.3890$ volts;
$v(5) = 0.1660$ volts;
$v(6) = 0.1661$ volts;
$v(7) = 0.1114$ volts.

Again, we have pretty good agreement with theory.

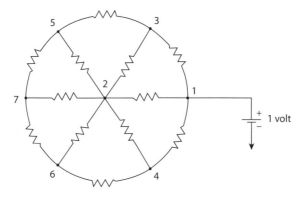

Figure 17.5. A pinwheel of resistors.

17.4 Symmetry, Superposition, and Resistor Circuits

Our "easy" solution of the 3-D resistor cube depended totally on the high degree of symmetry displayed by the cube and its connections to the voltage source. This no doubt accounts for much of the appeal of that problem to analysts, who generally place arguments based on symmetry high on their list of "beautiful" arguments.[9] The concept

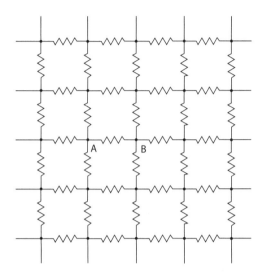

Figure 17.6. An infinite square grid of resistors.

of superposition ranks high on that list as well, and there is another famous resistor circuit problem that requires both concepts. This problem, which has been a favorite of electrical engineering and physics professors since its appearance in 1950,[10] is based on the infinite, two-dimensional square network of one-ohm resistors shown in Figure 17.6. The question is easy to state: what is the resistance between nodes A and B? The answer is certainly less than one ohm, but how do you account *exactly* for the infinity of resistors around the resistor that directly connects A and B?

The "easy" solution to this problem requires three simple, sequential steps.

Step 1: Imagine a one-ampere current source that injects current *into* the grid at node A. The current exits the grid along its "edge at infinity" (use your imagination with this!). By symmetry, the one ampere into A moves away from A by first splitting equally into four 1/4-ampere currents, one of which flows from A to B. This results in a 1/4-volt drop from A to B.

Step 2: Remove the one-ampere current source of Step 1 and hook another one up to the grid so that it injects current *into* the grid at infinity and removes it from the grid at node B. By symmetry, the one ampere removed at B flows into B as four 1/4-ampere currents, one of which is flowing from A to B. This results in a 1/4-volt drop from A to B, with the same polarity as in Step 1. (Important observation: we are, in Steps 1 and 2, treating both A and B as the "center" of the grid, and, since the grid is infinite in extent, we can do that; indeed, *any* node in an *infinite* grid can be thought of as the "center").

Step 3: Finally, leaving the one-ampere current source of Step 2 in place, reconnect the original one-ampere current source of Step 1 that injects current into node A. Then, by superposition, we can *add* the effects of the two currents when individually applied to get their combined effect when both are in place at the same time. That is, we have one ampere *into* A, one ampere *out* of B, *zero* current at infinity, and a total voltage drop of $1/4 + 1/4 = 1/2$ volt from A to B. So, 1/2 volt "causes" one ampere of current (or vice versa, if you wish), and therefore the resistance between A and B is 1/2 ohm. That's it!

This elegant solution did not appear along with the original 1950 statement of the infinite square grid of resistors, but rather in the solution manual to a 1966 physics textbook.[11] But this elegant solution works *only* because of the very high degree of symmetry in the problem. The original 1950 presentation actually solved the far more general problem of calculating the resistance between nodes A and B when A and B could be any pair of nodes in the infinite grid, not just adjacent ones. For example, if A and B are diagonally opposite nodes of one of the squares of the grid—that is, if A's coordinates are (0,0) and B's are (1,1)—then the resistance between A and B was given as $2/\pi$ ohms. If we call $R_{m,n}$ the resistance between nodes (0,0) and (m,n)—and so in the "easy" symmetrical case we analyzed earlier we have $R_{1,0} = \frac{1}{2}$ ohm—then the 1950 book gives the "Knight's move" resistance as $R_{1,2} = \frac{4}{\pi} - \frac{1}{2} = 0.77323\cdots$ ohms. Something of a puzzle still remained after 1950, however, because that solution involved the evaluation of a number of difficult integrations, and no discussion of how they were done was given.[12]

In 1972 Leo Lavatelli (1917–1998), a physicist at the University of Illinois, gave a beautiful derivation[13] of the $2/\pi$ result using only algebra and trigonometry. Thirteen years later the English mathematician Peter Trier (1919–2005) used two-dimensional Fourier series to express $R_{m,n}$ as a *double* integral.[14] Almost a decade later (1994)

Giulio Venezian, a physicist at Southeast Missouri State University, gave an explicit solution for the resistance between any two nodes of an infinite square grid of resistors in terms of one-dimensional integrals, and numerically evaluated them.[15] While Venezian's integrals don't look anything at all like the ones in the 1950 book (evaluated, as I've said, analytically by unexplained means), Venezian's results are in agreement. For example, Venezian's numerical integrations result in $R_{1,2} = 0.773$ ohms. The 1950 book gave the expressions for all $R_{0 \leq m \leq 3, 0 \leq n \leq 3}$, while Venezian's numerical values are for $R_{0 \leq m \leq 10, 0 \leq n \leq 10}$. By physical symmetry, we have $R_{\pm m, \pm n} = R_{m,n}$, $m, n \geq 0$, and $R_{m,n} = R_{n,m}$, which can be mathematically seen in Trier's double integral (see note 14 again).

Five years after Venezian's paper, another paper[16] appeared that, using Venezian's method, extended the original 2-D infinite resistor grid to 3-D and higher-dimensional infinite cubic lattices, as well as giving explicit evaluations of the resulting integrals in *analytical* form, using the powerful symbolic manipulation capability of the computer application *Mathematica*. The authors' code confirmed the known expressions $R_{1,0}$ $(= \frac{1}{2})$, $R_{1,1}$ $(= \frac{2}{\pi})$, and $R_{1,2}$ $(= \frac{4}{\pi} - \frac{1}{2})$; computer-generated expressions for $R_{0 \leq m \leq 5, 0 \leq n \leq 5}$ were given—for example, that $R_{2,4} = 6 - \frac{236}{15\pi} = 0.991924\ldots$ ohms and that $R_{3,5} = \frac{998}{35\pi} - 8 = 1.076379\ldots$ ohms, both in excellent agreement with Venezian's numerical evaluations.

CP. 17.1:

Show, analytically, that

$$\frac{(a+b)(c+d)}{a+b+c+d} = \frac{ac}{a+c} + \frac{bd}{b+d}, \quad a,b,c,d > 0$$

if

$$\frac{b}{a+b} = \frac{d}{c+d}.$$

CP. P17.2:

Consider again the infinite square grid of one-ohm resistors shown in Figure 17.6, but with the resistor joining nodes A and B removed. What then is the resistance between A and B?

CP. P17.3:

Take a look again, if you didn't solve it before, at CP. 3.5.

CP. P17.4:

In Discussion 3 you saw several examples of the analytical difficulties that an infinite number of resistors can cause. Here I'll illustrate for you how just *one* resistor can do the same. If we have a one-ohm resistor carrying current $i(t)$, then the instantaneous power is $p(t) = i^2(t)$, and the total energy dissipated (as heat) by the resistor is then given by $W = \int_{-\infty}^{\infty} p(t)\,dt = \int_{-\infty}^{\infty} i^2(t)\,dt$. And finally, by definition the current $i(t)$ is the rate at which electrical charge moves through the resistor, that is, $i(t) = \frac{dq}{dt}$, and so the total charge that passes through the resistor is $Q = \int_{-\infty}^{\infty} dq = \int_{-\infty}^{\infty} i(t)\,dt$. All this is just standard first-year electrical physics. Now, let's consider a specific $i(t)$. For c a positive constant, define

$$i(t) = \begin{cases} 0, & t < 0 \\ c^{-4/5}, & 0 < t < c \\ 0, & t > 0. \end{cases}$$

That is, $i(t)$ is a finite-valued *pulse* that is zero except for a finite length of time. The total charge transported through the resistor is

$$Q = \int_0^c c^{-4/5}\,dt = c^{-4/5} \cdot c = c^{1/5}.$$

Suppose now that we pick the value of c to be ever smaller, that is, we let $c \to 0$. Then the pulse-like current becomes ever briefer in duration and ever larger in amplitude. At the same time, $\lim_{c \to 0} Q(t) = 0$. And finally, since $i^2(t) = c^{-8/5}$ during the time when the pulse amplitude is non-zero, the dissipated energy $W = \int_0^c c^{-8/5}\,dt = c^{-8/5} \cdot c = c^{-3/5}$. So, $\lim_{c \to 0} W(t) = \infty$, which means the resistor should instantly vaporize because it's dissipating *infinite* energy in *zero* time! But how can that be, since in the limit as $c \to 0$ there is *no charge* transported through the resistor?

Notes and References

1. You can find Maxwell's words on p. 427 of volume 1 of the *Treatise*.
2. The quote is from Peter G. Doyle and J. Laurie Snell, *Random Walks and Electric Networks* (Carus Mathematical Monograph 22; Washington, D.C.: American Mathematical Association of America, 1984, p. 70). This seminal work includes an elegant discussion of Rayleigh's (this is the same Rayleigh who appeared in Discussion 14) monotonicity law—first stated by Rayleigh in 1871—in its Chapter 4. If you look back at the discussion about calculating the input resistance to the infinite circuit of Figure 3.9c, you'll see I've already invoked Rayleigh's monotonicity law in this book without calling it that.
3. Here I am following the notation of Monwhea Jeng, "Random Walks and Effective Resistances on Toroidal and Cylindrical Grids" (*American Journal of Physics*, January 2000, pp. 37–40).
4. This follows from the observation that $c_x = 0$ would require that every r_{xy} in (17.2) be infinite. But that would mean nodes x and node y are really not connected, contrary to assumption.
5. Raymond A. Sorensen, "The Random Walk Method for DC Circuit Analysis" (*American Journal of Physics*, November 1990, pp. 1056–1059).
6. The Wheatstone bridge—named for the English scientist Charles Wheatstone (1802–1875) who, in fact did not invent the circuit and freely admitted so—is a circuit for very accurately measuring the value of an unknown impedance. You can find its theory developed in any good electrical engineering or physics text that discusses electrical circuits.
7. Since data is a matrix, MATLAB requires all the rows of data to have the same length. Because all the nodes in a 3-D resistor cube each connect to the same number of other nodes, this requirement is automatically satisfied by this particular circuit. In circuits where that is not the case, however, all the shorter rows in data must be "padded out" with zeros to have their lengths equal that of the longest row. This keeps MATLAB happy, although the Monte Carlo code never actually looks at those padding zeros.
8. See my paper "A Simple Operator Approach for Analytic Solution of Linear Constant-Coefficient Difference Equations" (*IEEE Transactions on Education*, December 1968, pp. 234–239).
9. Here's a very pretty application of symmetry from pure mathematics. If you look in any good set of math tables you'll find the following definite

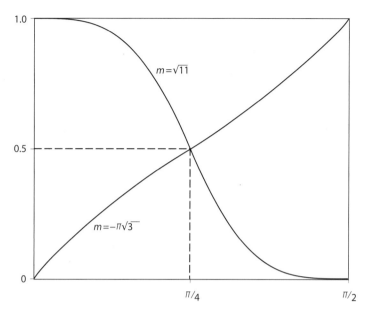

Figure 17.7. Symmetrical integrands.

integral: $\int_0^{\pi/2} \frac{dx}{1+\{\tan(x)\}^m} = \frac{\pi}{4}$. This holds for *any* real value of m, for example, $m = -\pi$, 0 (this special case should be obvious!) and $11,073,642.07$. How in the world, you might fairly ask, can such a wonderful result be established? If you try to prove it using any of the usual methods discussed in freshman calculus, then I think you'll find it a frustrating task. With symmetry, however, it's actually easy. In Figure 17.7 I've plotted the behavior of the integrand over the interval 0 to $\pi/2$ for two values of m picked at random, one negative and one positive ($-\pi/3$ and $\sqrt{11}$), and from those two curves, you might reasonably guess that the plots for *any* value of m will be symmetrical around the point $(\frac{\pi}{4}, \frac{1}{2})$. If that is so, then clearly the integrand divides the area of the box (equal to $\pi/2$) in half, and since the half that is beneath the integrand curve is the integral, then we immediately have our result! The symmetry for any m is easily established by noticing that, if true, it follows that if we write $f(x)$ for the integrand, then what we mean by saying $f(x)$ is symmetrical around $(\frac{\pi}{4}, \frac{1}{2})$ is just that $f(x) = 1 - f(\frac{\pi}{2} - x)$; that is, $f(x) + f(\frac{\pi}{2} - x) = 1$ for every value of x in the integration interval 0 to $\pi/2$. And, in fact, this *is* so because $f(x) + f(\frac{\pi}{2} - x) = \frac{1}{1+\{\tan(x)\}^m} + \frac{1}{1+\{\tan(\frac{\pi}{2}-x)\}^m} = \frac{1}{1+\{\tan(x)\}^m} + \frac{1}{1+\{\cot(x)\}^m}$
$= \frac{1}{1+\{\tan(x)\}^m} + \frac{1}{1+\frac{1}{\{\tan(x)\}^m}} = \frac{1}{1+\{\tan(x)\}^m} + \frac{\{\tan(x)\}^m}{1+\{\tan(x)\}^m} = \frac{1+\{\tan(x)\}^m}{1+\{\tan(x)\}^m} = 1$. That's it!

Symmetry, and the area interpretation of integration, are all we have used to show that $\int_0^{\pi/2} \frac{dx}{1+\{\tan(x)\}^m} = \frac{\pi}{4}$ for *any* real m.

10. B. van der Pol and H. Bremmer, *Operational Calculus* (Cambridge: Cambridge University Press, 1950 [2nd ed. 1955], pp. 371–372). The authors tell us in the book's opening that much of their writing was done during the Nazi occupation of the Netherlands during World War 2. Logic in the midst of madness.

11. E. M. Purcell, in the solutions manual for his book *Electricity and Magnetism* (New York: McGraw-Hill, 1966).

12. The integrals in the 1950 book are

$$R_{m,n} = \frac{1}{\pi} \int_{-1}^{1} \frac{1 - \cos\{m\cos^{-1}(u)\}e^{-|n|\cosh^{-1}(2-u)}}{(1-u)\sqrt{(1+u)(3-u)}} du.$$

I think just about everybody would disagree with the authors in note 10 who claimed this incredible integral can be "simply calculated" for given values of m and n. How they did it remains, I believe, a mystery to this day.

13. Leo Lavatelli, "The Resistive Net and Finite-Difference Equations" (*American Journal of Physics*, September 1972, pp. 1246–1257). An earlier paper extended the symmetry/superposition argument to other infinite, *non*-square resistor grids (triangular and honeycomb, for example) but was still limited to A and B being adjacent nodes; see Francis J. Bartis, "Let's Analyze the Resistance Lattice" (*American Journal of Physics*, April 1967, pp. 354–355).

14. P. E. Trier, "An Electrical Resistance Network and Its Mathematical Undercurrents" (*Bulletin of the Institute of Mathematics and Its Applications*, March/April 1985, pp. 58–60). Trier's double integral is the very pretty

$$R_{m,n} = \frac{1}{\pi^2} \int_0^\pi \int_0^\pi \frac{1 - \cos(mx+ny)}{2 - \{\cos(x)+\cos(y)\}} dx\, dy,$$

which clearly displays the symmetry $R_{m,n} = R_{n,m}$. Compare this expression for $R_{m,n}$ with the one in note 12.

15. Giulio Venezian, "On the Resistance Between Two Points on a Grid" (*American Journal of Physics*, November 1994, pp. 1000–1004).

16. D. Atkinson and F. J. van Steenwijk, "Infinite Resistive Lattices" (*American Journal of Physics*, June 1999, pp. 486–492).

The Big Noise

> Every sound shall end in silence, but the silence never dies.
> — Samuel Miller Hageman, "Silence" (1876)

18.1 An Interesting Textbook Problem

During the autumn quarter of my sophomore year (1959–1960) at Stanford I took Math 130, my first course in differential equations. The assigned textbook was Ralph Palmer Agnew's classic *Differential Equations*. I have kept my copy of Agnew all these past fifty years since, and as I write it is open on my desk in front of me to page 40. On that page is a problem I've always remembered because it described such a novel idea (to me) that I was immediately and forever fascinated by it. This will be our final discussion of the book (final serious physics discussion, that is—but don't overlook the "for fun" bonus discussion at the very end of the book!), and I think Agnew's problem presents just the right note for serving in that exit role. The book will end—metaphorically anyway—with a "big bang." Here's what Professor Agnew wrote:

> The pilot of a supersonic jet airplane wishes to make a big noise at point O by flying around O in a path such that all of the noise he makes is heard simultaneously at O. His *Mach number* is M [named after the famous German physicist Ernst Mach (1838–1916), whose philosophical writings greatly influenced Albert Einstein], which means his speed is Mc where c is the speed of sound, and $M > 1$ because the speed [of the jet airplane] is supersonic. Letting O be the origin of a plane with polar coordinates θ and r, and supposing that the pilot starts at time $t = 0$

from the point $\theta = 0$, $r = a$, and flies around O in the positive direction [that is, counterclockwise for an observer looking downward], find the simplest equation of the path.

There then followed a very brief sketch by Agnew for how to set up the analysis mathematically. I thought the whole idea of "the big noise" to be a hugely entertaining one. And why wouldn't I? I was, after all, a college sophomore!

So, I was greatly interested when, some years later, I came across a paper[1] in the *American Journal of Physics* that solved the problem all over again (no citation of Agnew's or any other textbook, for that matter, was given). Then, a bit later, yet another paper[2] appeared in the same journal to give what the author said was "perhaps a somewhat simpler derivation of the trajectory." The major contribution of that paper was not in the mathematics, however, but rather in its novel illustration of the *physics* of the problem. Instead of a jet airplane flying faster than the speed of sound, the "noise disturbance" was produced by the detonation of one end of a narrow strip of high explosive, with the detonation point then swiftly (7,300 meters/second) propagating along the strip. The strip was bent into the shape of the path that would have the shock front of the detonation, generated at every point along the strip, arrive at point O at the same time. The paper included a sequence of fascinating, high-speed camera pictures demonstrating the reality of the simultaneously converging shock fronts. That is, what that second paper described is a *shaped charge explosive lens* that generated a tremendous *impulsive* force at O, a force produced by focusing the shock waves generated by a traveling explosion occurring over an interval of time. The mathematics of that lens—that is, the curve of the explosive strip—is almost as fascinating as the explosion. I'll continue on from here with Agnew's airplane version, but the explosion math is the same.

18.2 The Polar Equations of the Big-Noise Flight

There are two general initial observations we can make about this problem without doing much mathematics. First, the entire flight, from start to the big noise finish, is of *finite* duration, equal to the time

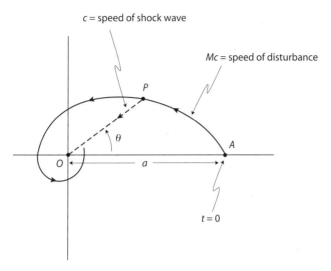

Figure 18.1. A moving disturbance an airplane traveling at speed Mc on and its shockwave traveling at speed c.

it takes the *initial* shock wave (generated at time $t = 0$ at point A in Figure 18.1) to reach O. If we denote that entire flying time by T, then

$$T = \frac{a}{c}. \tag{18.1}$$

Second, the airplane itself arrives at O at time $t = T$, along with all of the shock waves it has generated at each instant of time over the interval $0 \leq t \leq T$ (this includes the one *just* generated at time T). At each such instant the shock wave starts at the airplane's location, leaves the airplane, and then moves inward toward O at speed c. For all the shock waves and the airplane to arrive at O simultaneously, the airplane must be moving radially inward at speed c as well. Since we'll be working in polar coordinates (r, θ) as shown in Figure 18.1, we can then write

$$\frac{dr}{dt} = -c, \tag{18.2}$$

and, since $r(0) = a$, we have

$$r(t) = a - ct. \tag{18.3}$$

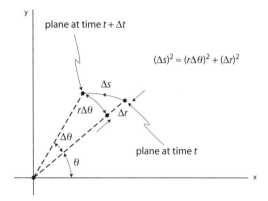

Figure 18.2. The arc-length Δs in polar coordinates.

Okay, that's the easy stuff. Now we are ready to do some serious mathematics.

The path of the airplane, assumed to take place entirely in a plane surface, is shown in Figure 18.1 as some sort of inward spiral. Suppose P is an arbitrary point on that spiral path at time t ($P = A$ at $t = 0$). Let's write t_{AP} as the time for the airplane to travel from A to P, and t_{PO} as the time for the shock wave produced at P to travel from P to O. Our problem, then, is that of determining what the airplane's path is so that

$$W = t_{AP} + t_{PO} \tag{18.4}$$

is a constant, the same for all points P on the spiral path. As I said before, it will be convenient to work in polar coordinates, and so we'll write the coordinates of P as (r,θ), where θ is as shown in Figure 18.1 and r is the length of OP. We have, then,

$$W = \frac{\text{arc-length of } AP}{Mc} + \frac{r}{c}. \tag{18.5}$$

To determine the arc-length of AP, take a look at Figure 18.2. There you'll see the airplane at time t at $(r + \Delta r, \theta)$, moving to $(r, \theta + \Delta\theta)$ at time $t + \Delta t$, along the airplane's inward spiral path through the "small" arc-length Δs. If *all* the delta quantities are "small," then the

Pythagorean theorem tells us that
$$(\Delta s)^2 \approx (r\Delta\theta)^2 + (\Delta r)^2,$$
and, in the limit as the delta quantities go to zero and become differentials, we can write
$$ds = \sqrt{(rd\theta)^2 + (dr)^2} = \sqrt{r^2 + \left(\frac{dr}{d\theta}\right)^2}\, d\theta. \tag{18.6}$$

Since arc-length is $\int ds$, then using (18.2) we can write (18.5) as
$$W = \frac{1}{Mc}\int_0^\theta \sqrt{r^2(u) + \left(\frac{dr}{du}\right)^2}\, du + \frac{1}{c}r, \tag{18.7}$$
where of course $r = r(\theta)$, and the u in (18.6) is simply a dummy variable of integration.

Now, here's the crucial observation: by definition, W is a constant. Thus, W is independent of P (that is, of θ). So, it must be that
$$\frac{dW}{d\theta} = 0.$$
Differentiating the right-hand side of (18.6) with respect to θ (take a look back at CP 5.1 for how to differentiate an integral), and setting the result equal to zero, we arrive at
$$\frac{1}{Mc}\sqrt{r^2 + \left(\frac{dr}{d\theta}\right)^2} + \frac{1}{c}\cdot\frac{dr}{d\theta} = 0,$$
or, with just a bit of algebra and separating variables, we have
$$\frac{dr}{r} = \pm\frac{d\theta}{\sqrt{M^2 - 1}}, \quad M > 1. \tag{18.8}$$

This last result is easy to integrate:
$$\int_a^r \frac{du}{u} = \pm \int_0^\theta \frac{du}{\sqrt{M^2 - 1}},$$
or
$$\ln(r) - \ln(a) = \ln\left(\frac{r}{a}\right) = \pm\frac{\theta}{\sqrt{M^2 - 1}},$$

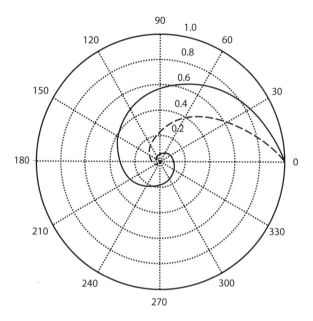

Figure 18.3. Two big noise flight paths.

or, finally,

$$r(\theta) = ae^{-\theta/\sqrt{M^2-1}}, \quad M > 1, \theta \geq 0, \tag{18.9}$$

where I've used the minus sign in the exponent because we know, *physically*, that $\lim_{\theta \to \infty} r(\theta) = 0$. Figure 18.3 shows the "big noise" flight path (*normalized* polar plots of r/a) for $M = 3$ (solid line) and $M = 1.5$ (dotted line).

As Agnew writes of (18.8),

> This path is a spiral. It is easy to imagine that a pilot could follow this spiral as long as r is sufficiently great. It is not so clear that the pilot and airplane could accomplish and survive the operation of making an infinite set of turns around [O to reach O] when $t = a/c$.

What does Agnew mean by this? He doesn't say, leaving it to his readers to explore the problem on their own, but I think it clear that he was thinking of the ever increasing acceleration experienced by the pilot and the airplane as they spiral inward to, and spin ever more rapidly around, O. To calculate this acceleration we will need to know both

r and θ as functions of *time*, not just r as a function of θ as given by (18.8). We already know $r(t)$, from (18.3), so let's now find $\theta(t)$.

Combining (18.3) and (18.8), we have

$$\frac{dr}{a-ct} = -\frac{d\theta}{\sqrt{M^2-1}},$$

or

$$\frac{d\theta}{dt} = -\frac{\sqrt{M^2-1}}{a-ct} \cdot \frac{dr}{dt}.$$

Using (18.2),

$$\frac{d\theta}{dt} = \frac{c}{a-ct}\sqrt{M^2-1} = \frac{1}{\frac{a}{c}-t}\sqrt{M^2-1},$$

or, using (18.1),

$$\frac{d\theta}{dt} = \frac{\sqrt{M^2-1}}{T-t}, \quad T = \frac{a}{c}, \quad M > 1, \quad (18.10)$$

where T is the (finite) duration of the big-noise flight. It is easy to integrate (18.10), with the initial condition of $\theta(0) = 0$:

$$\int_0^{\theta(t)} du = \sqrt{M^2-1} \int_0^t \frac{du}{T-u} = \theta(t),$$

or, with the change of variable $v = T - u$ (and so $dv = -du$),

$$\theta(t) = \sqrt{M^2-1} \int_T^{T-t} \frac{-dv}{v} = \sqrt{M^2-1} \int_{T-t}^T \frac{dv}{v} = \sqrt{M^2-1} \ln\left(\frac{T}{T-t}\right).$$

That is,

$$\theta(t) = \sqrt{M^2-1} \ln\left(\frac{1}{1-\frac{t}{T}}\right), \quad 0 \le t \le T = \frac{a}{c}, \quad M \ge 1. \quad (18.11)$$

Notice that if $M = 1$ (the airplane is not really supersonic) then $\theta(t) = 0$ for all t, that is, the airplane does *not* fly a *spiral path* around O but rather always moves straight in along the horizontal axis *directly toward* O. You can also see now, too, that even though the total flight time is finite for $M > 1$, the number of loops around O is infinite because $\lim_{t \to T} \theta(t) = \infty$ (this explains Agnew's "infinite set of turns" phrase). Now, with (18.3) and (18.11) in hand, we are ready to calculate the acceleration of the pilot and airplane as they fly a big-noise path.

18.3 The Acceleration on a Big-Noise Flight Path

To calculate the acceleration of a moving point in polar coordinates, in general, I'll take advantage of the vector nature of the complex exponential $e^{i\theta(t)}$, where $i = \sqrt{-1}$. The famous expansion formula of Euler says

$$e^{i\theta(t)} = \cos\{\theta(t)\} + i\sin\{\theta(t)\},$$

and the right-hand side of this expression is a vector of unit length making angle $\theta(t)$ at time t with the x-axis, directed outward from the origin. Since our moving point isn't always unit distance from the origin of our coordinate system (O in the big noise problem), we write the so-called *position vector* of the moving point as

$$\mathbf{p}(t) = r(t)e^{i\theta(t)}. \tag{18.12}$$

I'm using bold notation for $\mathbf{p}(t)$ to indicate it is a vector. Notice carefully that $r(t)$ is *not* a vector; it is the *length* of the vector $r(t)e^{i\theta(t)}$. It is the $e^{i\theta(t)}$ all by itself that provides the vector nature to $\mathbf{p}(t)$.

From (18.12) we can find the moving point's velocity and acceleration vectors, $\mathbf{v}(t)$ and $\mathbf{a}(t)$, respectively, by successive differentiations. For example, for $\mathbf{v}(t)$ we have

$$\mathbf{v}(t) = \frac{d\mathbf{p}(t)}{dt} = \frac{dr}{dt}e^{i\theta} + rie^{i\theta}\frac{d\theta}{dt},$$

or, more transparently,

$$\mathbf{v}(t) = \left(\frac{dr}{dt}\right)e^{i\theta} + \left(r\frac{d\theta}{dt}\right)ie^{i\theta}. \tag{18.13}$$

The first term on the right-hand side of (18.13) is a vector with length $\left|\frac{dr}{dt}\right|$ that is pointing in the same direction as $e^{i\theta}$ if $\frac{dr}{dt} > 0$ and in the direction opposite that of $e^{i\theta}$ if $\frac{dr}{dt} < 0$ (that is, radially inward). The second term on the right in (18.13) is a vector, too, but now we must remember that the vector nature of that term is due to $ie^{i\theta}$, not just $e^{i\theta}$. The presence of the factor i produces a unit vector that is rotated[3] 90 degrees counterclockwise from $e^{i\theta}$. So, the second term of (18.13) is a vector with length $|r\frac{d\theta}{dt}|$ that is pointing 90 degrees counterclockwise from $e^{i\theta}$ if $r\frac{d\theta}{dt} > 0$ and 90 degrees clockwise from $e^{i\theta}$ if $r\frac{d\theta}{dt} < 0$.

The first term of (18.13) is the *radial* velocity and the second term is the *tangential* velocity. The magnitude of their vector sum is the *speed* (a scalar) of the moving point, and it is instructive to calculate it for the big-noise problem. That is, the speed of the jet airplane is, using (18.1), (18.2), (18.3), and (18.10),

$$\sqrt{\left(\frac{dr}{dt}\right)^2 + \left(r\frac{d\theta}{dt}\right)^2} = \sqrt{c^2 + (a-ct)^2 \frac{M^2-1}{(T-t)^2}} = \sqrt{c^2 + c^2\left(\frac{a}{c}-t\right)^2 \frac{M^2-1}{(T-t)^2}}$$

$$= c\sqrt{1 + (T-t)^2 \frac{M^2-1}{(T-t)^2}} = c\sqrt{1 + M^2 - 1} = Mc,$$

a *constant* for all t which is, of course, correct (take a look again at Agnew's original presentation of the big-noise problem).

To find the acceleration of the moving point, we differentiate (18.13) and get

$$\mathbf{a}(t) = \frac{d\mathbf{v}(t)}{dt} = \left[\frac{d^2r}{dt^2}e^{i\theta} + \frac{dr}{dt}ie^{i\theta}\frac{d\theta}{dt}\right]$$

$$+ i\left[\frac{dr}{dt}e^{i\theta}\frac{d\theta}{dt} + r\left(ie^{i\theta}\frac{d\theta}{dt}\cdot\frac{d\theta}{dt} + e^{i\theta}\frac{d^2\theta}{dt^2}\right)\right],$$

or, collecting terms,

$$\mathbf{a}(t) = \left[\frac{d^2r}{dt^2} - r\left(\frac{d\theta}{dt}\right)^2\right]e^{i\theta} + \left[2\frac{dr}{dt}\cdot\frac{d\theta}{dt} + r\frac{d^2\theta}{dt^2}\right]ie^{i\theta}. \quad (18.14)$$

Again, we can calculate the magnitude of the acceleration—which is physically what most interests the pilot—from the component radial and tangential accelerations in (18.14). Thus,

$$|\mathbf{a}(t)| = \sqrt{\left[\frac{d^2r}{dt^2} - r\left(\frac{d\theta}{dt}\right)^2\right]^2 + \left[2\frac{dr}{dt}\cdot\frac{d\theta}{dt} + r\frac{d^2\theta}{dt^2}\right]^2}. \quad (18.15)$$

Since (18.3) tell us that $\frac{d^2r}{dt^2} = 0$ for the big-noise problem, and using our earlier result of (18.10) to calculate $\frac{d^2\theta}{dt^2}$ as

$$\frac{d^2\theta}{dt^2} = \sqrt{M^2-1}\left[-\frac{-1}{(T-t)^2}\right] = \frac{\sqrt{M^2-1}}{(T-t)^2},$$

we can use these and our earlier results for all the other terms in (18.15) to write

$$|\mathbf{a}(t)| = \sqrt{(a-ct)^2\frac{(M^2-1)^2}{(T-t)^4} + \left[-2c\frac{\sqrt{M^2-1}}{T-t} + (a-ct)\frac{\sqrt{M^2-1}}{(T-t)^2}\right]},$$

which, after just a bit of routine algebra, becomes the remarkably simple

$$|\mathbf{a}(t)| = \frac{Mc}{T-t}\sqrt{M^2-1}, \quad T = \frac{a}{c}, \; M \geq 1. \quad (18.16)$$

Suppose, as an example of what (18.16) can tell us, that the pilot starts his big noise flight at a distance of $a = 100,000$ feet from O, flying at one-and-a-half times the speed of sound ($M = 1.5$). As a fairly good approximation to the speed of sound, I'll use $c = 1,000$ feet/second, which means the entire flight time is $T = 100$ seconds. The big question, of course, is, can the pilot actually survive this flight? From (18.16) we see that the acceleration experienced by the pilot is

$$|\mathbf{a}(t)| = \frac{1,500\sqrt{1.25}}{100-t} \text{ feet/second}^2 = \frac{1,677}{100-t} \text{ feet/second}^2.$$

A military combat pilot can probably remain aware of his surroundings up to about seven gravities of acceleration (7 gees). Beyond that value blood flow to the brain is reduced below that necessary for a human to remain conscious, much less functional. This acceleration will be reached when

$$\frac{1,677}{100-t} = 7 \cdot 32.2,$$

or, solving for t, when $t = 92.6$ seconds. That looks like almost all of the flight—but looks are deceiving! From (18.3) the plane will, at that time, then be 7,400 feet from O and will have looped around O through an angle, from (18.11), of

$$\sqrt{1.25}\ln\left(\frac{1}{1-0.926}\right) = 2.91 \text{ radians.}$$

This is an angle of just 167 degrees, *less than one-half a single loop* around O. There are still an *infinity* of loops to go in the remaining 7.4 seconds of flight time. That is, when the flight is more than 92% finished in time, it has only just begun in terms of looping around O. If this was a Disney World ride, I think I'd take a pass.

CP. P18.1:

At what time does the pilot in the above analysis complete the first loop around O, and what is his acceleration at that moment? Assume both the plane and the pilot are indestructible.

Notes and References

1. Sherman L. Gerhard, "Shock-Wave Convergence Demonstrated by Surface Waves on Water" (*American Journal of Physics*, June 1967, pp. 509–513).

2. I. G. Clator, "Uniformly Converging Shock Waves" (*American Journal of Physics*, May 1970, pp. 660–661).

3. For much more on this, see my two books, *An Imaginary Tale: The Story of $\sqrt{-1}$* (1998, 2007), and *Dr. Euler's Fabulous Formula* (2006), both from Princeton University Press.

*SOLUTIONS
TO THE CHALLENGE
PROBLEMS*

Challenge Problem P.1

The hyperbolic functions are defined as follows:

$$\cosh(z) = \frac{e^z + e^{-z}}{2}, \ \sinh(z) = \frac{e^z - e^{-z}}{2}, \ \tanh(z) = \frac{\sinh(z)}{\cosh(z)}.$$

So,

$$1 - \tanh^2(z) = 1 - \frac{\sinh^2(z)}{\cosh^2(z)} = \frac{\cosh^2(z) - \sinh^2(z)}{\cosh^2(z)} = \frac{[\frac{e^z + e^{-z}}{2}]^2 - [\frac{e^z - e^{-z}}{2}]^2}{\cosh^2(z)}$$

$$= \frac{\frac{e^{2z} + 2 + e^{-2z}}{4} - \frac{e^{2z} - 2 + e^{-2z}}{4}}{\cosh^2(z)} = \frac{1}{\cosh^2(z)}.$$

Thus,

$$\sqrt{1 - \tanh^2(z)} = \frac{1}{\cosh(z)},$$

and so the right-hand side of our "great truth" is

$$\sum_{n=0}^{\infty} \frac{\cosh(z)\sqrt{1 - \tanh^2(z)}}{2^n} = \sum_{n=0}^{\infty} \frac{1}{2^n} = 1 + \frac{1}{2} + \frac{1}{4} + \cdots = 2.$$

Now, on the left-hand side of the "great truth" you should recognize that $\lim_{x \to \infty}(1 + x^{-1})^x = e$, and also that $\sin^2(y) + \cos^2(y) = 1$. Thus, the left-hand side is simply

$$\ln(e) + 1 = 1 + 1.$$

That is, our great truth is the undeniably true

$$1 + 1 = 2.$$

Challenge Problem P.2

$$I(t) = \int_0^{2\pi} \{\cos(\phi) + \sin(\phi)\} \sqrt{\frac{1 - \sqrt{t}\sin(\phi)}{1 - \sqrt{t}\cos(\phi)}} d\phi.$$

Notice that the radical in the integrand is real for all t in the interval 0 to 1, as for all ϕ both the numerator and the denominator are non-negative. Thus, $I(t)$ is purely real. Now, change variable to $\theta = \phi + \pi/4$. Then $d\theta = d\phi$ and we have

$$\cos(\phi) = \cos(\theta - \pi/4) = \cos(\theta)\cos(\pi/4) + \sin(\theta)\sin(\pi/4)$$
$$= \frac{1}{\sqrt{2}}\cos(\theta) + \frac{1}{\sqrt{2}}\sin(\theta) = \frac{1}{\sqrt{2}}[\cos(\theta) + \sin(\theta)],$$

and

$$\sin(\phi) = \sin(\theta - \pi/4) = \sin(\theta)\cos(\pi/4) - \cos(\theta)\sin(\pi/4)$$
$$= \frac{1}{\sqrt{2}}\sin(\theta) - \frac{1}{\sqrt{2}}\cos(\theta) = \frac{1}{\sqrt{2}}[\sin(\theta) - \cos(\theta)].$$

So,

$$I(t) = \int_{\pi/4}^{2\pi+\pi/4} \frac{2}{\sqrt{2}} \sin(\theta) \sqrt{\frac{1 - \sqrt{t}\frac{1}{\sqrt{2}}\{\sin(\theta) - \cos(\theta)\}}{1 - \sqrt{t}\frac{1}{\sqrt{2}}\{\cos(\theta) + \sin(\theta)\}}}\, d\theta,$$

or, writing $u = \sqrt{\frac{t}{2}}$,

$$I(u) = \sqrt{2} \int_{\pi/4}^{2\pi+\pi/4} \sin(\theta) \sqrt{\frac{1 + u\{\cos(\theta) - \sin(\theta)\}}{1 - u\{\cos(\theta) + \sin(\theta)\}}}\, d\theta$$

$$= \sqrt{2}\left[\int_0^{2\pi} + \int_{2\pi}^{2\pi+\pi/4} - \int_0^{\pi/4}\right].$$

By the periodicity of the sine and cosine functions, the integrand value over the second integration interval is, for every value of θ, *identical* to the integrand value over the third integration interval. That is, the last two integrals cancel. Thus,

$$I(u) = \sqrt{2} \int_0^{2\pi} \sin(\theta) \sqrt{\frac{1 + u\{\cos(\theta) - \sin(\theta)\}}{1 - u\{\cos(\theta) + \sin(\theta)\}}}\, d\theta,$$

where u varies from 0 to $1/\sqrt{2}$ as t varies from 0 to 1. Now, $\int_0^{2\pi} = \int_0^{\pi} + \int_{\pi}^{2\pi}$.

Since cos(θ) is symmetrical around π, while sin(θ) is antisymmetrical around π, we can write the π to 2π integral as an integral from 0 to π if we replace every cos(θ) with cos(θ) and every sin(θ) with −sin(θ). That is,

$$I(u) = \sqrt{2}\left[\int_0^\pi \sin(\theta)\sqrt{\frac{1+u\{\cos(\theta)-\sin(\theta)\}}{1-u\{\cos(\theta)+\sin(\theta)\}}}\,d\theta\right.$$

$$\left.-\int_0^\pi \sin(\theta)\sqrt{\frac{1+u\{\cos(\theta)+\sin(\theta)\}}{1-u\{\cos(\theta)-\sin(\theta)\}}}\,d\theta\right].$$

Or

$$I(u) = \sqrt{2}\left[\int_0^\pi \sin(\theta)\left\{\sqrt{\frac{1+u\{\cos(\theta)-\sin(\theta)\}}{1-u\{\cos(\theta)+\sin(\theta)\}}}\right.\right.$$

$$\left.\left.-\sqrt{\frac{1+u\{\cos(\theta)+\sin(\theta)\}}{1+u\{\sin(\theta)-\cos(\theta)\}}}\right\}d\theta\right].$$

Now, $\int_0^\pi = \int_0^{\pi/2} + \int_{\pi/2}^\pi$. Since sin(θ) is symmetrical around π/2, while cos(θ) is antisymmetrical around π/2, we can write the π/2 to π integral as an integral from 0 to π/2 if we replace every sin(θ) with sin(θ) and every cos(θ) with −cos(θ). That is,

$$I(u) = \sqrt{2}\left[\int_0^{\pi/2} \sin(\theta)\left\{\begin{array}{l}\sqrt{\frac{1+u\{\cos(\theta)-\sin(\theta)\}}{1-u\{\cos(\theta)+\sin(\theta)\}}}+\sqrt{\frac{1-u\{\cos(\theta)+\sin(\theta)\}}{1+u\{\cos(\theta)-\sin(\theta)\}}}\\ -\sqrt{\frac{1+u\{\cos(\theta)+\sin(\theta)\}}{1+u\{\sin(\theta)-\cos(\theta)\}}}-\sqrt{\frac{1-u\{\cos(\theta)-\sin(\theta)\}}{1+u\{\sin(\theta)+\cos(\theta)\}}}\end{array}\right\}d\theta\right].$$

This may look like we are taking our original tough-looking integral and making it look even more awful, but that's an illusion. Here's why.

If we call the first radical x, then you'll notice that the second radical is the reciprocal of the first one and so is $1/x$. Also, if we call the fourth radical y, then the third one is its reciprocal and so is $1/y$. What I'll do next is show that, over the entire interval of integration, the integrand is non-negative (and it is obviously not identically zero), so $I(u) > 0$, that is $I(t) > 0$. (The factor of sin(θ) in front of all the radicals is never negative over the integration interval, so we can ignore it.)

That is, I will show you that

$$x + \frac{1}{x} \geq y + \frac{1}{y},$$

that is, that

$$x^2 + 1 \geq xy + \frac{x}{y},$$

that is, that

$$x^2 y + y \geq xy^2 + x,$$

that is, that

$$x^2 y - xy^2 \geq x - y$$

that is, that

$$xy(x - y) \geq x - y.$$

Now, *if* $x \geq y$, then that last inequality says it must be true that $xy \geq 1$. If we can show that both of these conclusions are actually so, then our starting inequality $\left(\text{that is, } x + \frac{1}{x} \geq y + \frac{1}{y}\right)$ will in fact be true, and we'll be done. So, first, is $x \geq y$? The answer is yes, because that inequality is simply the assertion that

$$\sqrt{\frac{1 + u\{\cos(\theta) - \sin(\theta)\}}{1 - u\{\cos(\theta) + \sin(\theta)\}}} \geq \sqrt{\frac{1 - u\{\cos(\theta) - \sin(\theta)\}}{1 + u\{\sin(\theta) + \cos(\theta)\}}},$$

which is the assertion that

$$[1 + u\{\cos(\theta) - \sin(\theta)\}][1 + u\{\sin(\theta) + \cos(\theta)\}] \geq$$
$$[1 - u\{\cos(\theta) - \sin(\theta)\}][1 - u\{\cos(\theta) + \sin(\theta)\}],$$

which is the assertion that

$$1 + 2u\cos(\theta) - u^2\{\sin^2(\theta) - \cos^2(\theta)\} \geq$$
$$1 - 2u\cos(\theta) + u^2\{\cos^2(\theta) - \sin^2(\theta)\},$$

which is the assertion that

$$4u\cos(\theta) \geq 0,$$

which is obviously true for all θ in the integration interval 0 to $\pi/2$ (remember, u itself is non-negative). Second, is $xy \geq 1$? The answer is again yes, because that inequality is simply the assertion that

$$\sqrt{\frac{1+u\{\cos(\theta)-\sin(\theta)\}}{1-u\{\cos(\theta)+\sin(\theta)\}}} \cdot \sqrt{\frac{1-u\{\cos(\theta)-\sin(\theta)\}}{1+u\{\sin(\theta)+\cos(\theta)\}}} \geq 1,$$

which is the assertion that

$$1 - u^2\{\cos(\theta)-\sin(\theta)\}^2 \geq 1 - u^2\{\cos(\theta)+\sin(\theta)\}^2,$$

which is the assertion that

$$\{\cos(\theta)+\sin(\theta)\}^2 \geq \{\cos(\theta)-\sin(\theta)\}^2,$$

which is obviously true for all θ in the integration interval 0 to $\pi/2$ because, in that interval both $\cos(\theta)$ and $\sin(\theta)$ are never negative (the sum of two non-negative numbers is always at least as large as their difference). This completes the high school algebra and trigonometry proof that $I(t) > 0$ for all t in the interval 0 to 1. My discussion here is based on the paper by John D. Morgan III, "The Positivity of an Integral Connected with the Helium Atom Problem" (*American Journal of Physics*, February 1978, pp. 180–181). I have expanded on numerous points in Morgan's paper that, for purposes of journal publication brevity, Morgan had to skip over with little mention.

Challenge Problem P.1.1

From (1.8) we have the speed of the wall end of the pole as

$$v_y = -v_0 \cot(\theta).$$

Therefore, the acceleration is, in general, *before* breakaway

$$a = \frac{dv_y}{dt} = \frac{v_0}{\sin^2(\theta)} \cdot \frac{d\theta}{dt}.$$

From (1.14) we have the breakaway angle $\theta = \theta_c$ as

$$\theta_c = \sin^{-1}\left(\sqrt[3]{\frac{2}{3} \cdot \frac{v_0^2}{gL}}\right),$$

and so

$$\sin(\theta_c) = \left(\frac{2}{3} \cdot \frac{v_0^2}{gL}\right)^{1/3}$$

and

$$\sin^2(\theta_c) = \left(\frac{2}{3} \cdot \frac{v_0^2}{gL}\right)^{2/3}.$$

This says, at breakaway, the acceleration is

$$a = \frac{v_0 \frac{d\theta}{dt}}{\left(\frac{2}{3} \cdot \frac{v_0^2}{gL}\right)^{2/3}}.$$

From (1.11) we have, at breakaway,

$$\frac{d\theta}{dt} = -\frac{v_0}{L\sin(\theta)} = -\frac{v_0}{L\left(\frac{2}{3} \cdot \frac{v_0^2}{gL}\right)^{1/3}}.$$

So, at breakaway, the acceleration of the wall end of the pole is

$$a = -\frac{v_0^2}{L\frac{2}{3} \cdot \frac{v_0^2}{gL}} = -\frac{3}{2}g.$$

Challenge Problem P.2.1

Let D and T be the total bounce distance and time, respectively. Then,

$$D = h_0 + (c^2 h_0 + c^2 h_0) + (c^4 h_0 + c^4 h_0) + \cdots = h_0 + 2c^2 h_0 + 2c^4 h_0 + \cdots,$$
$$= h_0(1 + 2c^2 + 2c^4 + \cdots) = h_0(1 + 2S),$$

where

$$S = c^2 + c^4 + \cdots = c^2(1 + c^2 + \cdots) = \frac{c^2}{1 - c^2}.$$

So,
$$D = h_0\left(1 + \frac{2c^2}{1-c^2}\right) = h_0 \frac{1+c^2}{1-c^2}.$$

For my garage experiment, $c = 0.913$ and $h_0 = 36$ inches and so $D = 397$ inches. We also showed that the total time *from the first impact* until the n-th impact is
$$T_n = 2\sqrt{\frac{2h_0}{g}} c \frac{1-c^n}{1-c}.$$

So, as $n \to \infty$, this becomes (because $c < 1$) $T_\infty = \frac{2c}{1-c}\sqrt{\frac{2h_0}{g}}$. We have to add to this the initial drop time $\sqrt{\frac{2h_0}{g}}$ to get
$$T = \sqrt{\frac{2h_0}{g}} + \frac{2c}{1-c}\sqrt{\frac{2h_0}{g}} = \sqrt{\frac{2h_0}{g}}\left(1 + \frac{2c}{1-c}\right) = \frac{1+c}{1-c}\sqrt{\frac{2h_0}{g}}.$$

For my garage experiment $T = 9.5$ seconds.

Challenge Problem P.3.1

If you look back at Figure 3.10, you can see that if we remove the first two resistors, we are left with the original infinite network *flipped over*. Since that network is in parallel with the removed resistor "on a slant", we can immediately write
$$R = 1 + 1 \parallel R,$$
which is just (3.1) again with $R_1 = R_2 = 1$. So, $R = (1+\sqrt{5})/2$. So, even though the network of Figure 3.10 "looks different" from that of Figure 3.1, the defining equation for the input resistance is the same for both.

Challenge Problem P.3.2

We start with $V(x, y, z) = \frac{q}{4\pi\epsilon_0}\left[\int_{-\infty}^{\infty} \frac{dz'}{\sqrt{r^2 + (z-z')^2}} - \int_{-\infty}^{\infty} \frac{dz'}{\sqrt{R^2 + (z-z')^2}}\right]$. Let $u = z' - z$ (and so $du = dz'$).

Then,

$$V(x,y,z) = \frac{q}{4\pi\epsilon_0}\left[\int_{-\infty}^{\infty}\frac{du}{\sqrt{r^2+u^2}} - \int_{-\infty}^{\infty}\frac{du}{\sqrt{R^2+u^2}}\right]$$

$$= \frac{q}{4\pi\epsilon_0}\lim_{b\to\infty}\left[\int_{-b}^{b}\frac{du}{\sqrt{r^2+u^2}} - \int_{-b}^{b}\frac{du}{\sqrt{R^2+u^2}}\right].$$

From integral tables we have $\int \frac{du}{\sqrt{a^2+u^2}} = \ln\left[u+\sqrt{a^2+u^2}\right]$, and so

$$V(x,y,z) = \frac{q}{4\pi\epsilon_0}\lim_{b\to\infty}\left[\left\{\ln\left[u+\sqrt{r^2+u^2}\right]\right\}\Big|_{-b}^{b} - \left\{\ln\left[u+\sqrt{R^2+u^2}\right]\right\}\Big|_{-b}^{b}\right]$$

$$= \frac{q}{4\pi\epsilon_0}\lim_{b\to\infty}\left[\begin{array}{l}\ln\left[b+\sqrt{b^2+r^2}\right] - \ln\left[-b+\sqrt{b^2+r^2}\right]\\ -\ln\left[b+\sqrt{b^2+R^2}\right] + \ln\left[-b+\sqrt{b^2+R^2}\right]\end{array}\right]$$

$$= \frac{q}{4\pi\epsilon_0}\lim_{b\to\infty}\left[\ln\left(\frac{b+\sqrt{b^2+r^2}}{b+\sqrt{b^2+R^2}}\right) + \ln\left(\frac{-b+\sqrt{b^2+R^2}}{-b+\sqrt{b^2+r^2}}\right)\right]$$

$$= \frac{q}{4\pi\epsilon_0}\lim_{b\to\infty}\left[\ln\left(\frac{b+\sqrt{b^2+r^2}}{b+\sqrt{b^2+R^2}} \cdot \frac{-b+\sqrt{b^2+R^2}}{-b+\sqrt{b^2+r^2}}\right)\right].$$

Now, $\sqrt{b^2+R^2} = b\sqrt{1+\left(\frac{R}{b}\right)^2} = b\left[1+\frac{1}{2}\left(\frac{R}{b}\right)^2 - \frac{1}{8}\left(\frac{R}{b}\right)^4 + \cdots\right]$ and so $V(x,y,z) = \frac{q}{4\pi\epsilon_0}\lim_{b\to\infty}\left[\ln\left(\frac{b+b+\frac{1}{2}\frac{r^2}{b}-\cdots}{b+b+\frac{1}{2}\frac{R^2}{b}-\cdots} \cdot \frac{-b+b+\frac{1}{2}\frac{R^2}{b}-\cdots}{-b+b+\frac{1}{2}\frac{r^2}{b}-\cdots}\right)\right]$. Now, as $b \to \infty$ the first factor inside the log function goes to 1, and the second factor goes to R^2/r^2. So, $V(x,y,z) = \frac{q}{4\pi\epsilon_0}\ln\left(\frac{R^2}{r^2}\right)$ or, at last, we have the finite $V(x,y,z) = \frac{q}{2\pi\epsilon_0}\ln\left(\frac{R}{\sqrt{x^2+y^2}}\right)$. There is no z dependency because the electrically charged wire is infinitely long in the z direction, and there is nothing to physically distinguish one value of z from any other.

Challenge Problem P.3.3

Figure S3.11 shows an arbitrary node along the ladder, with node voltage $v(n)$. Kirchhoff's node current law says the current into the

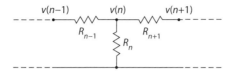

Figure S3.11. An arbitrary node of an infinite ladder.

node equals the current out of the node. So,

$$\frac{v(n-1) - v(n)}{R_{n-1}} + \frac{v(n+1) - v(n)}{R_{n+1}} = \frac{v(n)}{R_n}.$$

If we multiply each resistance in this expression by the same factor k, we will still have a valid statement of Kirchhoff's law *with the same node voltages*. All that changes are the individual resistor currents, which will be the original currents divided by k. In particular, the input current to the first section of the ladder will be divided by the factor k, and so we will have the original input voltage but an altered (by the factor k) input current. Thus, the input resistance changes (is divided) by the factor k.

Challenge Problem P.3.4

I don't know the answer to this question. If I calculate the sequence of input resistances R_n to finite-length versions of the infinite ladder of Figure 3.9c (that is, ladders with $n = 1, 2, 3$, and 4 stages) I get $R_1 = 2$, $R_2 = 1.9099$, $R_3 = 1.9019$, and $R_4 = 1.9090909$. From this flimsy "evidence" one might conjecture that the exact result is $R_c = 1.909090909\cdots = \frac{21}{11}$ ohms. But I can't prove that—it may well be wrong. Any help from readers will be greatly appreciated, and a reader derivation of R_c will receive prominent display and attribution in the future paperback edition of this book.

Challenge Problem P.3.5

See the solution to Challenge Problem P.17.3.

Challenge Problem P.4.1

When the rocket motor cuts off at $t = t_1$, let's write the rocket's height as h_1 and its (maximum) speed at that instant as v_1. From there it continues to rise upward to its maximum height (H)—let's say the time to go from h_1 to H is Δt—and then from H it falls back down to h_1 *in the same time interval* Δt. As it falls through height h_1 it is again moving at speed v_1. From h_1, on down to the ground the rocket's speed obviously continues to increase beyond v_1. On the upward portion of its flight, however, from zero height to h_1, the rocket's speed was always *no more* than v_1. So, it takes longer for the rocket to accelerate from zero to v_1 (from the ground to h_1) than to fall from height h_1 to the ground. Thus, $t_{up} > t_{down}$.

Challenge Problem P.4.2

Inserting $f(v) = kv$ into (4.4) and taking the unit speed as the terminal speed, we have

$$\int_0^{v_i} \frac{v\,dv}{1+v} = \int_0^{v_f} \frac{v\,dv}{1-v}.$$

Letting $u = 1 + v$ (and so $du = dv$) in the left-hand-side integral, and $u = 1 - v$ (and so $du = -dv$) in the right-hand-side integral, we have

$$\int_1^{1+v_i} \frac{u-1}{u}\,du = \int_1^{1-v_f} \frac{1-u}{u}(-du),$$

or

$$\int_1^{1+v_i} \left(1 - \frac{1}{u}\right)du = \int_{1-v_f}^{1} \left(\frac{1}{u} - 1\right)du,$$

or

$$\{u - \ln(u)\}\,|_1^{1+v_i} = \{\ln(u) - u\}\,|_{1-v_f}^{1},$$

or

$$(1+v_i) - \ln(1+v_i) - 1 = -1 - \ln(1-v_f) + (1-v_f),$$

or

$$(1+v_i) - (1-v_f) = \ln(1+v_i) - \ln(1-v_f) = \boxed{\ln\left\{\frac{1+v_i}{1-v_f}\right\} = v_i + v_f.}$$

From (4.3)

$$\tau = \frac{1}{2v_i}\left[\int_0^{v_i}\frac{dv}{1+v} + \int_0^{v_f}\frac{dv}{1-v}\right] = \frac{1}{2v_i}\left[\{\ln(1+v)\}|_0^{v_i} - \{\ln(1-v)\}|_0^{v_f}\right],$$

or, using the above boxed result,

$$\tau = \frac{\ln(1+v_i) - \ln(1-v_f)}{2v_i} = \frac{\ln\left\{\frac{1+v_i}{1-v_f}\right\}}{2v_i} = \frac{v_i + v_f}{2v_i} = \frac{1}{2}\left(1 + \frac{v_f}{v_i}\right).$$

Since $v_f < v_i$ for any physically plausible drag force law (including our linear one), then $\tau < 1$ for $f(v) = kv$.

Challenge Problem P.5.1

Starting with

$$v^2 = e^{-\alpha e^y}\int_\alpha^{\alpha e^y}\frac{e^z}{z}dz$$

we have, from how to differentiate a product and remembering that α is a constant,

$$\frac{d(v^2)}{dy} = \left\{\frac{d}{dy}\left(e^{-\alpha e^y}\right)\right\}\int_\alpha^{\alpha e^y}\frac{e^z}{z}dz + e^{-\alpha e^y}\frac{d}{dy}\left(\int_\alpha^{\alpha e^y}\frac{e^z}{z}dz\right)$$

$$= e^{-\alpha e^y}(-\alpha e^y)\int_\alpha^{\alpha e^y}\frac{e^z}{z}dz + e^{-\alpha e^y}\left\{\frac{e^{\alpha e^y}}{\alpha e^y}(\alpha e^y)\right\}$$

$$= -\alpha e^y e^{-\alpha e^y}\int_\alpha^{\alpha e^y}\frac{e^z}{z}dz + 1,$$

or
$$\frac{d(v^2)}{dy} = -\alpha e^y v^2 + 1.$$

At the start of the fall $v = 0$, and so
$$\frac{d(v^2)}{dy}\bigg|_{y=0} = +1.$$

Since the initial rate of change of v^2 is positive, v will increase as the fall continues (as will e^y—remember that $y = 0$ at the start of the fall and is measured in the *downward* direction and so *increases* as the fall progresses) until $\frac{d(v^2)}{dy} = 0$; this will obviously occur after a finite fall distance. The value of v at that instant is $v = V$. From this point on the speed must do one of three things: (1) continue to increase beyond V, (2) remain fixed at V, or (3) start to decrease from V. Possibility (1) can be rejected because

$$\frac{d(v^2)}{dy} = 2v\frac{dv}{dy} = -\alpha e^y v^2 + 1,$$

which says

$$\boxed{\frac{dv}{dy} = \frac{-\alpha e^y v^2 + 1}{2v},}$$

and, since both v and e^y are increasing, the boxed expression says $\frac{dv}{dy} < 0$ because the first term in the numerator of the right-hand side of the boxed expression is growing ever more negative during the fall. But if v is to be increasing beyond V, we must have $\frac{dv}{dy} > 0$, and we have a contradiction. Possibility (2) can also be rejected, because that says $\frac{dv}{dy} = 0$ while e^y continues to increase, and so the first term in the numerator of the right-hand side of the boxed expression is growing ever more negative during the fall and so, again, that says $\frac{dv}{dy} < 0$, and we have a contradiction. Therefore, what actually happens must be our final possibility, possibility (3), that is, v *decreases* from V. That says $\frac{dv}{dy} < 0$, which is not in an inescapable contradiction with the boxed expression, just so long as the increase in e^y more than compensates for the decrease in v. Thus, v increases from 0 to V (which is, in fact, the maximum speed $V_{\max}$) and thereafter decreases, just as shown in the curves of Figure 5.3.

Challenge Problem P.6.1

From (6.7) we have $x(t) = r_e \cos(\alpha t)$, $\alpha = \sqrt{\frac{4}{3}\pi G \rho}$. Thus, the anvil's speed is $|\frac{dx}{dt}| = r_e \alpha \sin(\alpha t)$. When the anvil arrives in Hell, located at $x = 0$ (which means $\alpha t = \frac{\pi}{2}$), its speed is $|\frac{dx}{dt}| = r_e \alpha$. A mass m on the surface of Earth has *weight* mg, where $mg = G\frac{mm_e}{r_e^2}$ and so $g = G\frac{m_e}{r_e^2} = G\frac{\frac{4}{3}\pi r_e^3 \rho}{r_e^2} = G\frac{4}{3}\pi r_e \rho$. So, $G\rho = \frac{g}{\frac{4}{3}\pi r_e}$ which gives the anvil's arrival speed as $r_e \sqrt{\frac{4}{3}\pi \frac{g}{\frac{4}{3}\pi r_e}} = \sqrt{gr_e}$. Plugging in numbers, the anvil arrives *in* Hell with speed $\sqrt{32.2 \cdot 3,960 \cdot 5,280}$ feet/second $= 25,947$ feet/second, which is ≈ 4.9 miles/second. This may be slower *or* faster than the speed of the proverbial "bat out of Hell," a calculation that remains out of the realm, at least for the time being, of mathematical physics as we know it.

Challenge Problem P.7.1

To calculate $\int_0^1 \int_0^1 \frac{dx\,dy}{1-x^2y^2}$ begin by defining D as $\boxed{D = \int_0^1 \int_0^1 \frac{dx\,dy}{1-xy} - \int_0^1 \int_0^1 \frac{dx\,dy}{1+xy}} = \int_0^1 \int_0^1 \frac{2xy\,dx\,dy}{1-x^2y^2}$. Then, changing variables to $u = x^2$ (and so $du = 2x\,dx$) and $v = y^2$ (and so $dv = 2y\,dy$), we have $D = \int_0^1 \int_0^1 \frac{2xy \cdot \frac{du}{2x} \cdot \frac{dv}{2y}}{1-uv} = \frac{1}{2} \int_0^1 \int_0^1 \frac{du\,dv}{1-uv}$, that is, $D = \frac{1}{2} \int_0^1 \int_0^1 \frac{dx\,dy}{1-xy}$ because u, v, x, and y are all, of course, dummy variables of integration. Next, define S as $\boxed{S = \int_0^1 \int_0^1 \frac{dx\,dy}{1-xy} + \int_0^1 \int_0^1 \frac{dx\,dy}{1+xy}} = 2\int_0^1 \int_0^1 \frac{dx\,dy}{1-x^2y^2}$. Then, adding the two boxed expressions, $D + S = 2\int_0^1 \int_0^1 \frac{dx\,dy}{1-xy} = \frac{1}{2} \int_0^1 \int_0^1 \frac{dx\,dy}{1-xy} + 2\int_0^1 \int_0^1 \frac{dx\,dy}{1-x^2y^2}$ and so, rearranging and dividing by 2, $\int_0^1\int_0^1 \frac{dx\,dy}{1-x^2y^2} = \int_0^1\int_0^1 \frac{dx\,dy}{1-xy} - \frac{1}{4}\int_0^1\int_0^1 \frac{dx\,dy}{1-xy} = \frac{3}{4}\int_0^1\int_0^1 \frac{dx\,dy}{1-xy}$. Now, as shown in the discussion the last integral on the far right is $\zeta(2)$, and so the answer to our first question is

$$\int_0^1\int_0^1 \frac{dx\,dy}{1-x^2y^2} = \frac{3}{4}\zeta(2) = \frac{3}{4}\cdot\frac{\pi^2}{6} = \frac{\pi^2}{8}.$$

To calculate $\int_0^1\int_0^1 \frac{dx\,dy}{1+xy}$, notice that the second boxed expression says

$$\int_0^1\int_0^1 \frac{dx\,dy}{1+xy} = 2\int_0^1\int_0^1 \frac{dx\,dy}{1-x^2y^2} - \int_0^1\int_0^1 \frac{dx\,dy}{1-xy} = 2\cdot\frac{\pi^2}{8} - \zeta(2) = \frac{\pi^2}{4} - \frac{\pi^2}{6},$$

and so $\int_0^1\int_0^1 \frac{dx\,dy}{1+xy} = \frac{\pi^2}{12}$.

Challenge Problem P.7.2

If you calculate the Fourier series expansion of $f(x) = x^3$ over the interval $-\pi$ to π, you should get

$$\boxed{x^3 = \sum_{n=1}^{\infty}\left\{\frac{12}{n^3} - \frac{2\pi^2}{n}\right\}\cos(n\pi)\sin(nx).}$$

If, for example, you set $x = 1$, then the left-hand side of the expression in the box is of course 1, while a numerical evaluation of the sum on the right-hand side, using the first one million terms, is $0.99999840332\ldots$. Notice, however, that if we set $x = \pi$, the summation is *not* π^3 but rather *zero* because of the $\sin(nx)$ factor—$\sin(n\pi) = 0$ for all integer n. This behavior is caused by the fact that a Fourier series expansion of a discontinuous function converges, *at a discontinuity*, to the average of the function values on each side of the discontinuity (and the average of $+\pi^3$ and $-\pi^3$ is, of course, zero). If it isn't clear why $f(x) = x^3$ is discontinuous at $x = \pi$, then you should sketch x^3 over the interval

$-\pi$ to π, and then make the periodic extension of that fundamental period to cover the entire x-axis. In fact, there is no value to which we can set x in the boxed expression that gives anything on the right-hand side of use, that is, anything that has $\zeta(3)$ in it.

Challenge Problem P.8.1

From (8.15), $R_m = \frac{v_0\sqrt{v_0^2+2gh}}{g}$, which is easily expanded to give $v_0^4 + 2ghv_0^2 - R_m^2 g^2 = 0$. This is quadratic in v_0^2, and so $v_0^2 = \frac{-2gh\pm\sqrt{4g^2h^2+4R_m^2g^2}}{2}$, or, as $R_m >> h$, $v_0 \approx \sqrt{R_m g}$ is the horizontal speed of the shell. Now, for point-blank fire, the time of flight T is the time for the shell to fall vertically to the ground through a distance of h, that is, $\frac{1}{2}gT^2 = h$, or $T = \sqrt{\frac{2h}{g}}$. So, $R_{pb} = v_0 T = \sqrt{R_m g}\sqrt{\frac{2h}{g}}$, or, at last, $R_{pb} = \sqrt{2R_m h}$. For $R_m = 20,000$ feet and $h = 4$ feet, $R_{pb} = \sqrt{2 \cdot 20,000 \cdot 4} = 400$ feet, which is surprisingly (I think) small. For the obvious purpose of point-blank fire, however, it is big enough.

Challenge Problem P.9.1

From (9.19) we have

$$S = \frac{m}{2k}\ln\left\{1 - \frac{kv_0^2}{4mg}[\ln(0.268) - 3.464]\right\} = \frac{m}{2k}\ln\left\{1 + \frac{4.78kv_0^2}{4mg}\right\}$$
$$= \frac{m}{2k}\ln\left\{1 + \frac{1.195kv_0^2}{mg}\right\}.$$

Now, as $k \to 0$, we have

$$S = \frac{m}{2k}\left(\frac{1.195kv_0^2}{mg}\right) = \frac{0.5975v_0^2}{g} = \frac{0.5975(146.7)^2}{32.2} = 399.3 \text{ feet},$$

which is in very good agreement with the computer value of 399.1 feet. (If you use more than three decimal places in the intermediate calculations—which is all I used on my hand calculator—as well as 146.667 ft/s for the initial speed of 100 mph instead of 146.7 ft/s, then the agreement is even better.)

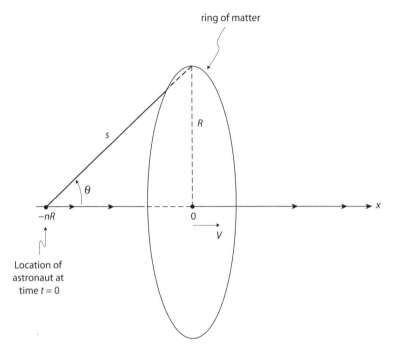

Figure S10.4. The geometry of the ring accelerator.

Challenge Problem P.10.1

Let the mass of the astronaunt be m and his speed be v. Then, from the geometry of the problem shown in Figure S10.4, we can write the gravitational force "felt" by the astronaunt that accelerates him along the x-axis toward the ring's center as $F = G\frac{m\mu}{s^2}\cos(\theta) = m\frac{d^2x}{dt^2}$, where $s = \sqrt{R^2 + x^2}$ and $\cos(\theta) = -\frac{x}{s} = -\frac{x}{\sqrt{R^2+x^2}}$. Thus, $\frac{d^2x}{dt^2} = -G\mu\frac{x}{(x^2+R^2)^{3/2}}$.

Since $v = \frac{dx}{dt} = \frac{dx}{dv}\cdot\frac{dv}{dt}$ and $\frac{d^2x}{dt^2} = \frac{dv}{dt}$, then $\frac{dv}{dt} = \frac{v}{\frac{dx}{dv}} = -G\mu\frac{x}{(x^2+R^2)^{3/2}}$, and so $v\,dv = -G\mu\frac{x}{(x^2+R^2)^{3/2}}\,dx$. Integrating indefinitely, with C the constant of integration, $\frac{1}{2}v^2 + C = -G\mu\int\frac{x}{(x^2+R^2)^{3/2}}\,dx$. Next, change variable to $u = x^2 + R^2$ (and so $dx = du/2x$). Then $\frac{1}{2}v^2 + C = -G\mu\int\frac{x\frac{du}{2x}}{u^{3/2}} = \frac{-G\mu}{2}\int\frac{du}{u^{3/2}} = \frac{-G\mu}{2}(-2u^{-1/2}) = \frac{G\mu}{\sqrt{u}} = \frac{G\mu}{\sqrt{x^2+R^2}}$. Since the astronaut starts from rest ($v = 0$) at $x = -nR$, we have

$$C = \frac{G\mu}{\sqrt{n^2R^2 + R^2}} = \frac{G\mu}{R\sqrt{1+n^2}}.$$

So,
$$\frac{1}{2}v^2 + \frac{G\mu}{R\sqrt{1+n^2}} = \frac{G\mu}{R\sqrt{1+\left(\frac{x}{R}\right)^2}},$$

or
$$v^2 = \frac{2G\mu}{R}\left[\frac{1}{\sqrt{1+\left(\frac{x}{R}\right)^2}} - \frac{1}{\sqrt{1+n^2}}\right].$$

We know $v = V$ at the ring's center ($x = 0$), and so

$$V = \sqrt{\frac{2G\mu}{R}\left(1 - \frac{1}{\sqrt{1+n^2}}\right)}.$$

If, for example, the ring has the mass and radius of the Earth, and (to pick a value at random) if $n = 3$, then

$$V = \sqrt{\frac{2 \cdot 6.67 \cdot 10^{-11} \cdot 5.98 \cdot 10^{24}}{6.37 \cdot 10^6}\left(1 - \frac{1}{\sqrt{10}}\right)} \text{ m/s}$$

$$= 9{,}254 \text{ m/s} \approx 5.7 \text{ miles/second}.$$

The maximum possible value for V is attained when n is very large, and is 11,190 m/s $\approx$ 6.9 miles/second. If $n = 0$ (that is, if the astronaut *starts* at rest at the ring's center), then our expression for V reduces to the obviously correct $V = 0$. To check the two claims made for Niven's ring, first recall that the acceleration at the ring's inner surface is, in feet/second^2 (where v is the ring rotation speed and r is the ring's radius), given by

$$\frac{v^2}{r} = \frac{(770 \cdot 5{,}280)^2}{(93{,}000{,}000 \cdot 5{,}280)} \text{ feet/second}^2 = 33.7 \text{ feet/second}^2$$

which is just slightly more than 1 g (32.2 ft/s^2), and that the area of the ring's inner surface (in units of the Earth's surface area) is (where w is the ring's width and R is the Earth's radius) given by

$$\frac{2\pi r w}{4\pi R^2} = \frac{93{,}000{,}000 \cdot 1{,}000{,}000}{2(3{,}960)^2} = 2{,}965{,}259,$$

which is pretty nearly the claimed factor of three million.

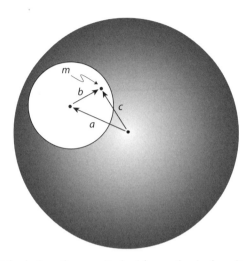

Figure S10.5. Calculating the gravity inside a spherical cavity.

Challenge Problem P.10.2

Figure S10.5 shows the geometry of the spherical Earth with a spherical cavity. A mass m is arbitrarily positioned inside the cavity. The vector from the center of the Earth to the center of the cavity is **a**, the vector from the center of the cavity to m is **b**, and the vector from the the center of the Earth to m is **c**. By the very definition of vector addition, we have **c** = **a** + **b**. Now, imagine first that there is no cavity. Then, the gravity force on m due to a solid Earth is, from the second superb theorem, due only to that part of the Earth no farther from the center of the Earth than $|\mathbf{c}|=c$, and that force is directed toward the center of the Earth (along **c**) (c is, of course, the *length* of **c**). This force is therefore the vector $\frac{G(\frac{4}{3}\pi c^3 \rho)m}{c^2}\left(-\frac{\mathbf{c}}{c}\right)$, where $-\frac{\mathbf{c}}{c}$ is the *unit* vector pointing along **c** toward the center of the Earth, that is, in the direction *opposite* that of **c**. Simplifying, this force is $-\frac{4}{3}G\pi\rho m$ **c**. Next, to account for the cavity, we "create" it by inserting into the Earth a sphere with negative density $-\rho$. The gravity force on m due to just this negative mass sphere is, again, because of the second superb theorem, due only to that mass no farther from the cavity's center than $|\mathbf{b}|=b$, where b is the length of **b**. But now, of course, that force is a *repulsive* force, directed away from the cavity's center along the direction of **b**. This force is the vector $\frac{G(\frac{4}{3}\pi b^3 \rho)m}{b^2}\left(\frac{\mathbf{b}}{b}\right) = \frac{4}{3}G\pi\rho m$ **b**.

The *total gravity* force on m is the sum of the two individual forces, which is $\frac{4}{3}G\pi\rho m(\mathbf{b}-\mathbf{c})$. But since $\mathbf{c}=\mathbf{a}+\mathbf{b}$ then $\mathbf{b}-\mathbf{c}=-\mathbf{a}$, and so the total gravity force on m is just $\frac{4}{3}G\pi\rho m(-\mathbf{a}) = \frac{4}{3}G\pi\rho ma\left(-\frac{\mathbf{a}}{a}\right)$, where $-\frac{\mathbf{a}}{a}$ is the unit vector from the center of the cavity to the center of the Earth. This is a force with constant magnitude $\frac{4}{3}G\pi\rho ma$, independent of the location of m within the cavity. There is no dependency on either the radius of the Earth or the radius of the cavity (I don't think this is at all obvious before doing the analysis). The magnitude of the force depends only on m, on the density of the Earth, and on how far apart the centers of the Earth and on the cavity are. The *direction* of the force is everywhere in the cavity parallel to the line joining the two centers, and directed to the Earth's interior (but not toward the center of the Earth except in the special case where m is at the center of the cavity).

Challenge Problem P.11.1

Let r, T_s, and T_m be the distance of the satellite from the Earth's center, the period of the satellite, and the period of the Moon, respectively. Then, from Kepler's third law, and using the given distance of the Moon from the Earth's center as 238, 850 miles, we have

$$\frac{T_m^2}{(238,850)^3} = \frac{T_s^2}{r^3}.$$

Since the satellite is geostationary we have $T_s = 1$ day, and since $T_m = 27.3$ days, then we can solve for r as

$$r = \left\{\frac{238,850^3 \cdot 1^2}{27.3^2}\right\}^{1/3} = 26,344 \text{ miles}.$$

The satellite is therefore $26{,}344 - 3{,}960 = 22{,}384$ miles above the surface of Earth.

Challenge Problem P.12.1

We can solve this problem with just a bit more generality than asked for, and then get our answers as special cases. Suppose the density of a planet increases linearly from ρ_0 at the surface to $n\rho_0$ at the center.

(Thus, $n = 2$ for Planet 1 and $n = 3$ for Planet 2.) Then, the density is $\rho(r) = \rho_0 \left[n - \frac{(n-1)r}{R} \right]$, where R is the radius of the planet and r is the distance from the center. The planet mass M is

$$M = \int_0^R 4\pi r^2 \rho(r)\, dr = 4\pi \rho_0 \int_0^R r^2 \left[n - \frac{(n-1)r}{R} \right] dr$$

$$= 4\pi \rho_0 \left[n \int_0^R r^2\, dr - \frac{(n-1)}{R} \int_0^R r^3\, dr \right]$$

$$= 4\pi \rho_0 \left[\frac{nR^3}{3} - \frac{(n-1)R^3}{4} \right] = 4\pi \rho_0 R^3 \left[\frac{n}{3} - \frac{n}{4} + \frac{1}{4} \right],$$

or $M = 4\pi \rho_0 R^3 \frac{n+3}{12} = \pi \rho_0 R^3 \frac{n+3}{3}$. Thus,

Planet 1 mass (with $n = 2$) $= \pi \rho_0 R^3 \frac{5}{3}$

Planet 2 mass (with $n = 3$) $= \pi \rho_0 R^3 \frac{6}{3}$

and so the percent increase in mass from Planet 1 to Planet 2 is

$$100 \frac{6-5}{5}\% = 20\%.$$

To calculate the percent increase in pressure P_c at the center of Planet 2 compared to the center pressure of Planet 1, I'll use the differential equation of static equilibrium of (12.14): $\frac{dP}{dr} = -g(r)\rho(r)$. We already have $\rho(r)$ from above, so all we need to write this differential equation is $g(r)$. The mass of the planet within a sphere of radius r is (with x as a dummy variable of integration) given by

$$M(r) = 4\pi \rho_0 \int_0^r x^2 \left[n - \frac{(n-1)x}{R} \right] dx$$

$$= 4\pi \rho_0 \left[n \int_0^r x^2\, dx - \frac{(n-1)}{R} \int_0^r x^3\, dx \right]$$

$$= 4\pi \rho_0 \left[\frac{nr^3}{3} - \frac{(n-1)r^4}{4R} \right] = 4\pi \rho_0 R^3 \left[\frac{nr^3}{3R^3} - \frac{(n-1)r^4}{4R^4} \right].$$

Since the gravitational force on a mass m at distance $r \leq R$ from the center of the planet is $F(r) = mg(r) = \frac{GmM(r)}{r^2}$, then $g(r) = \frac{GM(r)}{r^2}$, and so $g(r) = 4\pi \rho_0 G R^3 \left[\frac{nr}{3R^3} - \frac{(n-1)r^2}{4R^4}\right]$. Thus,

$$\frac{dP}{dr} = -4\pi \rho_0^2 G R^3 \left[n - \frac{(n-1)r}{R}\right]\left[\frac{nr}{3R^3} - \frac{(n-1)r^2}{4R^4}\right],$$

or, if we write $C = 4\pi \rho_0^2 G R^3$,

$$dP = -C\left[\frac{n^2 r}{3R^3} - \frac{n(n-1)r^2}{4R^4} - \frac{n(n-1)r^2}{3R^4} + \frac{(n-1)^2 r^3}{4R^5}\right] dr.$$

Integrating r from 0 (where $P = P_c$) to $r = R$ (where $P = 0$), we have

$$P_c = C \int_0^R \left[\frac{n^2 r}{3R^3} - \frac{n(n-1)r^2}{4R^4} - \frac{n(n-1)r^2}{3R^4} + \frac{(n-1)^2 r^3}{4R^5}\right] dr$$

$$= C\left[\frac{n^2}{6R} - \frac{n(n-1)}{12R} - \frac{n(n-1)}{9R} + \frac{(n-1)^2}{16R}\right]$$

or, after some simple algebraic reduction, $P_c = \frac{C}{R}\left[\frac{5n^2 + 10n + 9}{144}\right]$. So,

$$P_c(n=2) = \frac{C}{R} \cdot \frac{49}{144}$$

$$P_c(n=3) = \frac{C}{R} \cdot \frac{84}{144}$$

and therefore the percent increase in center pressure from Planet 1 to Planet 2 is

$$100 \frac{84 - 49}{49} \% = 71.4\%.$$

Challenge Problem P.12.2

With reference to Figure S12.7, let d denote the distance *along the curved surface* of the Earth between the points A and B. The angle θ

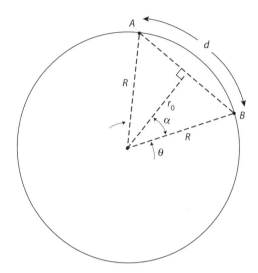

Figure S12.7. A tunnel through the Earth.

between the two radii to A and B is $\theta = 2\pi \frac{d}{2\pi R} = \frac{d}{R}$, where R is the radius of the Earth. Thus, the half-angle $\alpha = \frac{1}{2}\theta = \frac{d}{2R}$. Now, with r_0 as the distance from the center of the Earth to the center of the tunnel, $\frac{r_0}{R} = \cos(\alpha) = \cos(\frac{d}{2R})$, or $r_0 = R\cos(\frac{d}{2R})$. As the figure illustrates, the maximum depth, $maxd$, of the tunnel beneath the Earth's surface is

$$maxd = R - r_0 = R - R\cos\left(\frac{d}{2R}\right) = R\left[1 - \cos\left(\frac{d}{2R}\right)\right].$$

For A as New York City and B as either Boston or Philadelphia, $d \ll R$, and since $\cos(x) \approx 1 - \frac{1}{2}x^2$ for $x \ll 1$, then, to a good approximation,

$$maxd = R\left[1 - 1 + \frac{1}{2}\left(\frac{d}{2R}\right)^2\right] = \frac{d^2}{8R}.$$

For the New York City–Boston tunnel, with $R = 3,960$ miles and $d = 190$ miles,

$$maxd = \frac{(190)^2}{8 \cdot 3,960} \text{miles} = 1.14 \text{miles} = 6,017 \text{feet},$$

which is pretty deep, but may be not impossibly deep. For the New York City–Philadelphia tunnel, with $d = 85$ miles,

$$maxd = \frac{(85)^2}{8 \cdot 3{,}960} \text{ miles} = 0.228 \text{miles} = 1{,}204 \text{ feet},$$

which seems (to me) to be within the realm of the "reasonably possible".

Challenge Problem P.12.3

If we take the center of the Earth as our zero potential energy reference point, then a mass m on the surface of the Earth has a potential energy equal to the energy required to transport it (against the Earth's gravity force $F(r)$) from the center to the surface. This force is given by

$$F(r) = G\frac{M(r)m}{r^2} = G\frac{\frac{4}{3}\pi r^3 \rho m}{r^2} = \frac{4}{3}G\pi r \rho m$$

where of course $M(r)$ is the mass inside a sphere of radius r and ρ is the (assumed constant) density of the Earth. The transport energy is then

$$\int_0^R F(r)dr = \frac{4}{3}G\pi\rho m \int_0^R r\,dr = \frac{2}{3}G\pi\rho m R^2.$$

Now, if r_0 is the distance from the center of the Earth to the center of the tunnel (where the traveler's speed is maximum $= V_m$), we then have, by conservation of energy,

$$\frac{2}{3}G\pi\rho m R^2 - \frac{2}{3}G\pi\rho m r_0^2 = \frac{1}{2}mV_m^2,$$

where the left-hand side is the change in potential energy from the surface to the tunnel's midpoint and the right-hand side is the change in the kinetic energy. So,

$$V_m = \sqrt{\frac{4}{3}\pi G\rho(R^2 - r_0^2)}.$$

As shown in the previous solution, $r_0 = R - \frac{d^2}{8R}$ as long as $d \ll R$. So,

$$V_m = \sqrt{\frac{4}{3}\pi G\rho(R^2 - R^2 + \frac{d^2}{4} - \frac{d^4}{64R^2})} \approx d\sqrt{\frac{\pi G\rho}{3}}.$$

For the New York City–Boston tunnel,

$$V_m = 190\sqrt{\frac{\pi 6.67 \cdot 10^{-11}\frac{m^3}{kg \cdot s^2} \cdot 5,540\frac{kg}{m^3}}{3}} \text{ miles/second}$$

$$= 0.118 \text{ miles/second} = 425.5 \text{ mph},$$

which is far less (and far more reasonable) than is Goddard's speed. Since the maximum speed scales linearly with d, then for the New York City–Philadelphia tunnel we get the even less extreme speed of $V_m = 190$ mph. For a full-diameter trip we simply set $r_0 = 0$ and get

$$V_m = 2R\sqrt{\frac{\pi G \rho}{3}} = 2 \cdot 3,960\sqrt{\frac{\pi 6.67 \cdot 10^{-11}\frac{m^3}{kg \cdot s^2} \cdot 5,540\frac{kg}{m^3}}{3}} \text{ miles/second}$$

$$\approx 4.9 \text{ miles/second}.$$

Compare this result with the answer to Challenge Problem P.6.1.

Challenge Problem P.12.4

Write the acceleration of gravity as g. The mass falling from point A takes time T_A to fall a distance of $2R$. Thus, $2R = \frac{1}{2}gT_A^2$ or, $T_A^2 = 4R/g$. The mass at point B slides a distance of l. Since, as given in the hint, the triangle ABC is a right triangle, then if θ is the angle the wire (along which the sliding mass travels) makes with the horizontal, then θ is also the vertex angle of the triangle at A. Thus, $l/2R = \sin(\theta)$. Now, the acceleration of the mass along the wire is $a = g\sin(\theta) = l\,g/2R$. If the time for the slide is T_B, then $l = \frac{1}{2}aT_B^2 = \frac{1}{2} \cdot \frac{lg}{2R}T_B^2 = \frac{lgT_B^2}{4R}$. Since l cancels on both sides, then $T_B^2 = 4R/g$. But this is T_A^2. So, $T_B = T_A$.

Challenge Problem P.13.1

Figure 13.8 is reproduced below (as Figure S13.11a) with the subsquares labeled in bold as **a**, **b**, **c**, and so on. If we call the edge lengths of the **a**, **b**, and **c** subsquares x, y, and w, respectively, then we can immediately write the edge length of **d** as $y - w$, and the edge

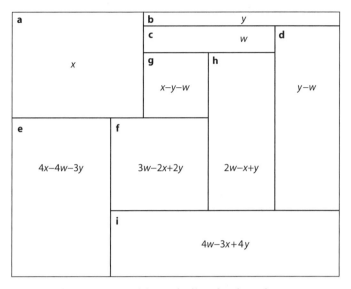

Figure S13.11a. Figure again (with symbolic edge lengths).

length of **g** as $x - y - w$, as shown. That tells us that the edge length of **h** is $w - (x - y - w) = 2w - x + y$, as shown. Then, comparing **c** and **h** with **d**, we see that $w + 2w - x + y = y - w$, that is, $x = 4w$. Continuing, from **g** and **h** we see that the edge length of **f** is $(2w - x + y) - (x - y - w) = 3w - 2x + 2y$, as shown. The edge length of **i** is the sum of the edge lengths of **f**, **h**, and **d**, that is, $(3w - 2x + 2y) + (2w - x + y) + (y - w) = 4w - 3x + 4y$, as shown. And finally, the edge length of **e** is obviously $(x + y) - (4w - 3x + 4y) = 4x - 4w - 3y$. Now, using the vertical constraints on the far left and right edges, we have $(x) + (4x - 4w - 3y) = (y) + (y - w) + (4w - 3x + 4y)$, which reduces to $8x = 7w + 9y$. Remembering that $x = 4w$, we have $32w = 7w + 9y$, or $25w = 9y$. An obvious solution of these two relations, $x = 4w$ and $25w = 9y$, is $y = 25$, $w = 9$, and $x = 36$. If these values are inserted into the expressions for the individual subsquare edge lengths we get the solution tiling, with no two sub-squares equal, shown in Figure S3.11b.

Challenge Problem P.13.2

Let's label the smallest subsquare of a tiling as S, which we know from the discussion in the text is *not* on the border. Following the hint, let's

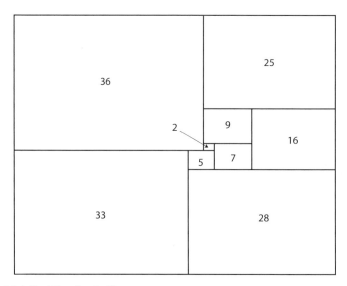

Figure S3.11b. The final tiling.

assume we have five subsquares surrounding S. To do this, we must have two (each larger than S, of course) sub-squares along one of S's edges, something like as shown in Figure S13.12a. Then, adding yet another subsquare (call it X)—which *must* go along one of S's vertical sides, because if X is along S's bottom side then no (larger than S) subsquare could fit along either vertical side of S—we have Figure S13.12b. And now you can see we are only one step away from failure. If we position yet another (larger than S) subsquare along the bottom side of S then we can't get a (larger than S) subsquare along the remaining vertical side of S, and vice versa. If we assume S has even more than five adjacent neighbors, then we obviously have a similar disaster. So, the only possible tiling is one with exactly one subsquare along each side of S (that is, *four* adjacent neighbors), as shown in Figure S13.12c.

Challenge Problem P.14.1

The MATLAB code river.m performs one million walks, each six days in duration. The fundamental idea behind the code is that, if a return to the river occurs on day x, then a return to the river occurs for *all*

CHALLENGE PROBLEM P.14.1 ---- **361**

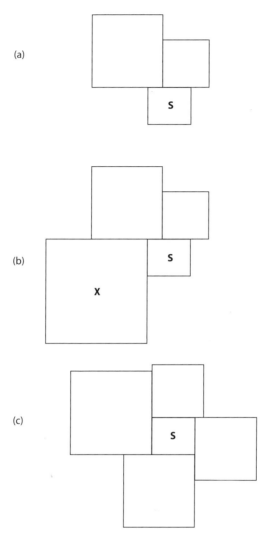

Figure S13.12. The neighbors of S.

walks of duration $x, x+1, x+2, \cdots$. Note carefully that a return to the river does *not* have to occur on the last day of a walk. Indeed, there may be multiple returns to the river during a given walk, but only the occurrence of the *first* return matters. In detail, then, here's how river.m works. Line 01 creates the six-element vector *riverret*, initially with all its elements set to zero. When the simulation ends *riverret(k)* will equal the number of times a journey of duration k days returned to the river. Line 02 simply sets the constant C to 2π; I did that here to avoid constantly multiplying 2 by π when the code gets down to work. The outermost loop, defined by lines 03 and 25, counts off the one million journeys. When that loop is entered, at line 04, to start each new simulated journey, the vector *angle* is loaded with six random numbers uniformly distributed from 0 to 2π. These are the angles, one after the other, that our traveler starts off with each new day; of course, the value of the first angle should be limited to the interval 0 to π, and that explains line 05. Line 06 generates the distance the traveler is from the river at the end of the first day, storing it as *distance(1)*, and then lines 07 through 09 fill in the rest of the elements of the vector *distance*; that is, *distance(k)* is the distance the traveler is from the river at the end of day k. Lines 10 through 20 now look at *distance* to determine if a return to the river has occurred (in line 13). This examination begins with *distance(2)*—notice that the variable *loop3* is initialized to 2 in line 11, and the variable *exit* is initialed to 0 in line 10—and continues until the first occurrence of a return. When (if) that happens, *exit* is set to 1 in line 14 (to terminate the *while* loop that is examining *distance*), and then all elements of *riverret* indexed on the current value of *loop3* up to 6 are incremented by 1 to record a return to the river for all journeys of durations *loop3* days to 6 days. If a first return does not occur for a journey of duration *loop3* days, however, then line 19 increments *loop3* by 1 (and lines 21 through 23 force *exit=1* to terminate the *while* loop if the code has examined all six elements of *distance*). This all happens all over again for a total of one million journeys, and then line 26 prints the code's estimates for the probability of a return to the river for journeys of length k days. When run, **river.m** produced the following estimates:

---

river.m

```
01    riverret=zeros(1,6);
02    C=2*pi;
03    for loop1=1:1000000
04        angle=C*rand(1,6);
05        angle(1)=angle(1)/2;
```

(continued)

(continued)

```
06      distance(1)=sin(angle(1));
07      for loop2=2:6
08          distance(loop2)=distance(loop2-1)
                +sin(angle(loop2));
09      end
10      exit=0;
11      loop3=2;
12      while exit==0
13          if distance(loop3)<=0
14              exit=1;
15              for j=loop3:6
16                  riverret(j)=riverret(j)+1;
17              end
18          else
19              loop3=loop3+1;
20          end
21          if loop3==7
22              exit=1;
23          end
24      end
25  end
26  riverret/1000000
```

Challenge Problem P.14.2

From (14.1) we have the pdf of the random variable **X** as

$$f_{\mathbf{X}}(x) = \begin{cases} \frac{2}{n}xe^{-\frac{x^2}{n}}, & x \geq 0 \\ 0, & x < 0. \end{cases}$$

That is, **X** is a Rayleigh random variable. To find the pdf of $\mathbf{Y} = \mathbf{X}^2$, we first find **Y**'s distribution function and then differentiate. So,

$$F_{\mathbf{Y}}(y) = P(\mathbf{Y} \leq y) = P(\mathbf{X}^2 \leq y) = P(-\sqrt{y} \leq \mathbf{X} \leq \sqrt{y})$$

duration (days)	probability of returning to the river
2	0.2500
3	0.3749
4	0.4534
5	0.5081
6	0.5490

Notice that the estimate for two days is pretty close to the estimate produced by the far less general code **crofton.m** given in Discussion 14 (section 14.4).

or, since **X** is never negative,

$$F_\mathbf{Y}(y) = P(0 \leq \mathbf{X} \leq \sqrt{y}) = \int_0^{\sqrt{y}} f_\mathbf{X}(x)\,dx = \int_0^{\sqrt{y}} \frac{2}{n}xe^{-\frac{x^2}{n}}\,dx.$$

This integral is actually doable, but why bother? We are, after all, just going to differentiate the result! So, let's differentiate the integral (using Leibnitz's formula) straightaway to get

$$f_\mathbf{Y}(y) = \frac{d}{dy}F_\mathbf{Y}(y) = \frac{2}{n}\sqrt{y}\,e^{-\frac{(\sqrt{y})^2}{n}} \cdot \frac{1}{2} \cdot \frac{1}{\sqrt{y}} = \frac{1}{n}e^{-\frac{y}{n}}.$$

That is,

$$f_\mathbf{Y}(y) = \begin{matrix} \frac{1}{n}e^{-\frac{y}{n}}, y \geq 0 \\ 0, y < 0 \end{matrix}$$

which is, indeed, the pdf of an exponential random variable, as was to be shown.

Challenge Problem P.15.1

The MATLAB code **cp15.m** estimates the pdf of a 21-step stretched walk with $s = 1.01$. The result is shown in Figure S15.6.

---

cp15.m

```
01    N=21;s=1.01;location=zeros(1,2097152);
02    for k=1:N
```

(continued)

(continued)

```
03          step(k)=s^(k-1);
04      end
05      for walk=1:2097152
06          x=0;
07          for k=1:N
08              if rand<0.5
09                  x=x+step(k);
10              else
11                  x=x-step(k);
12              end
13          end
14          location(walk)=x;
15      end
16      L=(s^N-1)/(s-1);
17      edges=linspace(-L,L,4648);
18      n=histc(location(:),edges);
19      pdf=n/20971.52;
20      plot(edges,pdf,'k')
21      axis([-24 24 0 .31])
22      xlabel('distance','FontSize',16)
23      ylabel('estimated pdf','FontSize',16)
24      title('Figure S15.6 - Stretched walk pdf for
                s=1.01','FontSize',16)
25      text(-2,.26,'N=21 steps','FontSize',16)
```

Challenge Problem P.16.1

Differentiating the expression given in the problem statement, $\int_0^z f_{\mathbf{Z}}(u)\, du = 1 - (1 - \frac{z^3}{R^3})^N$, we arrive at $f_{\mathbf{Z}}(z) = -N(1 - \frac{z^3}{R^3})^{N-1}\left(-\frac{3z^2}{R^3}\right)$,

or

$f_{\mathbf{Z}}(z) = \frac{3z^2 N}{R^3}(1 - \frac{z^3}{R^3})^{N-1}$, $0 \leq z \leq R$. Thus, $E(\mathbf{Z}) = \frac{3N}{R^3} \int_0^R z^3 (1 - \frac{z^3}{R^3})^{N-1}\, dz$.

Since
$$\left(1 - \frac{z^3}{R^3}\right)^{N-1} = \sum_{k=0}^{N-1} \binom{N-1}{k} \left(-\frac{z^3}{R^3}\right)^k,$$
then
$$z^3 \left(1 - \frac{z^3}{R^3}\right)^{N-1} = R^3 \frac{z^3}{R^3} \sum_{k=0}^{N-1} \binom{N-1}{k} (-1)^k \left(\frac{z^3}{R^3}\right)^k$$
$$= R^3 \sum_{k=0}^{N-1} \binom{N-1}{k} (-1)^k \left(\frac{z^3}{R^3}\right)^{k+1}$$

and so
$$E(\mathbf{Z}) = 3N \sum_{k=0}^{N-1} \binom{N-1}{k} (-1)^k \int_0^R \left(\frac{z^3}{R^3}\right)^{k+1} dz$$
$$= 3N \sum_{k=0}^{N-1} \binom{N-1}{k} (-1)^k \frac{1}{R^{3k+3}} \int_0^R z^{3k+3} dz$$
$$= 3N \sum_{k=0}^{N-1} \binom{N-1}{k} (-1)^k \frac{1}{R^{3k+3}} \left(\frac{z^{3k+4}}{3k+4}\right)\bigg|_0^R,$$

or, at last,
$$E(\mathbf{Z}) = \left[3N \sum_{k=0}^{N-1} \binom{N-1}{k} \frac{(-1)^k}{3k+4}\right] R.$$

Evaluating numerically, we have:

N	$E(\mathbf{Z})$
1	0.7500R
2	0.6429R
3	0.5786R
4	0.5431R
5	0.5007R
6	0.4743R
7	0.4528R
8	0.4347R
9	0.4191R
10	0.4056R

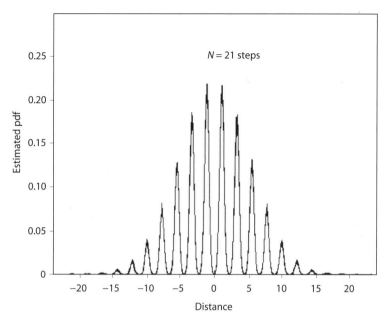

Figure S15.6. Stretched walk pdf for $s = 1.01$

Challenge Problem P.16.2

Let **W** be the random variable representing the distance of the *most distant* star from the center star; the distribution function of **W** is $F_\mathbf{W}(w) = P(\mathbf{W} \leq w)$. Thus, since $\frac{w^3}{R^3}$ is the probability *each* of the individual neighbor stars is within distance w of the center star—and so *all* the neighbor stars are within distance w of the center star (all the other neighbor stars must, by definition, be even closer than the *farthest* star)—then

$$F_\mathbf{W}(w) = \left(\frac{w^3}{R^3}\right)^N = \left(\frac{w}{R}\right)^{3N}.$$

For the most distant star to be no closer than $0.9R$, **W** must be in the interval $0.9R \leq \mathbf{W} \leq R$. The probability of that is

$$P(0.9R \leq \mathbf{W} \leq R) = F_\mathbf{W}(R) - F_\mathbf{W}(0.99R) = \left(\frac{R}{R}\right)^{3N} - \left(\frac{0.9R}{R}\right)^{3N}$$

$$= 1 - (0.9)^{3N}.$$

For this probability to exceed 0.99, that is, for $1 - (0.9)^{3N} > 0.99$, we have $0.01 > (0.9)^{3N}$, or, taking logs on both sides (to the base 10),

$$-2 > 3N \log_{10}(0.9) = -0.137N \quad \text{or} \quad N > 2/0.137 = 14.57,$$

which of course means that $N \geq 15$.

Challenge Problem P.17.1

If $\frac{b}{a+b} = \frac{d}{c+d}$, then $bc + bd = ad + bd$, and so $bc = ad$. This says that $b = \frac{ad}{c}$, and so $b + c = \frac{ad}{c} + c$. Thus,

$$\frac{(a+b)(c+d)}{a+b+c+d} = \frac{(a+b)(c+d)}{a + \frac{ad}{c} + c + d} = \frac{(a+b)(c+d)}{\frac{ac+ad+c(c+d)}{c}}$$

$$= \frac{(a+b)(c+d)c}{a(c+d) + c(c+d)} = \frac{(a+b)c}{a+c} = \boxed{\frac{ac}{a+c} + \frac{bc}{a+c}}.$$

Because $bc = ad$, then $bc + cd = ad + cd$, and so $c(b+d) = d(a+c)$, and therefore $bc(b+d) = bd(a+c)$, which says $\frac{bc}{a+c} = \frac{bd}{b+d}$. Using this in the box, we have $\frac{(a+b)(c+d)}{a+b+c+d} = \frac{ac}{a+c} + \frac{bd}{b+d}$ (*if* the given condition is true), and we are done.

Challenge Problem P.17.2

Call the required resistance R. We know that if we replace the missing one-ohm resistor, that is, if we put it in *parallel* with R, then we'll get the original, classic infinite 2-D resistor network with a resistance between A and B of $\frac{1}{2}$ ohm. So, $R \| 1 = \frac{1}{2} = \frac{R}{R+1}$ or $R + 1 = 2R$ or, at last, $R = 1$ ohm.

Challenge Problem P.17.3

We can solve the infinite honeycomb resistor circuit of Figure 3.11 in the same way we did the infinite square grid of Figure 17.6. Start by inserting one ampere into A and removing it at infinity. By symmetry, the current out of A splits equally into three currents of $\frac{1}{3}$ ampere each, and the $\frac{1}{3}$ ampere heading toward B then splits equally (again, by symmetry) into two currents of $\frac{1}{6}$ ampere each. So, the voltage drop from A to B is $\frac{1}{3}$ volt, and the voltage drop from A to C is $\frac{1}{3} + \frac{1}{6} = \frac{1}{2}$ volt. Next, insert one ampere into infinity and remove it at B. Making the same sort of symmetry argument, there will again be a $\frac{1}{3}$ volt drop from A to B. Then, using superposition, we conclude that one ampere into A and out at B results in a total voltage drop of $\frac{2}{3}$ volts. So, the resistance between A and B is $\frac{2}{3}$ ohm. If we remove the one ampere inserted at infinity from C, then the voltage drop from A to C is $\frac{1}{2} + \frac{1}{2} = 1$ volt, and so the resistance between A and C is one ohm.

Challenge Problem P.17.4

I'll put the best reader solutions I receive in the paperback edition of this book.

Challenge Problem P.18.1

As derived in the discussion, the acceleration of the airplane in the example is $\boxed{\frac{1,667}{100-t} \text{ ft/s}^2}$. If the plane has completed one loop then, from (18.11),

$$\theta(t) = 2\pi = \sqrt{1.25} \ln\left(\frac{1}{1 - \frac{t}{100}}\right) = \sqrt{1.25} \ln\left(\frac{100}{100-t}\right)$$
$$= \sqrt{1.25}\left[\ln(100) - \ln(100-t)\right].$$

Thus,

$$\ln(100-t) = \ln(100) - \frac{2\pi}{\sqrt{1.25}} = -1.014681599,$$

and so

$$100 - t = e^{-1.014681599} = 0.362517837.$$

This says the first loop is completed at time $t = 99.637$ seconds (not leaving much time for the infinity of loops yet to do!), and the acceleration at that instant is, from the boxed expression, $4,598.4$ ft/s$^2 = 142.8\,g$.

SPECIAL BONUS DISCUSSION

Do **not** Read before Reading Discussion 17

Electricity in the Fourth Dimension

> The other day we heard from a student that one of his friends was doing a lab project constructing a four-dimensional cube with one-ohm resistors and measuring the resistance across two opposite vertices. This prompted us to try to solve the problem analytically.
> — O. J. Tretiak and T. S. Huang, "Resistance of *N*-Dimensional Cube" (1965)

> "What's a tesseract?" "Didn't you go to school? A tesseract is a hypercube, a square figure with four dimensions to it. . . ."
> — Dialogue from Robert Heinlein's short story, "And He Built a Crooked House" (February 1941)

19.1 The Tesseract

Back in Section 17.3 I showed you how to analyze the three-dimensional resistor cube, made from one-ohm resistors, with the result that an applied voltage difference of one volt across two opposite vertices would result in a current into and out of the cube of 6/5 amperes. That is, the so-called *body diagonal* resistance of the cube is 5/6 ohm. You may not have even thought of it then—we take "seeing" in 3-D so much for granted—but one reason we could solve the problem so easily (once we had the symmetry and superposition trick in hand) is because we can *see*, that is we *know*, how to connect that cube's twelve resistors together. But how did the student in the above first quote know how to connect the thirty-two resistors (you'll see where that number comes from soon) of a 4-D cube—often called a *tesseract*—together? And how can you actually make a tesseract in the lab, anyway?

In this final, mostly for fun, appendix of the book I'll show you a nifty way to discover the connection topology of a tesseract (actually, the connection topology of a resistor cube of *any* dimension) using binary numbers, and then we can simply put that topology into the MATLAB computer code mccube.m of Section 17.3. First, however, you must shake off any psychological hang-ups you have about the fourth dimension. It isn't mysticism, it isn't science fiction,[1] and it isn't the inane babbling of pseudoscientific cranks. Mathematicians and physicists, in fact, often work in *infinite*-dimensional spaces. The *fourth* dimension is mere child's play in comparison. Our task, as beings living in a world with three space dimensions, of trying to understand the "structure" of a tesseract is entirely analogous to that of a being living in a world with two space dimensions trying to understand the "structure" of a cube. That task was, of course, the theme of Edwin Abbott's 1884 masterpiece *Flatland* (authored, so it was claimed at the time of the book's original publication, by the flatlander A. Square).[2] To see how both A. Square and we could each solve our respective problems, I'll begin in a space with zero space dimensions and then, from the only object that exists in that so-called null space—a point— we'll work our way up to the cube in four dimensions.

But first, before we do that, let me address one issue that might be on your mind. As one writer expressed it,[3] "Isn't time the fourth dimension? Time is sometimes treated mathematically as something like a fourth dimension, but time is "imaginary" in the mathematical sense of involving the square root of minus one.[4] The fourth dimension which I am about to discuss is a fourth *space* dimension, exactly like the three—length, width, and height—with which we are all familiar; and standing at right angles to all three, just as each of them stands at right angles to the other two." With time as the fourth dimension we have the famous four-dimensional *space-time* of Einstein, whose *curvature* reduced the apparently "occult" power of gravity as an action-at-a-distance force (which perplexed even the great Newton) to mere geometry. But that isn't the four-dimensional space we are discussing

object	points	line segments	squares	cubes
point	1	0	0	0
line segment	2	1	0	0
square	4	4	1	0
cube	8	12	6	1

here; our four-dimensional space is one with all four dimensions as dimensions of length.

Okay, now we can start.

We start with a point and move it a certain distance—we can, with no loss of generality, take it to be the unit distance—along our *first* space dimension. The result is a single line segment with *two* end points, which we'll call *vertices*. We follow this by a second motion, that is, we move the line segment a unit distance along our second space dimension, which is in a direction at right angles to our first motion. The result is a *square*, with four vertices and four line segments. What has happened is that our vertices have doubled, that is, each original vertex has duplicated itself. Also, our square has the original line segment, plus the original line segment in its new position, plus the two new line segments traced out by the motions of the original two vertices.

Next, we perform a third motion: we move the square a unit distance along our third space dimension which is at right angles to both of the first two motions. The result is a cube. Again, each of the square's four vertices duplicates itself, and so the number of vertices in a cube is eight. Also, the number of line segments is equal to the original number (4), plus the original line segments in their final positions (4), plus the new line segments traced out by the motions of the original four vertices (4), for a total of twelve line segments. The number of square surfaces (faces) is the original surface (1), plus the original surface in its final position (1), plus the four new square surfaces traced out by the motion of the original number of line segments (4), for a total of six square surfaces. We can summarize all of this as above, where we read across a row to see, for a given object, how many points, line segments, squares, etc. are in that object:

If you think over the previous discussion you should be able to see how each entry, in each row, is generated by the following two rules: (1) the number of points in a n-dimensional cube is 2^n and (2) all the other numbers in a row are twice the number directly above, plus the number directly to the left of the number just doubled. For example, the number of squares in a cube is $6 = 2 \cdot 1 + 4$. The value of these rules is that, using them, we can now construct the next row for a tesseract *without having to move a cube unit distance along a fourth space dimension at right angles to the first three dimensions.* (If you say you can visualize doing that, well, good for you—but I don't believe it!) The result, for the tesseract, as you can easily confirm, is that it has sixteen vertices, thirty-two line segments, twenty-four square surfaces, and eight cubes. And once you have the tesseract row, you could keep right on going to generate the row for the 5-D cube and so learn that it has thirty-two vertices, eighty line segments, eighty square surfaces, forty cubes, and ten tesseracts![5] And then you could do the 6-D cube, the 7-D cube, ... well, maybe I'm getting carried away here and we *won't* do that. Back to business.

Now we know that the four-dimensional resistor cube has thirty-two resistors (a resistor for each line segment that forms an edge of the tesseract), but that doesn't tell us how they are all connected together. For that, we turn next to binary numbers.

19.2 Connecting a Tesseract Resistor Cube

To see how to solve the connection problem, imagine you are A. Square in Flatland; how could he have figured out how a resistor cube in three dimensions is connected, even though he couldn't "see" a three-dimensional object? Each of the eight vertices (or nodes) of the resistor cube can be located in three-dimensional space by specifying its coordinates, which will of course require three numbers, one for each dimension. Because we have constructed our cube out of motions of unit length along the various dimensions, all at right angles to each other, these numbers will be either 0 or 1, and so the coordinates of the eight nodes are the eight three-bit binary groups from (0,0,0) to (1,1,1), that is, we are numbering the nodes from 0 to 7.

Now, here's the crucial observation: when we move from one node through a connecting resistor to the next node, we change just *one* of the bits in the initial binary group; for example, (0,0,0) is connected through a resistor to (0,0,1), through another resistor to (0,1,0), and through a third resistor to (1,0,0). Using this idea[6] on all eight nodes in the cube, A. Square could have constructed the following connection wire list for the 3-D cube:

node	connects	to	nodes		node	connects	to	nodes
(0,0,0)	(0,0,1)	(0,1,0)	(1,0,0)		0	1	2	4
(0,0,1)	(0,0,0)	(0,1,1)	(1,0,1)		1	0	3	5
(0,1,0)	(0,1,1)	(0,0,0)	(1,1,0)		2	3	0	6
(0,1,1)	(0,1,0)	(0,0,1)	(1,1,1)	⇒	3	2	1	7
(1,0,0)	(1,0,1)	(1,1,0)	(0,0,0)		4	5	6	0
(1,0,1)	(1,0,0)	(1,1,1)	(0,0,1)		5	4	7	1
(1,1,0)	(1,1,1)	(1,0,0)	(0,1,0)		6	7	4	2
(1,1,1)	(1,1,0)	(1,0,1)	(0,1,1)		7	6	5	3

where it should be clear that node $(0, 0, 0) = 0$ and node $(1, 1, 1) = 7$ are opposite nodes, that is, body diagonal nodes. This last wire list is still not quite in the proper form for use by mccube.m, however, because it violates a couple of that code's conventions. First, there can be no node 0; mccube.m assumes the numbering of the nodes starts with 1. And second, mccube.m wants to see nodes 1 and 2 as the body diagonal nodes (the nodes where the positive and negative terminals of the applied voltage source connect), not 0 and 7. Both of these transgressions are, fortunately, easy to fix. First, relabel 0 as 1 and 1 as 8. Second, relabel 7 as 2 and 2 as 7. That gives us the following final wire list:

node	connects	to	nodes
1	8	7	4
2	6	5	3
3	7	8	2
4	5	6	1
5	4	2	8
6	2	4	7
7	3	1	6
8	1	3	5

This wire list is what A. Square in Flatland could have constructed and, in fact, as Figure 19.1 shows, it does indeed correctly represent the connection of a 3-D cube. The nodes are not labeled as in Figure 17.3 (except for the convention of where nodes 1 and 2 have to be), but that just confirms what I stated in Discussion 17, that the labeling of all nodes other than 1 and 2 is arbitrary. Note *carefully* that A. Square does not need to actually draw Figure 19.1; I've included it here simply to illustrate that the binary method for constructing the cube's wire list does indeed work. All that matters, however, *all that A. Square needs*, is the wire connection list.

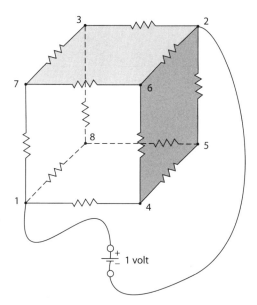

Figure 19.1. A square's 3-D cube.

Now, to answer the question that opened this discussion—what is the body diagonal resistance of a 4-D resistor cube?—for the analogous problem faced by A. Square with the 3-D cube, he could next have fed mccube.m the above final wire list and used the code to find the voltages at nodes 3, 4, and 7, and so have calculated the input current from the one-volt voltage source to the cube as

$$[1 - V(4)] + [1 - V(7)] + [1 - V(8)] = 3 - V(4) - V(7) - V(8).$$

The body diagonal resistance of a 3-D cube would then be

$$R_3 = \frac{1}{3 - V(4) - V(7) - V(8)}.$$

When I did this the result was $R_3 = 0.8336$ ohms, which compares well with the theoretical result of $5/6 = 0.83333$ ohms. Well, what A. Square could do[7] in Flatland to find R_3 we can do in our world to find R_4.

I'll begin as before, by writing the coordinates of the tesseract's sixteen vertices as four-bit binary code groups from $(0, 0, 0, 0)$ to $(1, 1, 1, 1)$. Then, again as before, we can construct a preliminary connection wire list by examining where each code group "goes" as we successively change just one bit. This results in

node	connects	to	nodes	
(0,0,0,0)	(0,0,0,1)	(0,0,1,0)	(0,1,0,0)	(1,0,0,0)
(0,0,0,1)	(0,0,0,0)	(0,0,1,1)	(0,1,0,1)	(1,0,0,1)
(0,0,1,0)	(0,0,1,1)	(0,0,0,0)	(0,1,1,0)	(1,0,1,0)
(0,0,1,1)	(0,0,1,0)	(0,0,0,1)	(0,1,1,1)	(1,0,1,1)
(0,1,0,0)	(0,1,0,1)	(0,1,1,0)	(0,0,0,0)	(1,1,0,0)
(0,1,0,1)	(0,1,0,0)	(0,1,1,1)	(0,0,0,1)	(1,1,0,1)
(0,1,1,0)	(0,1,1,1)	(0,1,0,0)	(0,0,1,0)	(1,1,1,0)
(0,1,1,1)	(0,1,1,0)	(0,1,0,1)	(0,0,1,1)	(1,1,1,1)
(1,0,0,0)	(1,0,0,1)	(1,0,1,0)	(1,1,0,0)	(0,0,0,0)
(1,0,0,1)	(1,0,0,0)	(1,0,1,1)	(1,1,0,1)	(0,0,0,1)
(1,0,1,0)	(1,0,1,1)	(1,0,0,0)	(1,1,1,0)	(0,0,1,0)
(1,0,1,1)	(1,0,1,0)	(1,0,0,1)	(1,1,1,1)	(0,0,1,1)
(1,1,0,0)	(1,1,0,1)	(1,1,1,0)	(1,0,0,0)	(0,1,0,0)
(1,1,0,1)	(1,1,0,0)	(1,1,1,1)	(1,0,0,1)	(0,1,0,1)
(1,1,1,0)	(1,1,1,1)	(1,1,0,0)	(1,0,1,0)	(0,1,1,0)
(1,1,1,1)	(1,1,1,0)	(1,1,0,1)	(1,0,1,1)	(0,1,1,1)

or

node	connects	to	nodes	
0	1	2	4	8
1	0	3	5	9
2	3	0	6	10
3	2	1	7	11
4	5	6	0	12
5	4	7	1	13
6	7	4	2	14
7	6	5	3	15
8	9	10	12	0
9	8	11	13	1
10	11	8	14	2
11	10	9	15	3
12	13	14	8	4
13	12	15	9	5
14	15	12	10	6
15	14	13	11	7

And again, because of the conventions of mccube.m, we must relabel 0 as 1 and 1 as 16, and also relabel 15 as 2 and 2 as 15; nodes 0 and 15, in our preliminary wire list, are opposite (body diagonal) nodes on the tesseract resistor cube, which must be (to keep mccube.m happy) labeled as nodes 1 and 2. This relabeling gives us our final wire list. (on next page)

Now, since node 1 is the positive terminal of an applied one-volt source, and since node 1 connects to nodes 4, 8, 15 and 16, then the input current to the tesseract is

$$[1 - V(4)] + [1 - V(8)] + [1 - V(15)] + [1 - V(16)]$$

and so the body diagonal resistance of a tesseract resistance cube made from one-ohm resistors is

$$R_4 = \frac{1}{4 - V(4) - V(8) - V(15) - V(16)}.$$

node	connects	to	nodes	
1	16	15	4	8
2	14	13	11	7
3	15	16	7	11
4	5	6	1	12
5	4	7	16	13
6	7	4	15	14
7	6	5	3	2
8	9	10	12	1
9	8	11	13	16
10	11	8	14	15
11	10	9	2	3
12	13	14	8	4
13	12	2	9	5
14	2	12	10	6
15	3	1	6	10
16	1	3	5	9

When mccube.m was given this final wire list the result was $R_4 = 0.6673$ ohms. How good an estimate is this? It can be shown,[8] using a symmetry argument that is a generalization of the one I used in Discussion 17 to calculate R_3, that the body diagonal resistance of an n-dimensional resistor cube made from one-ohm resistors is

$$R_n = \frac{1}{n!} \sum_{k=0}^{n-1} k!(n-1-k)! \text{ ohms.}$$

For $n = 3$, for example,

$$R_3 = \frac{1}{3!} \sum_{k=0}^{2} k!(2-k)! = \frac{2+1+2}{6} = \frac{5}{6} \text{ ohms,}$$

which agrees with our earlier result. And for $n = 4$,

$$R_4 = \frac{1}{4!} \sum_{k=0}^{3} k!(3-k)! = \frac{6+2+2+6}{24} = \frac{16}{24} = \frac{2}{3} = 0.6667 \text{ ohms.}$$

So, again, mccube.m has done well.

If you're looking for a way to kill an afternoon, I'll let *you* work through setting-up mccube.m for the resistor cube in the fifth dimension![9]

Notes and References

1. Science fiction has, of course, had a lot of fun with the fourth dimension. For an extended discussion on this, see my book, *Time Machines: Time Travel in Physics, Metaphysics, and Science Fiction* 2nd ed. (New York: Springer, 1999, in particular Chapter 2 ["On the Nature of Time, Spacetime, and the Fourth Dimension"], pp. 97–178).

2. Edwin Abbott (1838–1926) was an English educator and theologian whose book, while on the surface about mathematical geometry in a plane, is actually a rather pointed critique of snobby Victorian society (for example, the more sides a regular n-gon has, the higher in social standing is that -gon; regular ∞-gons or circles have the "ultimate" status). A similar literary experiment was published a few years later by H. G. Wells, in his even more famous and much darker "scientific romance" *The Time Machine* (1895). Wells's motivation for "inventing" a time machine was simply that he needed a gadget to get his hero into the far future. Then he could come back to the present and report on the ultimate fate of the Victorian split between the poor and the rich (who would become the Morlocks and the Eloi, respectively), and to illustrate the horrors that that age's smug optimism—that all was right with the world, and nothing needed to change—might produce. *Flatland* has had several imitators over the decades, some good and some not. I recommend the 1965 effort by the Dutch mathematical physicist Dionys Burger (1892–1987), *Sphereland*, supposedly written by A. Square's grandson, "A. Hexagon." (As explained in *Flatland*, a son always has one more side than does his father, and thus each new generation rises in social status simply as a birthright, and so that is, indeed, the correct name for a *grand*son.)

3. The procedure I'll follow with this part of the discussion is based on the short essay by Ralph Milne Farley, "Visualizing Hyperspace" (*Scientific American*, March 1939, pp. 148–149), from which comes the quotation. This essay is well written, but many of its ideas had been around for decades; see Henry P. Manning, *The Fourth Dimension Simply Explained*, first published in 1910 (reprinted by Dover in 1960), which contains popular essays submitted to a 1909 contest sponsored by *Scientific American*. "Farley" was the pseudonym used by Roger Sherman Hoar (1887–1963), a Harvard-trained lawyer who once served as the assistant attorney general of Massachusetts. He adopted his pen name, apparently, to keep his science fiction writing (which was voluminous) separate from his legal writing (also voluminous). The irony in this comes from the fact that science fiction is the genre where he should have used his real name–*that's* the stuff people remember today!

4. For what this means, see my book, *An Imaginary Tale: The Story of* $\sqrt{-1}$, (Princeton, N.J.: Princeton University Press, 1998, 2007, pp. 97–104).

5. If you initially balked at four-dimensional space, then you might think five-dimensional space to be *really* ludicrous, but that is not so. In 1921 the German physicist Theodor Kaluza (1885–1954) expanded the four-dimensional space-time of Einstein (3 space dimensions + 1 time dimension) to a five-dimensional space-time; he did that by adding a fourth space dimension. The result was that Maxwell's electromagnetic field equations were an automatic conclusion of Einstein's gravitational field equations, that is, Kaluza's five dimension space-time unified electricity and gravity! The obvious question, of course, is just where *is* that extra space dimension? In 1926 the Swedish physicist Oskar Klein (1894–1977) speculated that it's "curled up" in a tiny circle, a concept you can read more about in an essay by Bryce S. DeWitt, "Quantum Gravity" (*Scientific American*, December 1983). Since the 1920s the so-called *Kaluza-Klein theory* has mutated into today's controversial "string theories of everything" that require spaces with even more dimensions beyond the fifth.

6. This changing of a single bit as we move through one edge of the cube is called a *Hamming distance* of one, in honor of the American mathematician Richard Hamming (1915–1998). Hamming made great use of this simple idea to show how to generate binary codes that can both detect and self-correct random errors in the digital transmission of information. The self-correcting feature was long thought by many to be impossible until Hamming's insight of motion in an n-dimensional space showed it was not only possible, but not even difficult. When I taught during my 1981–1982 sabbatical year at the Naval Postgraduate School in Monterey, California, where Hamming was on the computer science faculty, he was treated almost as a god. I was far too shy to exchange even a single word with him—something I regret to this day—but I often saw him walking across campus with a student entourage trailing after him and hanging on every utterance.

7. Notice that A. Square could have actually constructed a 3-D resistor cube in Flatland. That is, a 3-D cube can be collapsed into a planar circuit, with no edge of the cube crossing another edge. It wouldn't "look like" a cube anymore (but rather like two squares, one inside the other with analogous corner nodes connected together), but from a connection point of view it would still be a cube. However, A. Square could not have measured R_3 in Flatland by connecting a voltage source to the body diagonal nodes and measuring the resulting source current. That's because one of the wires connecting the voltage source to the planar cube would have to cross a cube edge, and the

only way to do that (to avoid a short circuit) is to "hop over" the blocking edge by using the third dimension. We, however, *can* measure R_4 because, after we've collapsed the tesseract into three dimensions, there is no concern of edges or connecting wires intersecting one another. You can find the schematic of a collapsed 4-D resistor cube used in an ELECTRONICS WORKBENCH simulation, in my book, *The Science of Radio*, 2nd ed. (New York: Springer, 2001, p. 376). As a final comment on collapsing, in Robert Heinlein's story (see the second opening quotation), the "collapse" occurs in the opposite sense. That is, a California architect's house, originally built in our 3-D spatial world but with all the collapsed connections of a tesseract, "uncollapses" into the 4-D spatial world when it is shaken by an earthquake!

8. O. J. Tretiak and T. S. Huang, "Resistance of an N-Dimensional Cube" (*Proceedings of the IEEE*, September 1965, pp. 1271–1272). The authors were not cranks. Both were at MIT, a known hothouse, then and now, for spectacular electrical engineering projects.

9. The answer is $R_5 = \frac{8}{15} = 0.5333$ ohms.

Acknowledgments

As a mathematician might put it, the author of any book is necessary but not sufficient. It is my pleasure here to thank all those who, while quietly nameless elsewhere in *Mrs. Perkins's Electric Quilt*, made contributions that significantly and materially aided the emergence of a book from a sheaf of notes. My wonderful editor at Princeton, Vickie Kearn, got the project going by offering a contract after reading my proposal. Two reviewers of the first draft of the book made it better; one wished to remain anonymous, and so I am limited on this score to thanking Lawrence Weinstein, professor of physics at Old Dominion University. The staff I worked with at Princeton University Press, Debbie Tegarden, Anna Pierrehumbert, Carmina Alvarez-Gaffin, and Dimitri Karetnikov, were terrific. Marjorie Pannell of Chicago, the book's copyeditor, did a similarly outstanding job.

At home, my wife of forty-seven years, the lovely Patricia Ann, maker of beautiful quilts and dispenser of the common sense I sometimes lack, kept me on an even keel whenever I ran into a period of depression after my writing hit a speed bump. And when I needed a nonjudgmental pal, my three furry buddies, Heaviside (the immortal cat—she's twenty-three!), Maxwell, and Tigger, reminded me that sometimes all you really need to make the next technical breakthrough is a sleeping cat on your lap.

Paul J. Nahin
Lee, New Hampshire
January 2009

Index

Abbott, Edwin, 374, 382
Abe, Michio, 231
acceleration vector, 328–330
action-at-a-distance, 137–138, 165, 374
Agnew, Ralph Palmer, 321–322, 326, 328–329
Agrippa, Henry, 1
air drag, 46, 48, 120, 345; linear, 58, 64; quadratic, xx–xxii, 51–53, 55, 59, 62–81, 121–122
analytic continuation, 106
Apollo 11, 181
Archimedes' principle, 79
Aristotle, 64, 67
astronomical unit, 174, 180
asymptotic expansion, 265, 282. *See also* Stirling's formula

ballistics: with air drag, 120–135; no air drag, 107–119
baseball, 126–131
Bashworth, Francis, 120
BASIC, 306
Bellman, Richard, 94
Bentley, Richard, 137
Bernoulli, James, 245
Bernoulli, Johann, 120–121, 131
Bethe, Hans, 159
Beukers, Frits, 104
Big Bang, 169
Big Dig, 213
binary numbers, 272–273, 374, 376–377, 379, 383

Binet, Jacques, 284
binomial: coefficient, 258, 264; theorem, xix, 287, 292
blackbody radiation, 100, 105
black hole. *See* Michell, John
Bloch, Felix, 92
body-diagonal resistance (of resistor cube), 373, 378–381, 383–384. *See also* tesseract
Boltzman, Ludwig, 168
Bouwkamp, Christofel, 232
Boys, C. V., 164–167
brachistochrone, 214. *See also* gravity: tunnel
Brooks, Rowland, 218. *See also* BSST
Brown, Robert, 261–262
Brownian motion, 261–263, 266–268, 270, 281
BSST, 218–219, 224, 226, 229
Burger, Dionys, 382

Callisto (moon of Jupiter), 180
capacitor, 27–28, 42
Carroll, Lewis, 261
Cavendish, Henry, 139–140, 165–166, 182
center of mass, 175–176
central force, 138, 172
centripetal law (of wandering), 236
chain rule (of calculus), 2, 65, 73, 123, 125
Challenger (space shuttle), 70–71, 80
coefficient of restitution, 19
Columbia (spaceship), 181
commutation (of two limit operations), 35
complex zeros. *See* Riemann hypothesis

conservation laws: electric charge, xxviii, 215, 223; energy, xxiv, 19, 83–84, 165, 215, 223; momentum, xxiv
constraints (vertical and horizontal), 217, 223, 227
Conway, John, 229
Crab Nebula, 183
Crofton, Morgan, 239–240, 245–246, 257
curvature (radius of), 122, 132–135

dal Monte, Guidobaldo, 214
Dante, 92
Darwin, Charles, 158
De Moivre, Abraham, 284
derivative parabola, 117–118
Descartes, René. *See* vortex hypothesis
diffusion coefficient, 267, 282
dimensional analysis, 14–15, 17
Dirac, Paul, 283–284
distribution function, 253–254, 258, 287, 292, 296–297, 363, 366–367. *See also* probability density function
drag coefficient, 58–59
drunkard's walk, 239, 241, 269. *See also* gambler's ruin
Dudeney, Henry, 216, 219
Duijvestijn, Adrianus, 217–219, 232
dwarf star, 183
Dyson, Freeman, 36, 59

Eagle (lunar lander), 181
Earth: age of, 158–159, 169; atmosphere of, 72, 75–78, 80–81; density of, 87, 139, 193–194, 196–197, 200–203; in freshman physics, 192, 194, 199–200, 202; interior of, 195–203, 212; mass of, 140; orbital speed of, 84; radius of, 82, 86, 148–149, 168
eccentricity (of elliptical orbit), 174, 185
Einstein, Albert, 164, 262–263, 281–282, 321, 374, 383
Electronics Workbench (EWB), xxiv, 230–231, 384
energy: kinetic, 19, 45–46, 83, 161–162, 357; potential, 19–20, 45–46, 83, 158, 356–357
equation of state, 169
Euler, Leonhard, 89, 95–97, 99, 328
expected value, 286–287, 289–291, 365–366
exponential integral, 75

Farley, R. M., 382
Feynman, Richard, 26, 28–30, 32–37, 41
Fezandié, Clement, 204–205, 213
Fibonacci, 273–278, 284
Foucault, Léon, 188
Fourier: series, 95–96, 101–102, 104, 315, 348; transforms, 271
fourth dimension, 374, 376, 378–384. *See also* tesseract
Frost, Robert, 299, 303
functional equation, 165
fusion (nuclear), 158–159, 169

Galileo, 44, 60, 64, 67–68, 79, 112, 210–211, 214
gambler's ruin, 241–249
gamma function, 105, 295, 297–298
Gamow, George, 25
Gauss, C. F., 268
geometric series, 21, 244, 269
geostationary, 183, 353
Gernsback, Hugo, 213
Girard, Albert, 284
Glaisher, J. W. L., 239–240
Goddard, Robert, 204–205, 207, 210, 213, 358
golden ratio, 27, 273
gravitational binding energy, 150–152, 162
gravitational contraction, 157–162, 169
gravity: acceleration of, 18–23, 138, 186–188; onstant, 191, 193–195; inside spherical cavity, 163–164, 352–353; inside spherical shell, 141, 147, 163–164, 192; of circular ring, 141–143, 153–154; inverse square law of, 71, 82, 92, 138, 140, 148–149, 152–157, 165, 167–168, 172, 184, 191; of spherical shell, 143–146, 152–156, 167; tunnel, 204–210, 213–214, 355–358; universal constant of, 71, 83, 87, 138, 207. *See also* Newton
grazing orbit, 181–182

Hageman, Samuel Miller, 321
Halley, Edmond, 111, 139, 164, 166, 170, 191
Hamming, Richard, 383
Hardy, G. H., 103
Heaven. *See* Hell
Heaviside, Oliver, 42, 169
Heinlein, Robert, 373, 384

Hell (falling into), 82, 86–91
Hesiod, 82, 86–87, 90–92, 182
Hilbert, David, 25
Hoar, Roger Sherman. *See* Farley
Hofstader, Robert, 92
Homer, 36, 86
Hooke, Robert, 168, 170–172, 184, 188, 190–191

impedance, 27, 32–33, 35, 42
impulse function. *See* Dirac
inductor, 27–28, 42
infinite: honeycomb circuit, 41, 320, 368–369; hotel, 24–25; ladder circuits, 25–43, 342–343; regress, 227; square grid, 314–315; triangular grid, 320
integrating factor, 73–74
inverse square law. *See* gravity
isochronous, 208, 214. *See also* gravity tunnel
iterative. *See* recursion

Jacobian transformation, 283
Jeffreys, Harold, 195
Jupiter, 175, 180

Kalman, Dan, 104
Kaluza, Theodor, 383
Kant, Emmanuel, 136
Keill, John, 120–121, 132
Kelvin, Lord, 158–162, 169
Kepler's laws (of planetary motion), 171–172, 174–175, 177, 181–182, 184, 353
Kerouac, Jack, 233
Keynes, Lord, 136
King, Clarence, 158
Kirchhoff's laws, xxiv, xxviii, 25, 42, 222–223, 225, 229, 302, 309, 342–343
Klein, Oskar, 383
Kluyver, J. C., 260

Laplace transform, 42
Lavatelli, Leo, 315
laws of motion: first, xix; second, xx, 2, 8, 47, 51, 64, 70–71, 88, 109, 122, 160, 207
laws (physically plausible), 45, 47, 49–50, 52–53, 58, 62, 345
Leaning Tower of Pisa (experiment), 68
Lehmann, Inge, 195

Leibniz, Gottfried, 132–133
Leibniz's formula (for differentiating an integral), 78, 196, 325, 345, 364
Littlewood, John, 167
Loyd, Sam, 215–216
Lucretius, 281
Luzin, Nikolai, 217, 219

Mach, Ernst, 321
Maclaurin, Colin, 175
Mailer, Norman, 136
Mathematica, 101, 316
MATLAB, 16, 75, 121, 128, 231, 246–251, 259, 273, 281, 285, 289–291, 306–313, 318, 360–365, 374
Maxwell, J. C., 165, 300, 318, 383
Michell, John, 139, 165–166
Milton, John, 87
Mohrovičić, Andrija, 195
moment of inertia, 8–9
momentum, 47–48, 60
monotonicity law, 300–301, 318. *See also* Rayleigh
Monte Carlo, 245–257, 259, 271–272, 279–281, 289–291, 305–313
Moon: density of, 178; mass of, 178–179, 181; orbital speed of, 176; test, 148–149. *See also* Newton
Morón, Zbigniew, 217–219, 222
Munk, Max, 71, 75, 80

Nahin, Pat, 229–230
neutron star, 183
Newton's experiment, 188–191, 211
Newton, Isaac, 40, 88, 120–121, 132, 135–141, 147–149, 164, 167–168, 170–172, 374. *See also* gravity; laws of motion; superb theorems
Newton-Raphson formula, 60, 91
Nicholas of Cusa, 1
Niven, Larry, 163, 168, 351

Ohm's law, 223
oil drop experiment (Millikan), xiv–xv, 59
Oldham, Richard, 195
optimal launch angle, 111, 113, 116

Pascal, Blaise, 242
Pearson, Karl, 236–240, 246–249, 252, 254, 256, 259–260, 268

Pendray, George, 205
Perrin, Jean, 263
Pisano, Leonardo. *See* Fibonacci
polar coordinates, 322–328
position vector, 328
pressure (at the center of the Earth), 200–203. *See also* static equilibrium
Principia, 120, 138, 140–141, 147–148, 164, 166–168, 170, 172, 179–180, 184, 192, 195, 300. *See also* Newton; superb theorems
probability density function, 237, 250, 252–256, 267–268, 270–272, 280, 282, 363–364; (exponential), 256, 258; (Gaussian), 268–269, 364; (nearest neighbor); 286–288, 291–295; (Rayleigh), 237–239, 256–259, 363; (uniform), 271
product rule (for differentials), 123
proof (by contradiction), 231–232; (deductive and inductive), xxviii
pulsar, 182–183
Pythagorean theorem, 6, 8, 14–15, 17, 85–86, 133, 325

quilting, 215–216, 229–230

Ramanujan, Srinivasa, 103
random variable, 241, 252–255, 257–258, 286, 296, 363–364, 366
random walk, 233–284; geometric, 269–278; golden, 278–280, 301–305
Rayleigh, Lord, 17, 237, 260, 300. *See also* monotonicity law; probability density function
recursion, 29–32, 34, 60–61, 295, 298
resistor cube, 308–313
Reynolds, Osborne, 58, 79
Riemann hypothesis, 102, 107. *See also* zeta function
Roche limit, 156–157
Ross, Ronald, 233–237, 239, 263

Saturn, 175, 180. *See also* Roche limit; tidal force
Schawlow, Arthur, 105
schematic capture, 230
shock wave, 322–324
shot-put, 112–116
simple harmonic motion, 88
Smith, Cedric, 218. *See also* BSST

Sputnik 1, 182
static equilibrium (differential equation of), 200–201, 354–355. *See also* pressure (at the center of the Earth)
Stefan, Josef, 168
stochastic process, 263–269
Stokes, George, 58–59
Stone, Arthur, 218. *See also* BSST
Sprague, Roland, 217–219, 229
Stirling's formula, 264–265, 282
Strutt, John William. *See* Rayleigh
Sun: age of, 158; falling into, 82–86, 173–175; mass of, 82; energy out of, 157–158; radius of, 82
superb theorems, 86, 140–150, 157, 163, 168, 192, 205
supernova, 183
superposition, 313–315, 320, 369
symmetry, 313, 318–320, 369

Taurus (constellation), 183
temperature of Hell joke, 87, 92
terminal speed, 44, 51–54, 56, 60, 65–66, 69, 72, 75, 78, 344
tesseract, 374, 376, 378–381, 384
Thomson, William. *See* Kelvin, Lord
tic-tac-toe, 231–232
tidal force, 156–157, 168. *See also* Roche limit
topology (of electric circuits), 37, 309. *See also* wire connection list
torque, 9
Torricelli, Evangelista, 108
transcendental equation, 55, 60
transition probability, 303–304, 309–310
Trier, Peter, 315
trivial zeros. *See* Riemann hypothesis
tunneling time, 182. *See also* gravity: tunnels
Tutte, William, 215, 218, 229. *See also* BSST

Ulam, S. M., 258

Vee, Bobby, 22–23
velocity vector, 328–329
Venezian, Giulio, 316
Verne, Jules, 166, 204
von Helmholtz, Hermann, 157–158

von Neumann, John, 44–45, 58
vortex hypothesis, 137, 148, 164

Watson, G. N., 105
weight (of the planets), 138, 178–181
Weinstock, Robert, 184–185
Wells, H. G., 16
Wheatstone, Charles, 306, 318

white dwarf. *See* dwarf star
Wiechert, Emil, 195
Wiener, Norbert, 263
wire connection list, 376–381. *See also* topology

zeta function, 94–106, 284
z-transform, 42

Also by Paul J. Nahin

Oliver Heaviside (1988, 2002)

Time Machines (1993, 1999)

The Science of Radio (1996, 2001)

An Imaginary Tale (1998, 2007)

Duelling Idiots (2000, 2002)

When Least Is Best (2004, 2007)

Dr. Euler's Fabulous Formula (2006)

Chases and Escapes (2007)

Digital Dice (2008)